Reactor Design for Chemical Engineers

Reactor Design for Chemical Engineers

Edited by

J.M. WINTERBOTTOM
M.B. KING
University of Birmingham
Birmingham, UK

Routledge
Taylor & Francis Group

LONDON AND NEW YORK

First published 1983 by Routledge

2 Park Square, Milton Park, Abingdon, Oxfordshire OX14 4RN
52 Vanderbilt Avenue, New York, NY 10017

Routledge is an imprint of the Taylor & Francis Group, an informa business

First issued in paperback 2019

A catalogue record for this book is available from the British Library

ISBN 978-0-367-39995-5

Typeset by PureTech India Ltd, Pondicherry
http://WWW.puretech.com

Publisher's Note
The publisher has gone to great lengths to ensure the quality of this reprint
but points out that some imperfections in the original may be apparent.

Contents

3 Kinetics of homogeneous reactions and of reactions on solid catalyst surfaces   103
M.B. KING and J.M. WINTERBOTTOM

4 Simple reactor sizing calculations   180
J.M. WINTERBOTTOM and M.B. KING

5 Non-ideal flow in chemical reactors and the residence time distribution
J. BOURNE

Contributors

J. Bourne

Visiting Professor in Chemical Engineering, University of Birmingham, Edgbaston, Birmingham BT15 2TT, UK

R. Crane

Eutech Engineering Solutions Limited, PO Box 99, Pavilion 99, Byland Way, Belasis Hal Technology Park, Billingham, Cleveland TS23 4YS, UK

A.N. Emery

Senior Lecturer, School of Chemical Engineering, The University of Birmingham, Edgbaston, Birmingham BT15 2TT, UK

E.M. Holt

RT&E Department, ICI Katalco, PO Box 1, Billingham, Cleveland TS23 1LB, UK

G.J. Kelly

RT&E Department, ICI Katalco, PO Box 1, Billingham, Cleveland TS23 1LB, UK

F. King

RT&E Department, ICI Katalco, PO Box 1, Billingham, Cleveland TS23 1LB, UK

M.B. King

School of Chemical Engineering, University of Birmingham, Edgbaston, Birmingham BT15 2TT, UK

R.M. Nedderman

University Lecturer, Department of Chemical Engineering, University of Cambridge, Pembroke Street, Cambridge CB2 3RA, UK

K. Thayanithy

School of Chemical Engineering, University of Birmingham, Edgbaston, Birmingham BT15 2TT, UK

J.M. Winterbottom

Reader in Catalytic Reaction Chemistry, University of Birmingham, Edgbaston, Birmingham BT15 2TT, UK

Preface

Central to virtually every chemical plant is a chemical reactor or system of reactors. This book is concerned with their design. It is intended to provide a forward looking introduction to the topic and is written with undergraduate students in Chemical Engineering primarily in mind. However, it is hoped that the book will also be useful to other readers.

The first chapters are concerned with the relevant stoichiometry, thermo-dynamics and reaction kinetics and their application to simple reactor design calculations. Subsequent chapters, written by industrial practitioners, provide insight into the important and overlapping areas of catalyst design and catalytic reactor design.

Three phase reactors, fluidised bed reactors and bioreactors are described in the final section of the book.

The editors wish to thank the many generations of undergraduate and postgraduate students who have, in their various ways, added to this book. They also thank most warmly the contributors and those who have checked the manuscript or proofs.

J.M.W.
M.B.K.

1 Introduction
J.M. WINTERBOTTOM

1.1 The need for good reactor design

In recent years, the chemical reactor has been elevated to the position of being arguably the most important unit to optimise within a chemical processing plant. Until the early 1970s, downstream separation processes were regarded as the key to profitable operation of a chemical plant. While every effort was made to obtain both high selectivity and conversion to the desired product within the reactor, the desired product purity was achieved largely by well-designed downstream separation operations.

The increased cost and less certain availability of feedstocks, along with legislation to protect the environment, has now led to a drive towards 'clean technology', with the objectives of more efficient processing and waste minimisation. The latter aspect is not only economically important but also contributes to a determined effort to ensure that harmful material is not present in effluent streams leaving chemical plant. The above objectives are better achieved by improvements in reactor technology to reduce the amount of waste material produced than by the installation of retrofit downstream units to remove low-value by-products and other waste from effluent streams, although, as a last resort, this may also be necessary.

The new objectives are centred around clean technology synthesis and are namely: (i) to give as far as possible only the desired product; (ii) to reduce by-products and pollution to zero or very small proportions; (iii) to intensify processing to give improved economics; and (iv) to ensure that the process products are used in an environmentally friendly manner with emphasis on recycling and re-use.

Progress has been achieved towards meeting these objectives and this has led in recent years to increased interest in the reactor and reaction chemistry in comparison with downstream processing. Two factors which have led to this have been (i) improvement in reactor design combined with the avail-ability of precise kinetic data and powerful modelling and computational techniques and (ii) the design and availability of novel and highly selective catalysts. Catalyst design, and some achievements in this area, is discussed in Chapter 6. In the modern chemical industry the majority of processes depend upon the use of catalysts and it is likely that an even higher propor-tion of new processes will do so.

A review of the principal reactor types currently in use follows: more quantitative discussions of the design and characteristics of many of these will be found in Chapter 4 and subsequent chapters.

1.2   Reactor types

There are three main types of chemical reactor namely:

(a) batch
(b) continuous
(c) semi-batch or semi-continuous.

These general reactor types can be divided into a number of subcategories according to the nature and manner of processing the feedstocks, as follows:

(a) single phase: gas, liquid or solid
(b) multi-phase, which can be described overall as fluid–solid but may encompass combinations of:
  (i) gas–liquid
  (ii) liquid–liquid
  (iii) gas–solid (catalysed or uncatalysed)
  (iv) liquid–solid (catalysed or uncatalysed)
  (v) gas–liquid–solid (catalysed or uncatalysed).

A large number of reactions involving solids are catalytic. In such cases, the catalytic solid is not consumed by the reaction and, ideally, can be re-used for a considerable period of time.

Batch and semi-batch reactors operate in various degrees of unsteady state. Batch reactors are relatively easy to describe since the variable and unsteady operating parameter is the concentration and all material is contained within the reactor during operations. Simple kinetic equations can therefore be used to describe the course of reaction. In the case of semi-batch reactors, however, some materials which are charged to the reactor remain there for the course of the reaction, but others may be added and/or removed continuously.

Continuous reactors, as their name implies, operate under steady-state conditions, feedstock and product being added and removed respectively at a steady rate. The only exception to this is at start-up and shut-down of the reactor.

Reactors operated continuously can be represented by two extremes of flow regime. The first of these is the 'plug flow' reactor (PFR) for which there is no mixing of fluid elements as they pass through the reactor. In this case there is a continuous concentration gradient for the various species from inlet to outlet. The other extreme is that represented by the 'perfectly mixed' reactor (PMR). In this case, the reactor contents are so well mixed that there is no concentration gradient within the reactor and the effluent concentration is identical to the concentration within the tank. In practice, by careful design both ideal flow patterns can be closely approached. A very close approximation to 'plug flow' is achieved in many of the tubular reactors used in the heavy chemicals industry while a close approximation to 'perfect mixing' is achieved in the continuous stirred tank reactors (CSTRs) described in Section 1.4.1.1. In both the above cases, approach to ideal behaviour can be sufficiently close to be useful for design purposes. Of course, some reactors exhibit flow regimes that are intermediate between the two ideals of the PFR and PMR. These reactors can be described as non-ideal, and the effects of non-ideal flow on conversion will be discussed in Chapter 5.

1.3 Batch reactors

Batch reactors process one batch of material at a time in a closed system. Typical types are shown in Figure 1.1. They have the advantage of flexibility in that they can be used to produce a variety of chemicals, i.e. they are not dedicated to a single product. They are used for fine chemical, pharmaceutical and polymer production on a relatively small scale. An additional advantage of batch reactor is that of relatively small capital investment. Nevertheless, their design and operation are being improved to meet the needs of clean technology operations, and the Buss reactor (shown later in Figure 1.5) is one example where improvements in batch reactor technology have led to significantly enhanced performance. The important operational parameters in the case of batch reactors are good mixing and heat transfer. The need for this is well illustrated by considering the case of a gas–liquid–solid catalysed exothermic reaction. Good mixing is particularly important to give a homogeneous dispersion of all species and to ensure good mass and heat transfer between the phases. Provision of good external heat exchange may also be necessary. Various batch reactor designs can be seen in Figure 1.1(a)–(d).

The most appropriate design for the stirrers in these reactors depends on the nature of the fluid phases requiring mixing. Figures 1.2 and 1.3 show typical turbine-type and wide-radius agitators, while Figure 1.4 illustrates several 'Archimedes' screw' and marine propeller-type agitators. Figure 1.4 also shows the types of circulation obtained from these and other stirring systems, and in a number of situations it is necessary to use baffles to control vortex formation. Turbine-type agitators can be used in liquids with viscosity up to about 10 Pa s[1]. Wide-radius or spiral agitators are required for the very viscous liquids with viscosities greater than this.

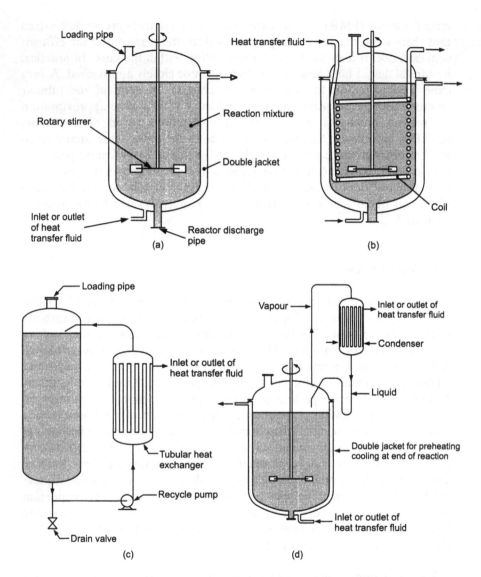

Figure 1.1 Some typical types of batch reactor: (a) batch reactor with double jacket; (b) batch reactor with double jacket and internal coil; (c) batch reactor with external heat exchanger on circulation loop; (d) batch reactor with cooling by vapour phase condensation and recycle. (From Trambouze *et al.* (1988), p. 98; reprinted by permission of Gulf Publishing Co., Editions Technip, 27 Rue Ginoux, Paris.)

Another method of inducing good mixing and heat transfer involves the use of closed loop circulation. Figure 1.1(c) shows a simple closed loop circulation reactor while Figure 1.5(a) and 1.5(b) show highly efficient three-phase reactors, the former being the Buss reactor [2] and the latter the

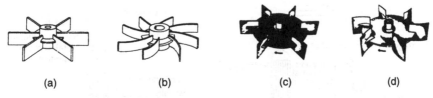

Figure 1.2 Turbine type agitators: (a) turbine with six flat blades; (b) turbine with eight curved blades; (c) turbine with six flat blades mounted on disc; (d) turbine with six curved blades mounted on disc. (From Trambouze *et al.* (1988), p. 540; reprinted by permission of Gulf Publishing Co., Editions Technip, 27 Rue Ginoux, Paris.)

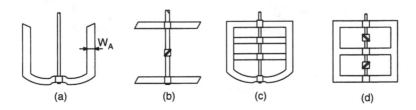

Figure 1.3 Wide-radius agitators: (a) simple anchor; (b) large inclined paddles (three mounted at 90°); (c) reinforced anchor; (d) frame with large inclined paddles mounted perpendicular to the frame. (From Trambouze *et al.* (1988), p. 542; reprinted by permission of Gulf Publishing Co., Editions Technip, 27 Rue Ginoux, Paris.)

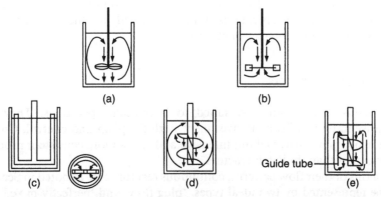

Figure 1.4 Types of circulation obtained in batch reactors using various agitation systems: (a) marine propeller (axial flow); (b) turbine (radial flow); (c) anchor (peripheral flow); (d) Archimedes' screw (axial flow); (e) Archimedes' screw with guide cylinder (axial flow with peripheral recirculation). (From Trambouze *et al.* (1988), p. 543; reprinted by permission of Gulf Publishing Co., Editions Technip, 27 Rue Ginoux, Paris.)

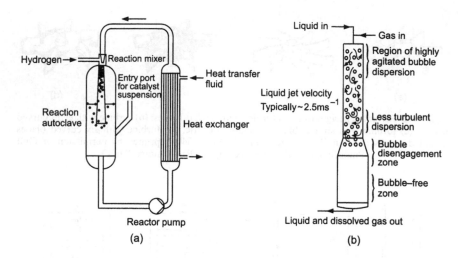

Figure 1.5 Three-phase reactors: (a) the Buss reactor; (b) the Cocurrent Downflow Contactor Reactor (CDCR).

Cocurrent Downflow Contactor Reactor (CDCR) [3]. These reactors are similar in many ways, the Buss reactor using a venturi device to mix gas with liquid and solid, while the CDCR employs a simple small orifice. Reference will be made to some of the configurations illustrated above in later chapters. Some of the design features of batch reactors are equally applicable to other well-mixed reactors, including the continuous stirred tank reactor (CSTR).

Typical reactions carried out in batch reactors are nitrations, sulphonations, hydrogenations, alkylations and polymerizations. They represent relatively small-scale operations typically suited to the fine chemical pharmaceutical and polymer industries.

1.4 Continuous reactors

Continuous reactors are best suited to large-scale operations. They are expensive in terms of capital investment both for plant and control. On the other hand, they are less labour intensive and give a more consistent product specification than do batch reactors.

In terms of their flow patterns, continuous reactors can, as noted in Section 1.2, be represented by two ideal types, 'plug flow' and 'perfectly mixed'. In 'plug flow' reactors (PFRs) the flow behaviour is 'ideal' in that there is no back mixing of fluid elements as material passes through the reactor. This results in a steadily decreasing concentration of each of the reactants as it passes from inlet to outlet. The second extreme, as far as flow characteristic is concerned, the PMR, is characterised by complete mixing of fluid as it enters the reactor such that the effluent concentration is identical to that in any part of the reactor. In practice, both 'plug flow' and 'perfect mixing' are usually attained to quite a high degree in tubular and continuous stirred tank reactors (CSTRs) respectively. However, a certain degree of back-mixing occurs in tubular reactors even with turbulent flow, while laminar flow can result in severe back-mixing. Mixing effects will be discussed further in Chapter 5, since non-ideal flow patterns can sometimes produce considerable deviations from conversions calculated using the 'ideal reactor' models.

1.4.1 Well-mixed continuous reactors

1.4.1.1 Continuous stirred tank reactors. Continuous stirred tank reactors (CSTRs) in their simple form (Figure 1.6) are in many ways very similar to stirred batch reactors, though in the case of the CSTR provision is made for the continuous introduction of feed and removal of product. They can be used as individual units or a set of such units can be joined together to form a chain of stirred tanks. In this situation an approach to 'plug flow' behaviour will be obtained. A configuration of this type is shown in Figure 1.7. Alternatively, a cascade of stirred units may be constructed within a single shell (Figure 1.8). The reactor selected, and the temperature control achievable within it, may be critical in terms of reaction selectivity, as discussed in Chapter 4. Reactors of the agitated well-mixed type can be operated either homogeneously or heterogeneously using a solid catalyst.

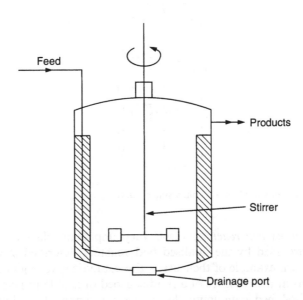

Figure 1.6 A basic CSTR unit with mechanical stirring. *Hatched areas*: Baffles (4 at 90°). These may be used to control vortex formation; alternatively, the stirrer may be mounted off-centre (as in Figure 1.7) or fixed counterblades may be used.

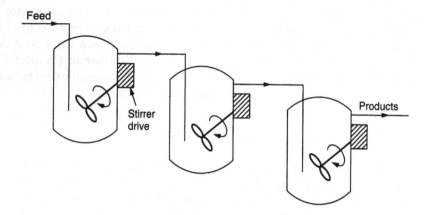

Figure 1.7 Cascade of individual CSTR units with propeller-type stirrers.

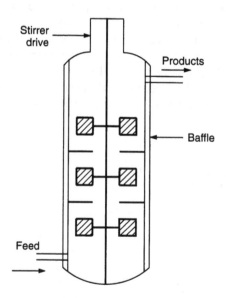

Figure 1.8 Cascade of stirred units within a single tank.

1.4.1.2 Fluidised bed reactors. One very important class of well-mixed reactors is provided by the fluidised bed reactors described in Chapter 8. The best-known example of their use is the catalytic cracking of long-chain hydrocarbons in the presence of a fluidised bed of zeolite to give gasoline.

The fluidised bed containing the catalyst is linked to a fluidised bed regenerator, which 'de-cokes' the used catalyst and returns it to the reactor, as shown in Figure 1.9. Similar types of reactor are used in a number of

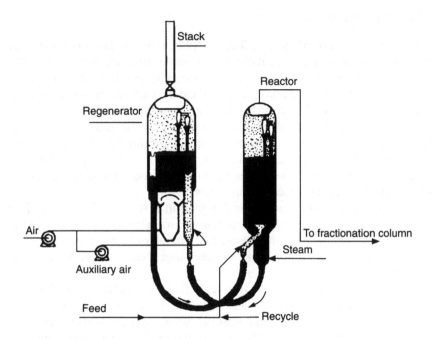

Figure 1.9 Fluidised bed reactor with fluidised bed catalyst regeneration unit used for catalytic cracking (Exxon Model IV). (From Trambouze *et al.* (1988), p. 432; reprinted by permission of Gulf Publishing Co., Editions Technip, 27 Rue Ginoux, Paris.)

oxidation processes which require the good heat transfer and temperature control features which a fluidised bed reactor can provide (Chapter 8).

Examples of processes carried out in various forms of well-mixed reactor involving catalysts are given in Table 1.1.

Table 1.1 Examples of processes carried out in well-mixed reactors

Reactor type	Process
Stirred batch	Hydrogenation of triglycerides
	Nitrocompounds
	Butyne-1, 4-diol
	Fatty nitriles
Continuous stirred tank	Fischer–Tropsch synthesis
	Triglyceride hydrogenation
Fluidised bed	Catalytic cracking
(i) Gas phase	Fischer–Tropsch synthesis
(ii) Gas/liquid phases	Hydrodesulphurization of residua
	Hydrocracking
Entrained catalyst (gas phase)	Catalytic cracking

1.4.2   Tubular reactors

Tubular reactors are often described as 'plug flow' reactors (PFRs) since their flow characteristics normally approximate quite closely to the 'plug flow' model. In a few instances the reactor may simply consist of an empty tube, but it will usually contain a bed of catalyst or occasionally a bed of inert material intended to improve the heat transfer.

The PFR is to be found in a large number of oil and secondary oil-processing operations and in heavy chemical production generally. Well-known examples are ammonia synthesis, platinum reforming of naphtha, steam reforming of methane and/or naphtha, hydrodesulphurisation of gas oil and hydrocracking, to select only a few.

Plug flow reactor systems can be designed in a variety of configurations which either provide close approximations to isothermal or adiabatic operation or which operate in a non-adiabatic and non-isothermal mode. (The last of these possibilities requires less heat exchange than is required for adiabatic operation.)

The isothermal operation of large-scale fixed bed reactors can be quite difficult to achieve because of heat exchange problems in large beds. Nevertheless, some temperature control is often required, both for exothermic and endothermic reactions if the required conversions and (for endothermic systems) reaction rates are to be achieved and also if the catalyst (if any) is to remain within its optimum operating range. For example, reactions such as ammonia synthesis, while exothermic, are reversible and favoured by (i) high pressure and (ii) low temperature. Hence some degree of heat exchange is essential in order to obtain viable and acceptable conversions. Table 1.2 gives examples of reactions in fixed bed and moving bed reactors which are either carried out adiabatically or require some degree of heat exchange.

1.4.2.1   Single-bed tubular reactors.

There are many examples of single-bed reactors. When designing such reactors it is frequently necessary to overcome problems arising from excessive temperature changes due to the heat of reaction and/or pressure drops. One method for preventing the development of excessive temperature changes in a continuous bed is exemplified in the ICI Quench Converter for ammonia synthesis [4]. In this case the catalyst bed is divided by 'lozenge' distributors which are placed at various points down the bed and through which quench gas in the form of cool reaction gas is injected thus cooling the bed.

The above technique is an example of 'cold shot' cooling, which is further discussed in Section 2.2.2.1.

In some circumstances the development of excessive pressure drop can limit the bed depth in longitudinal flow reactors. One way of overcoming this problem is to use a radial flow reactor. In reactors of this type the

Table 1.2 Examples of processes carried out in fixed bed reactors

Reactor type	Process
Adiabatic	
Single-fluid phase	SO_2 to SO_3
	Methanol synthesis
	Platinum reforming
	Ethyne hydrogenation to ethene
	Hydrodesulphurisation of naphtha
Two-fluid phases	Gas oil hydrodesulphurisation
	Hydrocracking
	Lube oil hydrorefining
With heat exchange	
Single-fluid phase	Steam reforming of CH_4/naphtha
	NH_3 synthesis
	Methanol synthesis
	Ethylene oxide synthesis
	Phthalic anhydride synthesis
	Fischer–Tropsch synthesis
Two-fluid phases	Selective hydrogeneations
	e.g. long-chain unsaturated fatty acids

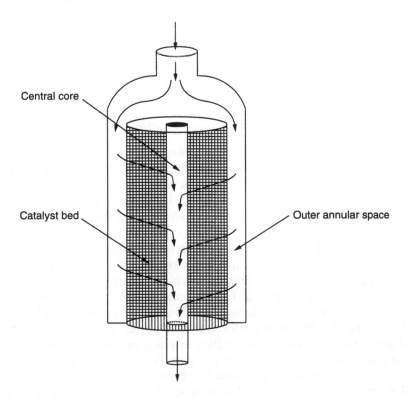

Figure 1.10 Radial flow reactor (schematic).

catalyst bed is cylindrical with a central hollow core formed by a porous mesh. The bed is located within an annular space from which feed gas can enter the bed through a porous support at all points along the length of the cylinder. Reaction occurs as the gas flows radially across the bed to the central core from which the resultant mixture is ducted. (The reverse flow is also possible.) The principle on which this system works is shown schematically in Figure 1.10.

Several reactors used for ammonia synthesis and also for methanol synthesis operate on this principle or modificators of it [4]. (The methanol synthesis and ammonia synthesis reactions have several common features: in each case the reaction is reversible and high conversions are favoured by high pressure and low temperature.) The ICI steam-raising unit is a typical reactor used for methanol synthesis. In this case [4] the necessary heat exchange is effected by heat exchange tubes embedded in the catalyst bed. The heat removed is used to raise steam.

1.4.2.2   Multiple-bed reactors.   Oxidation reactions tend to be strongly exothermic and it is necessary to cool the reactor fluids more effectively than can be achieved within a single catalyst bed. Multiple-bed reactors are normally used to overcome the problems associated with this high reaction exothermicity. For example, in the oxidation of sulphur dioxide a series of catalyst beds is used with intermediate cooling between the beds. Various layouts are in use, one of which is shown in Figure 2.17. In contrast to the oxidation reactions considered above, dehydrogenation reactions, such as the steam reforming of naphtha

$$C_nH_{2n+2} + nH_2O \rightarrow nCO + (2n + 1)H_2 \qquad (1.1)$$

are strongly endothermic. Adiabatic reactors would not be suitable in most of these cases because the temperature decrease across a catalyst bed of sufficient depth to produce the desired conversion would be such as to reduce the reaction rate virtually to zero. Hence reactions of this kind are carried out in multi-tubular reactors housed within furnaces.

1.4.2.3.   Moving-bed reactors.   An ever-present problem in the use of catalytic reactors is deactivation due to coke laydown. This problem has already been referred to in Section 1.4.1 as occurring in fluidised bed reactors for catalytic cracking. In that case the solution was continuous recirculation of the catalyst through a fluidised bed regenerator. The problem also arises in the catalytic reforming of naphtha (a mixture of paraffins and cyclic paraffins) to form aromatics. A platinum–rhenium catalyst supported on acidic alumina is used for this purpose. The reaction is a combination of dehydrogenation accompanied by isomerisation and cyclisation. It is endothermic and carried out at low pressure (1–3 bar) and 480–530°C. The modern configuration consists of three reactors, stacked one above the other. The catalyst and feed pass concurrently under gravity in plug flow, and the coked catalyst is passed through a regenerator and returned to the top of the first reactor. This enables the catalyst to be used for a considerable period of time with very small losses. This is essential since platinum is a very expensive metal and only small losses can be tolerated.

1.4.2.4 Reactors employing monolithic catalysts. The catalysts discussed up to this point have been in particulate form, typical particle diameters ranging from 20 microns in slurries to 3–13 mm in fixed beds. In addition, a small number of reactors employ monolithic (i.e. non-particulate) catalysts, usually in the form of a gauze. The catalyst material used to reduce the emissions of carbon monoxide, nitrogen oxides and hydrocarbons from car exhausts is in this form. A large-scale use of monolithic catalysts occurs in the reactors used for oxidising ammonia to nitric oxide in the production of nitric acid. These reactors employ large platinum–rhodium gauzes as the monolithic catalyst. A pad of these gauzes is normally supported on a large open mesh of high-chromium steel and typical gauze diameters range from 3 to 5 metres [4]. The preparation of these pads and the changes in morphology which occur as the gauzes pass from a virgin to a reacted/activated state are described in reference [4].

1.5 The importance of catalysts

It is clear from the survey of reactor types given in this chapter that a very large number of processes involve catalysts. It is not surprising, therefore, that the development of catalytic – and particularly heterogeneous catalytic – reactors (and the catalysts which they contain) accounts for much of the research and development work on chemical reactors currently in progress.

The use of catalysts in large-scale and other chemical processes has proceeded concurrently with more than 100 years of research and development. The range of operating temperatures of the catalysts used today extends from ambient to very high temperatures (about 1000 K for steam reforming). Not only does the presence of a catalyst permit reactions to occur under much milder conditions than their homogeneous counterparts, but they also facilitate the selective production of a desired product or products.

The aims of the catalyst designer may be seen as:

(1) to reproduce for the types of reaction which occur in the heavy chemicals industry the degree of selectivity which, under very different conditions and for very different reaction types, is achieved by the biocatalysts (or enzymes) which have been tailored by nature to give specific products and
(2) to outperform these natural catalysts in terms of activity.

Good progress has been made towards achieving these aims with the design of zeolites, catalysts for the production of asymmetric molecules and also some highly sophisticated and selective homogeneous catalysts. Some of these aspects will be discussed in Chapter 6.

References

1. Trambouze, P., van Landeghem, H. and Wauquier, J.P. (1988) *Chemical Reactors Design/ Engineering/Application*, Gulf Publishing Co., Editions Technip, Paris.
2. Greenwood, T.S. (1986) *Chemistry and Industry*, 3, 94.
3. Boyes, A.P., Chugtai, A., Khan, Z., Raymahasay, S., Sulidis, A.T. and Winterbottom, J.M. (1995) *J. Chem. Tech. Biotechnol.*, **64**, 55.
4. Twigg, M.V. (ed.) (1996) *Catalyst Handbook*, 2nd edition, Manson, London.

2 Reaction stoichiometry and thermodynamics

K. THAYANITHY and M.B. KING

2.1 Introduction, nomenclature and concepts

2.1.1 Homogeneous and heterogeneous reactions, homogeneous and heterogeneous catalysis

Although some of the reactions considered in this book are truly *homogeneous* in that they take place within a single phase (normally a fluid phase), many are *heterogeneous*. A heterogeneous reaction involves more than one phase and normally takes place at or near the interfaces between these phases. Typical phase combinations encountered in reactors have been listed in Section 1.2. Many reactions involve solid and fluid phases. In some cases (as in the combustion of coal in air described in Chapter 8) components from both these phases are used up in the reaction. A more complex but very important example of a heterogeneous reaction involving solid and fluid phases is that which occurs when fluid reactants combine on the surfaces of a solid catalyst to form fluid products. This form of catalysis, which for obvious reasons is termed *heterogeneous catalysis*, is very commonly encountered in the heavy chemicals industry. The catalyst affects the rate of reaction but is not used up in the reaction and does not affect the equilibrium. If several reactions are thermodynamically possible, a good catalyst will selectively accelerate the one which gives the desired product or products. An introduction to the kinetics and mechanisms of heterogeneous catalysis is given in Chapter 3 and examples of catalytic reactor layouts have already been given in Chapter 1.

Homogeneous catalysis can also be of considerable importance, particularly for the biological reactions considered in Chapter 10. Many of these reactions are homogeneous in the liquid phase, the reactants, the products and the catalyst (in this context described as an 'enzyme') all being soluble in this phase.

2.1.2 Reversible and 'irreversible' reactions

A truly 'irreversible' reaction would continue until at least one of the reactants (the limiting reactant) had been completely used up (see Section 2.1.4.4). Although it is possible that no reaction is completely irreversible, many (including the oxidation of hydrocarbons) are sufficiently nearly so to

be regarded as such for all practical purposes. A reaction which closely
conforms to the definition of irreversibility is the explosive reaction of
hydrogen with oxygen which is discussed in Section 3.1.3.

$$2H_2 + O_2 \rightarrow 2H_2O \tag{2.1}$$

So-called 'reversible' reactions, on the other hand, do not approach completion at the pressures and temperatures of interest due to the existence of an
appreciable back reaction in which the products interact to form the initial
reactants. As the reaction proceeds, the rate of the back reaction increases
due to increase in the product concentrations while the rate of forward
reaction falls due to the fall in the reactant concentrations (see Section
2.4.1 and Chapter 3). Eventually the back and forward reaction rates
become equal and the net reaction rate falls to zero. An example of a
reaction of this type, which will be considered later in the present chapter,
is the catalysed oxidation of sulphur dioxide.

$$2SO_2 + O_2 \rightleftharpoons 2SO_3 \tag{2.2}$$

The reverse arrow in equation (2.2) (and in similar cases) indicates that the
back reaction can be significant and that the overall reaction must be
regarded as 'reversible'.

2.1.3   Reaction stoichiometry and stoichiometric coefficients

Reaction stoichiometry is concerned with the relative amounts of each of the
components participating (either as reactants or products) in a chemical
reaction and with mass balance calculations based on this information.
Equation (2.2), for example, gives the **overall stoichiometry** for the oxidation
of SO_2. It tells us that, as the net result of the reaction, 2 molecules of SO_2
and 1 molecule of O_2 are used up in forming 2 molecules of SO_3. The
stoichiometric coefficients (ν) for each component give the number of molecules of that component participating in stoichiometric equations such as
(2.1) and (2.2). The convention usually adopted is to regard the coefficients
for the components being used up (i.e. the reactants) as negative while those
for the products formed are positive. Following this convention, the stoichiometric coefficients ν_{SO_2}, ν_{O_2} and ν_{SO_3} for SO_2, O_2 and SO_3 in equation
(2.2) are $-2, -1$ and $+2$ respectively.

   The above sign convention enables stoichiometric equations such as (2.1)
or (2.2) to be expressed in a concise algebraic form. Regarded as an algebraic
equation, equation (2.2), for example, may be written as,

$$0 = 2SO_3 - O_2 - 2SO_2$$
$$= (\nu_{SO_3})SO_3 + (\nu_{O_2})O_2 + (\nu_{SO_2})SO_2$$
$$= \sum_i \nu_i M_i \tag{2.3}$$

M_i is here a shorthand way of writing 'molecules of chemical species i'. ν_i is the stoichiometric coefficient for species i. In the above example i may be either SO_3, O_2 or SO_3. The stoichiometric coefficients can usually (though not invariably) be expressed as small integers. The simple molecular arguments outlined in Chapter 3 suggest that this should be true for single-stage reactions.

It should be noted, however, that equations such as (2.1) and (2.2) give **the overall stoichiometry** of the reaction and may in fact be giving the end results of quite complex reaction sequences. Equation (2.1), for example, tells us that the net result of the hydrogen/oxygen reaction is that two molecules of hydrogen are used up for every molecule of oxygen which reacts, and that two molecules of water are then formed. However, quite a complex branched chain reaction mechanism is believed to be involved in bringing about this end result.

Although stoichiometric equations are usually expressed, where possible, in terms of integers, it may not always be convenient to do so. There is no reason, for example, why the overall stoichiometry for the oxidation of SO_2 (equation (2.2)) should not be expressed as

$$SO_2 + \frac{1}{2}O_2 \rightleftharpoons SO_3 \qquad (2.4)$$

if this proves to be more convenient (see Example 2.1). (Equation (2.4) is obtained by dividing the stoichiometric coefficients in equation (2.2) through by 2). It should be noted, however, that the numerical values of some of the properties of the reaction (such as the equilibrium constant which will be discussed in Sections 2.4.1.1 and 2.4.3.1) depend on the numerical values attributed to the stoichiometric coefficients. The way in which the reaction is written should therefore be stated when numerical values for properties such as the equilibrium constant or standard heat of reaction are given.

2.1.4 The mole and molar mass and their use in stoichiometric calculations

When carrying out chemical reaction calculations it is convenient, for obvious reasons, to work in terms of moles rather than the very large numbers of elementary entities such as molecules, atoms, ions and radicals which may be present. The number of moles is equal to the number of the elementary entities considered divided by Avogadro's number (**N**). For example, the number of moles n of a given compound is equal to the number of molecules (N) divided by Avogadro's number (**N**)

$$n = \frac{N}{\mathbf{N}} \qquad (2.5)$$

while the number of moles of a given type of ion is equal to the number of ions of this type again divided by the Avogadro number. The discussion below will largely be in terms of molecules and atoms but it should be remembered that the same concepts apply to ionic species and radicals.

Formally, the mole has been defined as the amount of substance which contains as many elementary entities as there are atoms in 0.012 kg of carbon-12. The elementary entity must be specified and may be an atom, a molecule, an ion, a radical, etc. In terms of the above definitions, Avogadro's number is found experimentally to be $6.02 \times 10^{23} \, \text{mol}^{-1}$.

It is, of course, a consequence of equation (2.5) that the number of moles of a given compound participating in a chemical reaction is proportional to the number of molecules which participate. Equation (2.2), for example, tells us that, as the net result of the reaction shown, 2 moles of SO_2 and 1 mole of O_2 will be used up in forming 2 moles of SO_3.

By analogy with equation (2.3) (which referred to molecules), reaction equations can be expressed in a generalised form in terms of moles as follows:

$$0 = \sum_i \nu_i m_i \tag{2.6}$$

where m_i is here a shorthand way of writing 'moles of chemical species i'. ν_i is the stoichiometric coefficient for each species and, as in equation (2.3), is taken to be negative for the reactants and positive for the products. Taking the oxidation of SO_2 written as in (2.4) as an example, the stoichiometric coefficients ν_{SO_2}, ν_{O_2} and ν_{SO_3} for SO_2, O_2 and SO_3 are $-1, -\frac{1}{2}$ and $+1$ respectively. Equation (2.6) as applied to this reaction then reads,

$$1 \text{ mole of } SO_3 - 1 \text{ mole of } SO_2 - \frac{1}{2} \text{ mole of } O_2 = 0$$

2.1.4.1 Calculation of the number of moles of a compound from its mass. The number of moles of a substance is obtained by dividing the mass (m) of that substance by its molar mass (**M**)

$$n = \frac{m}{\mathbf{M}} \tag{2.7}$$

m and **M** in equation (2.7) should of course be expressed in consistent mass units. It is a consequence of the way in which the mole is defined that the molar mass (**M**) of a compound when expressed in grams mol^{-1} is numerically equal to the relative molecular mass (RMM) of that compound (RMM is the currently preferred term for 'molecular weight'). It then follows that:

$$n = \text{mass of substance in grams/RMM}.$$

The RMM for a substance may be calculated from its molecular formula and a table of relative atomic masses in the usual way. Taking **carbon monoxide** for example:

REACTION STOICHIOMETRY AND THERMODYNAMICS

RMM $= 12.01 + 16.00 = 28.01$

Molar mass in g $= 28.01\,\mathrm{g\,mol^{-1}}$

Molar mass in kg $= \mathbf{M}_K = 0.02801\,\mathrm{kg\,mol^{-1}}$

Number of moles of CO contained in 2.0 kg $= \dfrac{2.0}{0.02801} = 71.4$

It should be noted that, in primary SI units, the molar mass $\mathbf{M}_K$ is expressed in kg, not g. It is not therefore numerically equal to RMM but to RMM/1000.

$$\mathbf{M}_K = \frac{\mathrm{RMM}}{1000} \tag{2.8}$$

In equations in which it is important that all terms should be expressed in SI units, it is of course $\mathbf{M}_K$ given by equation (2.8) which should be used (see Problem 3.4). As a very simple example of the use of the mole in chemical reaction calculations, we calculate below the mass of carbon dioxide which is produced by complete oxidation of 12 kg of carbon monoxide. The overall stoichiometry of the reaction is given by:

$$2CO + O_2 \rightarrow 2CO_2$$

$$\text{2 moles} \quad \text{1 mole} \quad \text{2 moles}$$

This equation tells us that 1 molecule of oxygen is required for the complete oxidation of two molecules of carbon monoxide, i.e. from equation (2.5) 1 mole of oxygen is required for the complete oxidation of 2 moles of carbon monoxide, 2 moles of carbon dioxide being then produced.

Number of moles of CO to be oxidised $= 12.0/0.02801 = 429$

$= $ number of moles of CO_2 produced

The relative molecular mass (RMM) of CO_2 is 44.01. So, from equation (2.8),

Molecular mass of $CO_2 = 0.044\,\mathrm{kg\,mol^{-1}}$

and, from equation (2.7),

Mass of CO_2 produced $= 429 \times 0.04401 = 18.9\,\mathrm{kg}$

2.1.4.2 Calculation of number of moles from gas volumes and molar volumes: range of validity of perfect gas laws. Gas phase molar volumes are frequently required in reactor stream calculations (for example, for converting molar flow rates to volumetric flow rates and vice versa). In many cases the molar volumes can be calculated with sufficient accuracy using the perfect gas laws. According to these laws, the molar volume V_m^{PG} of a perfect gas is given by

$$V_m^{PG} = R_m T / P = 8.3144 T / P \tag{2.9}$$

where T is the absolute temperature in K, P is the pressure in $N\,m^{-2}$ (Pa) and V_m^{PG} is the molar volume in $m^3\,mol^{-1}$ and 8.3144 is the molar gas constant expressed in SI units ($m^3\,Pa\,mol^{-1}K^{-1}$ or $J\,mol^{-1}K^{-1}$). The molar volume of a perfect gas at STP (0°C and 1.0133×10^5 Pa) is

$$V_m^{PG}(STP) = 0.022414\,m^3 \tag{2.10}$$

and the number of moles n present in an ideal gas or gas mixture is given by

$$n = [V(STP)/(0.022414)]$$
$$= PV/(8.3144T) \tag{2.11}$$

where $V(STP)$ is the volume in m^3 occupied by the gas or gas mixture reduced to STP and V is the volume occupied at pressure P (Pa) and temperature T (K).

Deviations from the perfect gas laws can conveniently be expressed in terms of the compressibility factor Z, which is unity for ideal gases.

$$Z = (PV_m)/(R_m T) \tag{2.12}$$

where P is the pressure (in $N\,m^{-2}$ in SI units), V_m is the gas molar volume ($m^3\,mol^{-1}\,K^{-1}$ in SI units) and R_m is the molar gas constant ($N\,m\,mol^{-1}\,K^{-1}$ in SI units).

At near ambient temperatures, these laws are normally obeyed by gaseous components to within 3 or 4% at pressures up to ambient or a little above. At the higher temperatures which can be encountered in chemical reactions, the pressure range over which this degree of accuracy is maintained may be considerably extended, particularly for the lighter components. Consider, for example, the components in the 'shift' reaction

$$CO + H_2O \rightleftharpoons CO_2 + H_2 \tag{2.13}$$

This catalysed reaction, which is typically carried out in the range 400 to 500°C, forms one stage in the production of ammonia from natural gas and air. To integrate this stage satisfactorily into the overall process, it is usually carried out at a pressure of about 30 bar. Even at this somewhat elevated pressure, the molar volumes of the reactant and product components (including the water) all fall within 3 to 4% of the 'ideal gas' values, (The compressibility factors for CO, H_2O, CO_2 and H_2 at 400°C and 30 bar are 1.02, 0.96, 1.00 and 1.01 respectively [1,2].)

In a situation such as the above, the reactant/product mixtures would be expected to show deviations from ideality similar to or rather less than those shown by reactants and products (as pure components) at the same conditions.

Although directly determined data for the substances considered should be used when these are available (as in the above example), a good first guide to the range over which gas phase near-ideality prevails can in many cases be obtained from the familiar compressibility factor charts and tables [3] both for pure components and (using pseudo-critical constants) for the mixtures also.

These charts and tables give the compressibility factor as a function of reduced pressure $(P_r = P/P_c)$, reduced temperature $(T_r = T/T_c)$ and (in many cases) the acentric factor ω [3]. T_c and P_c in the above equations are the critical temperature and pressure of the substance. (In the case of mixtures these may be replaced by 'pseudo-critical' constants which, for approximate purposes, are frequently taken to be the molar averages of the constituent critical temperatures, critical pressures and acentric factors respectively). The acentric factor for pure substances provides a measure of the deviation of the substance behaviour from that of non-polar substances such as Ne or Ar which have spherical molecules. This parameter is listed in many data banks [3,] alongside the critical constants.

Unfortunately, this approach has several limitations. For example, it should not be applied to fluids in which hydrogen bonding occurs and it cannot be applied to the very light substances, hydrogen and helium, unless pseudo-critical parameters are employed in place of the true critical constants [3]. Also the use of pseudo-critical parameters in mixture calculations is found to give poor accuracy if the ratio of the pure component critical

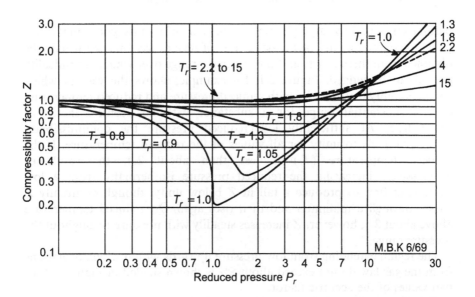

Figure 2.1 Compressibility factor $(Z = PV/RT)$ as a function of reduced pressure $(P_r = P/P_c)$ with reduced temperature $(T_r = T/T_c)$ as parameter (acentric factor approximately 0.25).

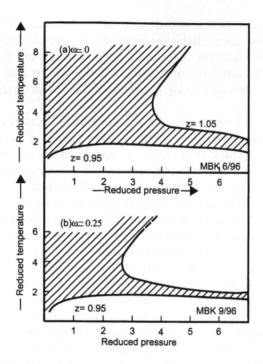

Figure 2.2 Region (shown shaded) for which compressibility factor falls between 0.95 and 1.05 (a) for substance (such as argon) with an acentric factor of approximately 0; (b) for substance (such as carbon dioxide) with an acentric factor of approximately 0.25).

parameters is unduly large [3]. Nevertheless, a sufficiently good qualitative guide to compressibility factor behaviour of reactor systems can usually be obtained from these charts and tables for the range over which near-ideality prevails to be judged. Figure 2.1, for example, shows the way in which compressibility factor varies with reduced pressure and reduced temperature over a substantial range for a substance with an acentric factor of about 0.25. At sufficiently low reduced pressure the compressibility factor (Z) tends to unity for all values of the reduced temperature. If the reduced temperature (T_r) is less than about 2.4, the result of progressively increasing the reduced pressure isothermally from a low value is at first to produce a fall in Z below unity, though eventually Z passes through a minimum and then rises again. At reduced temperature above about 2.4, however, Z increases steadily with pressure throughout the pressure range.

The region extending from low pressure upwards within which deviations from the gas law do not exceed about 5% is shown shaded in Figure 2.2 for two values of the acentric factor.

This figure is intended to provide a rough first guide only for substances with acentric factors between zero and 0.25. However, it is seen that **at**

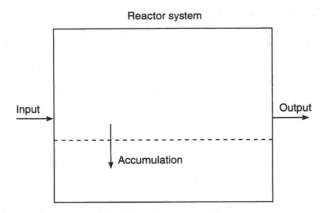

Figure 2.3 Illustrating equation (2.14) as applied to total mass and the masses of the constituent elements.

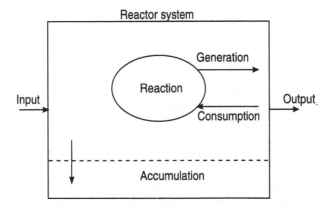

Figure 2.4 Illustrating equation (2.15) as applied to the masses and the numbers of moles of the individual chemical compounds.

reduced temperatures above 2.0, the shaded area becomes extensive, particularly at zero acentric factor. Even at the higher acentric factor shown, the boundary of the shaded area does not fall below a reduced pressure of 2.75 which, for a substance with a typical critical pressure of about 40 bar, would correspond to a reactor pressure slightly in excess of 100 bar.

The above notes are intended only to provide a qualitative picture of the ranges of conditions under which the ideal gas laws are or are not approximately valid. When more detailed calculations are required, procedures involving equations of state are to be preferred (particularly for mixtures) and the reader is referred to more specialised texts (see reference [3], for example).

2.1.4.3   Mass and molar balances.   It is a consequence of the law of conservation of mass that, when a purely chemical reaction takes place, the total
mass of the products which are produced is equal to the total mass of the
reactants which are used up. Likewise the mass (and the number of atoms) of
each element in the reactants which are consumed is equal to the mass (and
the number of atoms) of that element in the products which are produced.
However, the number of moles of products formed is not necessarily the
same as the number of moles of products used up.

   Many of the reactors which we shall be considering are constant flow
reactors with an appreciable hold up. Applying the principle of conservation
of mass to flow systems, it follows that the total mass of material leaving any
section of a plant (such as a reactor) in a given time should be equal to the
mass entering minus any accumulation which has taken place within that
section. The same is true (with the exception of nuclear reactions) for any of
the elements constituting the chemical compounds entering and leaving. The
chemical compounds themselves, however, may be either generated or used
up within the reactor section of a plant.

   The total mass and the mass of the elements constituting the compounds
thus follow the equation

$$\text{Output} = \text{Input} - \text{Accumulation} \tag{2.14}$$

The masses (and the number of moles) of the chemical compounds entering
and leaving the section of a plant will follow a different relationship

$$\text{Output} = \text{Input} - \text{Accumulation} + \text{Generation} - \text{Consumption} \tag{2.15}$$

For a reactant, the generation term is zero, while for a product the consumption term is zero.

   In the steady state (as in a continuous flow reactor for which the inlet
and outlet conditions have been maintained constant over a long period) no
accumulation of material takes place and therefore, under these conditions,
equations (2.14) and (2.15) reduce to (2.16) and (2.17) respectively.

Total mass balance and elemental masses in steady state:

$$\text{Output} = \text{Input} \tag{2.16}$$

Individual component mass balances in steady state:

$$\text{Output} = \text{Input} + \text{Generation} - \text{Consumption} \tag{2.17}$$

As an example, consider the reaction

$$2CO + O_2 \rightarrow 2CO_2$$

and suppose that this reaction takes place in a continuous flow reactor in the
steady state. An idealised situation is considered where the CO and O_2 are
fed in stoichiometric proportions (2 moles of CO for every mole of O_2) and

the reaction proceeds to completion. The relative molecular mass of CO is 28 while that of O_2 is 32 and that of CO_2 is 44. The following calculations are based on an input of 2 moles of CO.

(a) Consider the **masses** involved in the reaction:

The total mass of reactants entering
$= (2 \times 28) + (1 \times 32) = 56 + 32 = 88\,g$
The total mass of products leaving $= (2 \times 44) = 88\,g$
That is:
The total mass of reactants = The total mass of products

(b) Consider the **elements** involved in the reaction:

The C in the reactants $= 2$ g atoms $=$ the C in the product $= 2$ g atoms
The O in the reactants $= 4$ g atoms (2 from CO and 2 from O_2) $=$ the O_2 in the product $= 4$ g atoms
That is:
The number of atoms of each element in the reactants
= the number of atoms of that element in the product

(c) However, if the **compounds** in the reaction are considered:

3 moles of reactants (2 moles of CO and 1 mole of O_2) produce only 2 moles of the product
Therefore, there is a decrease in the number of moles of compounds after the reaction. This is because 3 molecules of fairly simple diatomic substances react to form 2 triatomic molecules.

It is also instructive to examine the material balance equation (2.15) as applied to each component in this reaction.

CO balance:

Input $= 2$ moles, generation and output $= 0$ moles, consumption $= 2$ moles

Output = Input + Generation − Consumption − Accumulation
$0 = 2 + 0 − 2 − 0$

O_2 balance:

Input $= 1$ mole, generation and output $= 0$ moles, consumption $= 1$ mole

Output = Input + Generation − Consumption − Accumulation
$0 = 1 + 0 − 1 − 0$

CO_2 balance:

Input and consumption $= 0$ moles, generation $= 2$ moles, output $= 2$ moles.

Output = Input + Generation − Consumption − Accumulation
$2 = 0 + 2 − 0 − 0$

2.1.4.4   Collected definitions of further terms used in chemical calculations: limiting reactant, conversion and yield.

(a) Stoichiometric proportions

When the reactants are present in the same relative amounts as those in which they occur in the reaction equation, they are said to be in stoichiometric proportions. For example, in the reaction $SO_2 + \frac{1}{2}O_2 \rightleftharpoons SO_3$, the ratio between the moles of sulphur dioxide and the moles of oxygen is (1.0 / 0.5) = 2.0. Therefore, the stoichiometric proportion of SO_2 to O_2 in the above reaction is 2 to 1.

(b) Limiting reactant

This is the reactant which would be used up first if the reaction proceeded sufficiently. Consider a mixture initially containing 1 mole of SO_2 and 1 mole of O_2 undergoing the reaction $SO_2 + \frac{1}{2}O_2 \rightleftharpoons SO_3$. Sulphur dioxide will be the limiting reactant, because even if it is all reacted some O_2 would still remain unreacted.

(c) Excess reactant

Reactants other than the limiting reactant are termed excess reactants: some of these would remain even if all the limiting reactant were consumed. In the example of the reaction between 1 mole of SO_2 and 1 mole of O_2, oxygen is the excess reactant because 0.5 mole of oxygen would be left behind unreacted even if all the SO_2 were converted to SO_3.

$$\% \text{ Excess} = \frac{\text{moles present} - \text{moles required for complete reaction}}{\text{moles required for complete reaction}} \times 100$$

(2.18)

So, in the example considered,

$$\% \text{ Excess oxygen} = \frac{1.0 - 0.5}{0.5} \times 100 = 100\%$$

(d) Conversion

This is the fraction (or the percentage) of the limiting reactant that is used up in any way during the reaction. Thus,

$$\% \text{ Conversion} = \frac{\text{moles of the limiting reactant used up}}{\text{total moles of the limiting reactant in the feed}} \times 100$$

(*Note*: The words 'in any way' cover the case where by-products are formed as well as the required products.)

   If the limiting reactant is designated reactant A, the fractional conversion (X_A) is given by,

$$X_A = \frac{\text{moles of limiting reactant } (A) \text{ used up}}{\text{total moles of limiting reactant in 'feed'}}$$

(2.19)

The conversion of the limiting reactant is a useful single measure of the progress of a reaction which will be used extensively in later sections. When this parameter is specified, the numbers of moles of all components in a batch reactor, for example, can be calculated from the numbers fed in before the reaction started and the stoichiometry of the reaction (in this case the 'feed' is the material fed to the reactor before initiation of the reaction (i.e. when $X_A = 0$)). Likewise the number of moles of all components leaving a given part of a *steady-state flow reactor system* over a given time interval can be calculated from the numbers of moles of unreacted feed entering the system over the same time interval if the conversion of the material leaving the reactor section is known. The equations developed in the two cases are formally identical. When deriving these equations the reaction equation is written in the form

$$(-\nu_A)A + (-\nu_B)B + \ldots \rightarrow (\nu_C)C + (\nu_D)D + \ldots \qquad (2.20)$$

where $A, B, \ldots$ are reactants and $C, D, \ldots$ are products. (The convention is followed, as in equation (2.6) of counting the reactant stoichiometric coefficients as negative numbers.) A is here the limiting reactant and X_A *is its fractional conversion.*

The case of the batch reactor will be developed first. n_{A0} moles of A are present before the reaction starts (i.e. when $X_A = 0$). Also present initially are n_{B0} moles of B and n_I moles of inerts. When the conversion of A has risen to X_A, the moles of A, B, C, D, and I will be n_A, n_B, n_C, n_D and n_I respectively. From the definition of fractional conversion given above,

$$\text{Moles of } A \text{ used up} = (n_{A0} X_A)$$

From the stoichiometry of the reaction,

$$\text{Moles of } B \text{ used up} = n_{A0} X_A \left(\frac{-\nu_B}{-\nu_A}\right) = \left(\frac{\nu_B}{\nu_A}\right) n_{A0} X_A$$

$$\text{Moles of } C \text{ formed} = \left(\frac{\nu_C}{-\nu_A}\right) n_{A0} X_A$$

$$\text{Moles of } D \text{ formed} = \left(\frac{\nu_D}{-\nu_A}\right) n_{A0} X_A$$

$$n_A = (\text{initial moles} - \text{moles used up}) = n_{A0}(1 - X_A) \qquad (2.21)$$

$$n_B = (\text{initial moles} - \text{moles used up}) = n_{B0} - \left(\frac{\nu_B}{\nu_A}\right) n_{A0} X_A \qquad (2.22)$$

$$n_C = (\text{initial moles} + \text{moles formed}) = n_{C0} - \left(\frac{\nu_C}{\nu_A}\right) n_{A0} X_A \qquad (2.23)$$

$$n_D = (\text{initial moles} + \text{moles formed}) = n_{D0} - \left(\frac{\nu_D}{\nu_A}\right) n_{A0} X_A \qquad (2.24)$$

$$n_I = n_{I0} \qquad (2.25)$$

n_{C0} and n_{D0} will normally be zero, i.e. no products will be present in the feed to the reactor.

In general for a component i,

$$n_i = n_{i0} - \left(\frac{\nu_i}{\nu_A}\right) n_{A0} X_A \tag{2.26}$$

where $n_{i0} = 0$ for products. The stoichiometric coefficient for an inert is zero.

The case of the steady-state flow system will now be considered. It will be supposed that $\underline{n}_A, \underline{n}_B, \ldots \underline{n}_C, \underline{n}_D, \ldots, \underline{n}_I, \ldots, \underline{n}_i$ are now the numbers of moles of components $A, B, \ldots C, D, \ldots, I, \ldots, i$ leaving such a system during a period when $\underline{n}_{A0}, \underline{n}_{B0}, \ldots \underline{n}_{C0}, \underline{n}_{D0}, \ldots \underline{n}_{I0}, \ldots \underline{n}_{i0}$ moles of these components enter.

Applying the steady-state mass balance equation (2.17) to component A:

Moles of A leaving $(\underline{n}_A)$ = moles of A entering $(\underline{n}_{A0})$ − moles of A used up in reactor

or, from equation (2.19):

$$\underline{n}_A = \underline{n}_{A0} - \underline{n}_{A0} X_A = \underline{n}_{A0}(1 - X_A) \tag{2.21a}$$

Likewise, for reactant B:

Moles of B leaving $(\underline{n}_B)$ = moles of B entering $(\underline{n}_{B0})$
− moles of B used up in reactor

From the stoichiometry of the reaction (equation (2.20)):

(Moles of B used up) = (moles of A used up) × $\left(\dfrac{\nu_B}{\nu_A}\right)$

So,

$$\underline{n}_B = \underline{n}_{B0} - \left(\frac{\nu_B}{\nu_A}\right) \underline{n}_{A0} X_A \tag{2.22a}$$

$$\underline{n}_C = \underline{n}_{C0} - \left(\frac{\nu_C}{\nu_A}\right) \underline{n}_{A0} X_A \tag{2.23a}$$

$$\underline{n}_D = \underline{n}_{D0} - \left(\frac{\nu_D}{\nu_A}\right) \underline{n}_{A0} X_A \tag{2.24a}$$

$$\underline{n}_I = \underline{n}_{I0} \tag{2.25a}$$

and in general for a component i:

$$\underline{n}_i = \underline{n}_{i0} - \left(\frac{\nu_i}{\nu_A}\right) \underline{n}_{A0} X_A \tag{2.26a}$$

Equations (2.21a) to (2.26a) are formally identical to (2.21) to (2.26). The only differences are that the moles initially fed to the reactor are replaced by the moles entering the reactor in a given time interval and that the moles of each component present in the reactor when the conversion is X_A are replaced by the moles of each component leaving the reactor in the given time interval when the conversion is X_A.

(e) Yield
Two different definitions for yield may be found in the literature; one is based on the quantity of the reactants which react and the other is based on the total amount of reactants in the feed. These definitions are:

$$\text{(i) } \% \text{ Yield} = \frac{\text{moles of desired product actually produced}}{\substack{\text{moles of desired product that could have been produced} \\ \text{for the given conversion if side reactions were absent}}} \times 100$$

$$(2.27)$$

$$\text{(ii) } \% \text{ Yield} = \frac{\text{moles of desired product actually produced}}{\substack{\text{moles of the desired product that could have been produced} \\ \text{for 100\% conversion if side reactions were absent}}}$$

$$\times 100 \qquad (2.28)$$

Example 2.1   A substance A reacts with another substance B to produce the desired compound C by the reaction $2A + B \rightarrow C$. In addition, the substance A can also decompose to give two other substances D and E by the side-reaction, $A \rightarrow D + E$. A feed containing 50 moles of A and 50 moles of B is reacted to produce a products stream which contains the following: $A = 4$ moles, $B = 32$ moles, $C = 18$ moles, $D = 10$ moles, $E = 10$ moles.
(a) What is the conversion?
(b) What is the yield by definition (i)?
(c) What is the yield by definition (ii)?

Solution
(a) Conversion
Since conversion is always based on the limiting reactant, this should be identified first. The reaction equations are:

$$2A + B \rightarrow C \qquad \qquad \text{①}$$
$$A \rightarrow D + E \qquad \qquad \text{②}$$

The substance A must be the limiting reactant because, even if all of the substance A were reacted in the main reaction ①, about 25 moles of substance B would still remain unreacted. By defunction,

$$\% \text{ Conversion} = \frac{\text{moles of the limiting reactant used up}}{\text{total moles of the limiting reactant in the feed}} \times 100$$

The amount of limiting reactant present in the feed = 50 moles and the amount of limiting reactant remaining in the products stream = 4 moles.

$$\therefore \text{ Moles of limiting reactant used up} = (50 - 4) = 46 \text{ moles}$$
$$\therefore \text{ \% Conversion} = (46/50) \times 100 = 92\%$$

(b) Yield by definition (i)

$$\% \text{ Yield} = \frac{\text{moles of desired product actually produced}}{\substack{\text{moles of desired product that could have been produced} \\ \text{for the given conversion if side reactions were absent}}} \times 100$$

The moles of the desired product actually produced = 18 moles, the amount of limiting reactant present in the feed = 50 moles, and the amount of limiting reactant remaining in the products stream = 4 moles.

$$\therefore \text{ Moles of limiting reactants used up} = (50 - 4) = 46 \text{ moles}$$

If all the 46 moles of A reacted to produce C:

The amount of C that would have been produced is $(46/2) = 23$ moles

$$\therefore \text{ \% Yield (by definition (i))} = (18/23) \times 100 = 78.3\%$$

(c) Yield by definition (ii)

$$\% \text{ Yield} = \frac{\text{moles of desired product actually produced}}{\substack{\text{moles of the desired product that could have been} \\ \text{produced for 100\% conversion if side reactions are absent}}} \times 100$$

The moles of the desired product actually produced = 18 moles and the amount of limiting reactant present in the feed = 50 moles.

$$\therefore \text{ Maximum amount of } C \text{ that could have been produced by reaction } ①$$
$$= (50/2) = 25 \text{ moles (because 2 moles of } A \text{ give 1 mole of } C)$$

$$\therefore \text{Yield (by definition (ii))} = (18/25) \times 100 = 72\%$$

Alternatively, use may be made of the result

$$\% \text{ Conversion} = \frac{\text{yield by definition (ii)}}{\text{yield by definition (i)}} \times 100$$

giving (as before),

$$\text{Yield (by definition (ii))} = \{(92 \times 78.3/100)\} = 72\%$$

(The above result follows at once from the definitions of the yield and conversion (equations (2.19), (2.27), (2.28)) when it is remembered that the moles of desired product produced under specific conditions is proportional to the moles of the limiting reactant used up and that all the limiting reactant would be used up for 100% conversion. Under these conditions, the constant of proportionality is given by the reaction stoichiometry.)

2.2. First stages in reactor design: material balance calculations

One of the first stages in reactor system design is normally to carry out preliminary material and energy balances around the reactor or reactors involved, bearing mind the overall plant requirements and, where appropriate, the equilibrium characteristics of the reactions taking place. At this stage reasonable assumptions are made about the approach to equilibrium and a flow diagram is produced. Preliminary sizing calculations involving consideration of the kinetics of the reactions usually follow later and may well lead to a rethink of the flow diagram produced in the first stage. The present chapter is concerned solely with material balances and chemical equilibrium. Kinetics and reactor sizing are the subjects of Chapters 3 and 4.

When carrying out material balance calculations, a systematic approach to the problem is necessary. The following **three** simple steps may be followed,

1. **Draw a simple flow diagram of the process:** *This should show the flows in and out, and the accumulation (if any). Boxes or circles are adequate at this stage to indicate equipment with joining lines for the flow streams.*
2. **List all the relevant information:** *List the flow rates, compositions, temperatures, pressures, chemical reactions and physical changes. This information can be shown on the above flow diagram.*
3. **Choose a basis for calculation:** *The basis is a convenient reference chosen by you for the calculation and it may be either a quantity of material entering or leaving, or a certain period of operation. It is very important that the basis is stated clearly in all calculations.*

The material balance equations may now be used to find the unknown flows and compositions. Full understanding of the application of the above steps can only be obtained by practice. To explain the principles some examples of material balance and stoichiometry are given below. The word 'stoichiometry' incidentally was derived from the two Greek words *stoikeon* (basic constituent) and *metrein* (to measure), and it deals with the combining masses of elements and compounds.

2.2.1   Simple reactor systems with no 'split-flow' features

Example 2.1   In the contact process for the manufacture of sulphuric acid, molten sulphur is burnt with air in a sulphur burner to produce SO_2 which is then oxidised in a catalytic reactor to SO_3. If the sulphur is burnt at the rate of 20 kg per minute, how much SO_3 is produced and what is the minimum air requirement? (The relative atomic masses are S = 32, O = 16). Air can be taken to contain 21 mol% O_2 and 79 mol% N_2 and its relative molecular mass is 29).

Solution
In order to obtain good conversion of SO_2 to SO_3 substantial excess air is used in practice. With careful design of the catalytic converter it is known that conversions as high as 98% can then be achieved. In order to obtain a preliminary estimate of the amount of SO_3 produced by the given sulphur input, conversion of the S to SO_2 and of the SO_2 to SO_3 will be regarded as complete.

$$S + O_2 \rightarrow SO_2$$

$$SO_2 + \frac{1}{2}O_2 \rightarrow SO_3$$

Although substantial excess air would be required in practice to achieve the desired conversion, it is convenient to calculate in the first instance the air requirement in a hypothetical process in which air and sulphur were fed in stoichiometric amounts (see Section 2.1.4.4) and in which, nevertheless, the conversion to SO_3 was virtually complete. The air requirement calculated in this way is called the minimum air requirement. In this hypothetical process the excess air is zero.

Basis: 20 kg of sulphur to burner

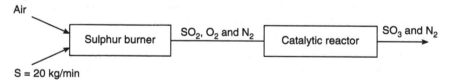

Diagram 2.1

The reactions are:

$$S + O_2 \rightarrow SO_2 \quad \text{and} \quad SO_2 + \frac{1}{2}O_2 \rightarrow SO_3$$

It is best to use the sulphur balance for this calculation because 1 mole of sulphur gives 1 mole of SO_2 which in turn gives 1 mole of SO_3. The 20 kg of sulphur in the feed = (20/32) = 0.625 kmol. This will produce the same number of kmol of SO_2, which in turn will produce the same number of kmol of SO_3. Thus, the amount of SO_3 produced should be equal to 0.625 kmol $= [0.625 \times \{32 + (16 \times 3)\}] = 50$ kg of SO_3 The air provides the necessary O_2 to form SO_2 and to oxidise it to SO_3. The reaction equations suggest that for every kmol of sulphur at least 1.5 kmol of O_2 would be required (i.e. 1 kmol to form SO_2 and $\frac{1}{2}$ kmol to form SO_3). Therefore, Minimum oxygen required $= 1.5 \times 0.625 = 0.937$ kmol.

This corresponds to 21 mol% (mol% and vol% are the same for **perfect gases**) of the minimum air required. So,

$$\text{Minimum air required} = 0.937/0.21 = 4.46 \text{ kmol or } 129.4 \text{ kg}$$

Example 2.2 Synthetic ammonia may be produced by the reaction $N_2 + 3H_2 \rightleftharpoons 2NH_3$. The reaction is reversible and it does not approach completion. If the mole ratio of hydrogen to nitrogen entering the reactor is 4 to 1 and the mole ratio of the hydrogen to nitrogen leaving the reactor is 4.25 to 1, how many ton moles per day of gases must enter the reactor to produce 300 tons per day of ammonia?

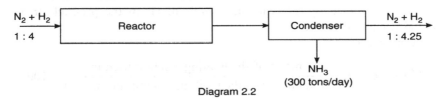

Diagram 2.2

Note: Unused reactants are recycled (see Section 2.2.2.3 and Chapter 5).

Solution
Basis: 1 mole of nitrogen entering the reactor

The forward reaction is $N_2 + 3H_2 \rightarrow 2NH_3$

$\therefore$ Amount of hydrogen entering the reactor $= 4$ moles

Let the ammonia formed $= \mathbf{A}$ moles, the amount of N_2 gas leaving the reactor $= \mathbf{N}$ moles and the amount H_2 gas leaving the reactor $= \mathbf{H}$ moles.

The mole ratio of H_2 gas to N_2 gas leaving the reactor is given as 4.25 to 1. Therefore,

$$\text{H/N} = 4.25.$$

Apply, N_2 balance (in moles). Since there is no accumulation of N_2 in the reactor, input = output.

N_2 entering reactor $= N_2$ leaving the reactor $+ N_2$ used up in forming ammonia

$\qquad 1 = \qquad\quad N \qquad\quad + \qquad\qquad A/2$

Similarly, a H_2 balance for the reaction should give:

H_2 entering reactor = H_2 leaving the reactor + H_2 used up in forming ammonia

$$4 = \qquad H \qquad + \qquad 3A/2$$

The nitrogen and hydrogen balances, and the ratio $H/N = 4.25$, give three equations for the three unknowns A, N and H. Solving the equations simultaneously gives $A = 0.4$, $N = 0.8$ and $H = 3.4$.

So, for 1 mole of nitrogen entering the reactor, 0.4 mole of ammonia is actually produced.

The required ammonia production $= (300/17) = 17.7$ ton moles day^{-1}. Therefore, the amounts of nitrogen and hydrogen required are:

$$\text{Nitrogen} \ = (1/0.4) \times 17.7 = 44 \text{ ton moles day}^{-1}$$

$$\text{Hydrogen} = (4/0.4) \times 17.7 = 177 \text{ ton moles day}^{-1}$$

$$\text{Total} \qquad = 221 \text{ ton moles day}^{-1}$$

If required, the conversion and yield for the reaction can also be calculated. The limiting reactant in this particular case is nitrogen, because even if all the N_2 were consumed by the reaction, 1 mole of H_2 in the product stream would remain unreacted.

(a) **Conversion**

$$\% \text{ Conversion} = \frac{\text{moles of the limiting reactant used up}}{\text{total moles of the limiting reactant in the feed}} \times 100$$

that is,

$$\% \text{ Conversion} = \{(1 - 0.8)/1\} \times 100 = 20\%$$

(b) **Yield (by definition (i))**

$$\% \text{ Yield} = \frac{\text{moles of desired product actually produced}}{\begin{array}{c}\text{moles of desired product that could have been produced}\\ \text{for the given conversion if side-reactions were absent}\end{array}} \times 100$$

that is,

$$\% \text{ Yield} = (0.4/0.4) \times 100 = 100\%$$

(c) **Yield (by definition (ii))**

$$\% \text{ Yield} = \frac{\text{moles of desired product actually produced}}{\begin{array}{c}\text{moles of desired product that could have been produced}\\ \text{for 100\% conversion if side-reactions were absent}\end{array}} \times 100$$

that is,

$$\% \text{ Yield} = \{0.4/(2 \times 1)\} \times 100 = 20\%$$

Example 2.3   Acetone can be produced by the combined oxidation and dehydrogenation reactions of isopropanol. The reactions forming acetone are as follows:

$$(CH_3)_2CHOH + \tfrac{1}{2}O_2 \rightarrow CH_3COCH_3 + H_2O \qquad ①$$
$$(CH_3)_2CHOH \rightarrow CH_3COCH_3 + H_2 \qquad ②$$

The isopropanol can also be completely oxidised to CO_2 and H_2O as shown below:

$$(CH_3)_2CHOH + 4\tfrac{1}{2}O_2 \rightarrow 3CO_2 + 4H_2O \qquad ③$$

The ratio of air to isopropanol vapour in the mixture entering the reactor is 2 moles of air to 1 mole of isopropanol, the conversion of isopropanol in the reactor is 90% and the yield of acetone is 98%, based on the isopropanol which reacts (by definition (i)). In the reactor, equal amounts of acetone are formed by the oxidation and dehydrogenation reactions. The gases leaving the reactor are cooled and scrubbed with water, which removes the acetone and the unreacted isopropanol. The acetone and isopropanol are separated from the solution by distillation and *the isopropanol is re-used as feed to the reactor*. Calculate:

(a) the composition (in mol%) of the gases leaving the reactor
(b) the amount of isopropanol required to produce 1000 kg of acetone.

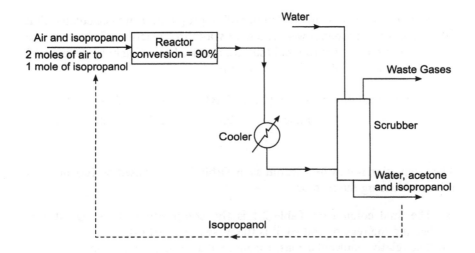

Figure 2.5 Schematic flow diagram of plant for production of acetone.

Solution
Basis: 100 moles of the mixture of air and isopropanol entering the reactor
On this basis, the amount of isopropanol entering the reactor $= \frac{1}{3} \times 100$ $= 33.3$ moles. The conversion of isopropanol is 90%, so the amount of isopropanol reacted *by all three reactions* is:

$$0.9 \times 33.3 = 30.0 \text{ moles}$$

Yield of acetone (by definition (i)) is given as 98%.
   Acetone is produced only by the first two reactions and, by each of these reactions, 1 mole of acetone is produced from 1 mole of isopropanol. Therefore, the maximum amount of acetone that could be produced from 30.0 moles of isopropanol $= 30.0$ moles. Hence

$$\text{Amount of acetone actually produced} = 30 \times 0.98 = 29.4 \text{ moles}$$

Equal amounts of acetone are formed by the oxidation and dehydrogenation reactions, so the amount of acetone formed by each of the two reactions is $(29.4/2) = 14.7$ moles.

Oxidation reaction : $\quad (CH_3)_2CHOH + \frac{1}{2}O_2 \rightarrow CH_3COCH_3 + H_2O$ ①
$\qquad\qquad$ moles : $\qquad\qquad$ 14.7 $\qquad$ 7.35 $\qquad\quad$ 14.7 $\qquad\quad$ 14.7

Dehydrogenation reaction : $\quad (CH_3)_2CHOH \rightarrow CH_3COCH_3 + H_2$ ②
$\qquad\qquad\qquad$ moles : $\qquad\qquad$ 14.7 $\qquad\qquad$ 4.7 $\qquad$ 14.7

These two reactions use up 29.4 moles of isopropanol. It was calculated that 30.0 moles of isopropanol were reacted by all three reactions. Therefore, the number of moles of isopropanol that are oxidised completely by reaction ③ should be equal to $(30 - 29.4) = 0.6$ moles.

Complete oxidation reaction : $(CH_3)_2CHOH + 4\frac{1}{2}O_2 \rightarrow 3CO_2 + 4H_2O$
$\qquad\qquad\qquad$ moles: $\qquad\quad$ 0.6 $\qquad\qquad$ 2.7 $\qquad$ 1.8 $\quad$ 2.4

③

The calculation is now best set out as in Table 2.1, still based on 100 moles of mixture entering the reactor.

(a) The final column on Table 2.1 is the composition of the gas stream leaving the reactor and so, it is the answer to part (a).
(b) The relative molecular mass (or molecular weight) of acetone is 58, so

$$1000 \text{ kg of acetone} = (1000/58) = 17.25 \text{ kmol}$$

Table 2.1

Component	Entering the reactor	Change in reaction	Products moles	Products mol %
Isopropanol	33.30	−30.0	30	2.7
Oxygen	14.0 (66.7 × (21/100))	−7.35 − 2.7 = −10.05	3.95	3.2
Nitrogen	52.7 (66.7 × (79/100))		52.70	42.8
Acetone	0	+14.7 + 14.7 = 29.4	29.40	23.9
Water	0	+14.7 + 2.4 = 17.1	17.1	13.9
Hydrogen	0	+14.7	14.70	12.0
Carbon dioxide	0	+1.8	1.80	1.5
Totals	100		122.95	100.0

From Table 2.1, 29.4 moles of acetone are produced from a feed of 33.30 moles of isopropanol. But, the 3.33 moles of isopropanol in the outlet stream are **recovered and re-used in the feed**. So,

Amount of fresh isopropanol required $= (33.30 − 3.30) = 30.0$ moles

Therefore,

Amount of isopropanol required to produce 17.25 kmol of acetone
$$= 17.25 \times (30.0/29.4) = 17.6 \, \text{kmol}$$

The relative molecular mass of isopropanol $= 60$. Therefore,

$(17.6 \times 60) = 1056 \, \text{kg}$ of isopropanol is required to produce
1000 kg of acetone

2.2.1.1   Some problems.
1. Chemical engineering principles can be applied in biochemical and minerals engineering processes, as well as in the more traditional chemical industries. This problem illustrates the use of mass balance calculations in (a) a biochemical engineering process and (b) a minerals engineering process
(a) Ethyl alcohol (C_2H_5OH) may be produced by the fermentation of carbohydrates using yeast. The carbohydrates are present in the feedstock as disaccharides with a typical formula $C_{12}H_{22}O_{11}$. The amount of carbohydrate present in the feedstock is 91 mass% and the rest is waste material. This feedstock is mixed with water before yeast is added for fermentation. 61 mass% of the carbohydrate is fermented by yeast to produce ethyl alcohol and carbon dioxide by the fermentation reaction

$$C_{12}H_{22}O_{11} + H_2O \rightarrow 4CO_2 + 4C_2H_5OH.$$

The mass ratio of ethyl alcohol to water at the end of fermentation is 1 to 9. The volume occupied by 1 kmol of a perfect gas at 0°C and 760 mm Hg is 22.4 m^3.

(i) How much feedstock and water are required in the initial mixture to produce 100 kg of ethyl alcohol?

(*Ans.* Feedstock = 335 kg, Water = 910 kg)

(ii) If all the carbon dioxide produced were removed as a gas at 30°C and 750 mm Hg pressure, what volume would the CO$_2$ occupy (perfect gas behaviour may be assumed)?   (*Ans.* 55 m^3)

(b) Copper is produced from copper concentrate (a mineral which may be represented by the formula CuFeS$_2$) by reaction with oxygen in a furnace. The reactions in the smelting zone of the furnace are:

$$2CuFeS_2 \rightarrow Cu_2S + 2FeS + S \quad \text{and} \quad S + O_2 \rightarrow SO_2$$

and the reactions in the converting zone of the furnace are:

$$Cu_2S + O_2 \rightarrow 2Cu + SO_2 \quad \text{and} \quad 2FeS + 3O_2 \rightarrow 2FeO + 2SO_2$$

(i) How many kmols of air are theoretically required for the complete reaction of 100 kmols of copper concentrate (air contains 21 volume % Oxygen and 79 volume % nitrogen)?   (*Ans.* 1190 kmol)

(ii) If twice the theoretical quantity of air is used, what is the composition of the gas leaving the furnace if all the copper concentrate reacts?

(*Ans.* O$_2$ = 10.7 mol%, N$_2$ = 80.7 mol% and SO$_2$ = 8.6 mol%)

2. Methanal (CH$_2$O: usually called formaldehyde in industry) is produced by a combination of oxidation and dehydrogenation of methanol over silver gauze catalyst. The reactions are:

$$CH_3OH + \tfrac{1}{2}O_2 \rightarrow CH_2O + H_2O \qquad \text{①}$$
$$CH_3OH \rightarrow CH_2O + H_2 \qquad \text{②}$$

The methanol can also be completely oxidised by the following reaction,

$$CH_3OH + 1\tfrac{1}{2}O_2 \rightarrow CO_2 + 2H_2O \qquad \text{③}$$

The feed to the reactor is an air + methanol mixture, containing 40 mol% methanol. After the reaction products leave the reactor the formaldehyde and any unreacted methanol are removed by condensation and scrubbing with water. The gases remaining are vented to the atmosphere, and the plant can be controlled by analysis of these gases. After removal of all the water from a sample of these gases the analysis was hydrogen = 22.1 mol%; nitrogen = 74.2 mol%; oxygen = 0.1 mol%; carbon dioxide = 3.6 mol%. Calculate:

(a) The percentages of the methanol entering which are reacted by each of the three reactions and which remain unreacted. Air can be taken to be 21 mol% oxygen and 79 mol% nitrogen.

(*Ans.* Reacted by ① = 45.4% of the feed; reacted by ② = 35.3%; reacted by ③ = 5.8% and unreacted = 13.5%)
(b) The conversion of methanol   (*Ans.* 86.5%)
(c) The yield of formaldehyde, based on the methanol which reacts
(*Ans.* 93.4%)
(d) The cost of producing formaldehyde (per 100 kg), if methanol costs £10.00 per 100 kg and the cost of operating the plant is £11.00 per 100 kg of formaldehyde produced. The unreacted methanol leaves with the formaldehyde in solution (it helps to prevent decomposition of the formaldehyde) and so it is not re-used as feed. The relative molecular masses of methanol = 32 and formaldehyde = 30.
(*Ans.* £24.21 per 100 kg)

2.2.2 Material balances for reactor systems with by-pass, recycle and purge

2.2.2.1 Systems with by-pass: 'cold shot' cooling.

A by-pass stream is one which skips one or two stages in a process and (sometimes after heat exchange or other treatment) re-enters the main stream at a later stage. A by-pass is used, for example, in 'cold shot' cooling. This is a method of cooling the reactant/product mixture leaving a catalyst bed in which a mildly exothermic reaction has occurred by injecting cold feed into it before the mixture enters the next catalytic bed (i.e. it is a method of 'interstage cooling'). This cooling method eliminates the need for expensive interstage heat transfer equipment (Figure 2.6). An obvious disadvantage

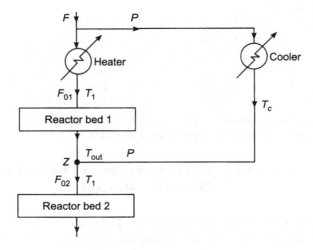

Figure 2.6 'Cold shot' cooling.

is the loss in reactor conversion resulting from the injection of fresh feed and for this reason it is of limited applicability. For the method to work well it is necessary for the required temperature reduction ($T_{out} - T_1$ in Figure 2.6) to be substantially less than the excess of the bed outlet temperature over the temperature of the injected feed {i.e.$(T_{out} - T_1) \ll (T_{out} - T_c)$}

Example 2.4 The feasibility is being examined of carrying out an exothermic reaction between reactants A and B in a series of catalyst beds with 'cold shot' intercooling between the beds. The general layout envisaged is as shown in Figure 2.6. The intercooling is necessary to retain the beds of catalyst within the recommended operating temperature range (Section 2.4.3.4) and to maintain favourable conversion. The inlet to the first bed is at 300°C and reactant A is the 'limiting reactant'. The feed mixture consists of 20 mol% of A and 80 mol% of B and the total flow rate at F in Figure 2.6 is 5000 kmol/h. The proposed bed depth is such that 65% of the component A entering the first bed is converted within the bed. Before passing the outlet stream from bed 1 to the next bed, it is proposed to return the stream temperature to 300°C by injecting fresh feed cooled to 40°C. By taking a heat balance and taking account of the known heat of reaction and the heat capacity data for the system it has been estimated that it would be necessary to by-pass 15% of the initial feed to obtain the necessary degree of cooling. Calculate the conversion of component A as it enters the second reactor bed and also the flow rate of this component.

Solution
Basis = 1 hour operation
Total feed (F) entering the reactor system = 5000 kmol
Reactant A in the feed (F) entering = $0.20 \times 5000 = 1000$ kmol
Feed (P) flowing around the by-pass = $0.15 \times 5000 = 750$ kmol
Feed (F_{01}) entering the first reactor bed = $0.85 \times 5000 = 4250$ kmol
Reactant A entering first reactor bed = $0.20 \times 4250 = 850$ kmol
Reactant A leaving first reactor bed = $0.35 \times 850 = 297.5$ kmol
Reactant A flowing around by-pass = $0.20 \times 750 = 150$ kmol
Taking material balance on component A at point Z

Reactant A entering bed 2 = kmol of A from reactor 1 + kmol of A in by-pass
= $297.5 + 150 = 447.5$ kmol

Conversion of A as it enters second bed = $[(1000 - 447.5)/1000] = 55.3\%$
(Had it not been for the cold shot cooling this would have been 65%)
Flow rate of A entering second reactor bed = 447.5 kmol h^{-1}.

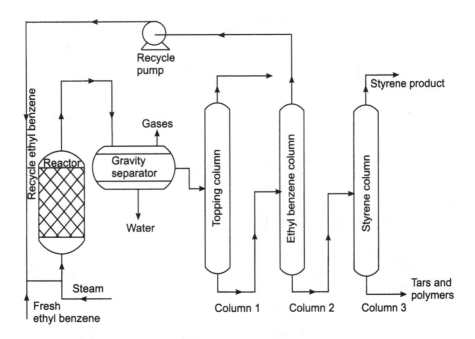

Figure 2.7 Schematic diagram of plant for production of styrene.

2.2.2.2 Systems with recycle. A recycle stream is one which returns some (or most) of the material leaving a unit (such as a reactor with separator) to the inlet side of the unit for further processing. For example, in the manufacture of styrene from ethyl benzene (Figure 2.7) the unreacted ethyl benzene is largely recovered as top product from column 2 and is returned to join the fresh ethyl benzene feed to the reactor.

Example 2.5 Methanol is produced from carbon monoxide and hydrogen by the reaction $CO + 2H_2 \rightleftharpoons CH_3OH$. The feed to the reactor consists of hydrogen and carbon monoxide with the stoichiometric ratio of H_2 to CO (i.e. 2 to 1, in moles). Part only of the reaction mixture entering the reactor reacts as it passes through. The remaining reactants are recycled after removal by condensation of the methanol produced by the reaction (Figure 2.8). The reactor/recycle system shown may be taken to be operating at steady-state conditions and the composition of the recycle stream is the same as that of the feed. The reaction conditions are such that only 15% of the CO entering the reactor reacts in a single pass. Calculate the ratio of the moles of recycle to the moles of fresh feed when the plant is running continuously.

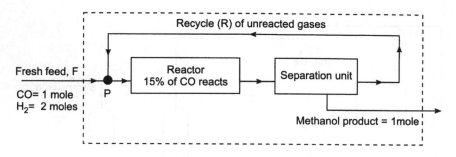

Figure 2.8 System with recycle.

Solution

Basis: 1 mole CO in the fresh feed

Moles of H_2 in the fresh feed = 2 moles

Moles of methanol product leaving = 1 mole

Let R be the total number of moles of CO and H_2 in the recycle. Since the ratio of the moles of H_2 to the moles of CO in the recycle stream is the same as that in the fresh feed (i.e. 2 to 1),

$$\text{Moles of CO in the recycle} = \{[1/(1+2)] \times R\} = (R/3) \text{ moles}$$

Overall CO balance for reactor:

$$\text{Moles of CO entering the reactor} = \text{CO in fresh feed} + \text{CO in the recycle}$$
$$= [1 + (R/3)] \text{ moles}$$

From conversion given:

$$\text{Amount of CO reacted in the reactor} = (15/100) \times [1 + (R/3)] \text{ moles}$$

This amount of CO produces $\{0.15[1 + (R/3)]\}$ moles of CH_3OH. This is the product.

$$\therefore \{0.15[1 + (R/3)]\} = 1$$

So, $R = 17$ moles; therefore:

$$\text{Moles of recycle/moles of feed} = 17/3 = 5.67$$

2.2.2.3 Use of purge streams.   A purge stream is one which is continuously bled off from a system with a recycle to prevent the build-up of unwanted material which is entering the system but for which there is no other significant outlet. The unwanted material usually takes the form of non-reacting impurities (or 'inerts') present in the feed to the reactor, though it can sometimes be other extraneous material added during a process [4]. An example of feed impurities in a reactor/separator system with recycle is provided by the feed (nitrogen + hydrogen) fed to reactors for ammonia synthesis (see Example 2.7). The nitrogen is obtained from air and contains about 1 mol% inert gases (mainly argon). If the hydrogen is produced by reacting natural gas with steam, this will probably result in the presence of other inerts also. If an outlet were not provided for argon and other inerts, their amounts would increase and eventually they would affect the performance of the reactor. Therefore, in practice, some of the recycled material is discharged (i.e. purged) from the recycle stream.

The use of a purge invariably involves waste of some of the feed material and this wastage increases with the 'inert' content of the feed and hence with the purge rate. Purification of feedstock(s) can, where practicable, reduce this form of wastage. A good discussion of waste minimisation in separation and recycle systems is given in reference [4].

Example 2.6 As in Example 2.5, methanol is produced by the reaction $CO + 2H_2 \rightleftharpoons CH_3OH$ in a reactor/separator unit with a recycle. In this case, however, the feed contains 33.3 mol% CO, 66.6 mol% H_2 and 0.1 mol% of inert material and a purge is provided on the recycle line to prevent the build-up of the inerts in the reactor (Figure 2.9). The fresh feed is mixed with the purged recycle and the mixture is fed to the reactor. 15% of the H_2/CO mixture is converted to methanol in a single pass and virtually all the methanol produced (150 kmol h^{-1}) is recovered from the unreacted gases by condensation. The composition of the purge stream has been measured and found to be 33.0 mol% CO and 66.0 mol% H_2 and 1.0 mol% inerts. Calculate the fresh feed and purge rates.

Solution
Basis = 1 hour operation (or 150 kmol of methanol produced)

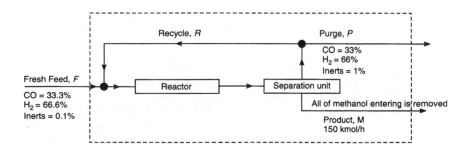

Figure 2.9 Flow diagram of methanol plant with recycle and purge. (The broken envelope lines enclose the region on which overall mass balances are carried out in the example.)

Overall CO balance:

$$CO \text{ in fresh feed} = CO \text{ in purge} + CO \text{ in product}$$
$$0.333F \quad = \quad 0.33P \quad + \quad 150 \qquad \text{①}$$

Inerts balance:

$$\text{Inerts in fresh feed} = \text{Inerts in purge} + \text{Inerts in product}$$
$$0.001F \quad = \quad 0.01P \quad + \quad 0 \qquad \text{②}$$

From which, $F = 500 \text{ kmol h}^{-1}$ and $P = 50 \text{ kmol h}^{-1}$

Example 2.7 Ammonia is produced by the catalytic reaction $N_2 + 3H_2 \rightleftharpoons 2NH_3$. In a reactor with recycle (see Figure 2.10), a fresh feed containing 24.57 mol% N_2, 75.16 mol% H_2 and 0.27 mol% argon is mixed with the recycle (this contains most of the gases that remain unconverted by the reaction) and fed to the reactor. 17% of the N_2 in the mixture entering the reactor is converted to ammonia, virtually all of which is removed from the exit gases by condensation. To prevent the build-up of argon, some of the unreacted gases in the recycle are purged to the atmosphere. The amount of the purge is controlled by maintaining the amount of argon in the exit gases from the separator at 4.1 mol%. Calculate the amount and composition of the purge and the recycle streams. Calculate also the purge to recycle and the recycle to feed ratios.

Solution
Let the mole fraction of hydrogen in the exit gases from the condenser and also in the recycle and purge streams be x, the mole fraction of nitrogen be y and the mole fraction of argon be z. The purge is controlled to give $z = 0.041$. Then,

$$x + y + z = 1.0$$
$$x + y = 1.0 - 0.041 = 0.959 \qquad \text{①}$$

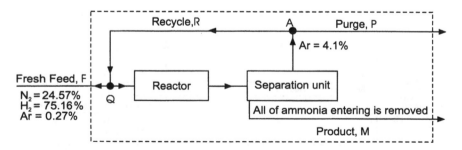

Figure 2.10 Flow diagram of ammonia converter showing recycle and purge streams. (The broken envelope lines enclose the region on which overall mass balances are carried out in the example.)

Basis = 100 kmol of fresh feed (*Note: This is NOT the feed to reactor*)
Overall mass balance: H balance:

H in the fresh feed = H in the product + H in the purge stream $\qquad$②
$$2 \times 75.16 \quad = \quad 3 \times M \quad + \quad 2 \times P \times x$$

N balance:

$$2 \times 24.57 \quad = \quad 1 \times M \quad + \quad 2 \times P \times y \qquad ③$$

Argon balance:

$$1 \times 0.27 \quad = \quad 0 \quad + \quad P \times 0.041 \qquad ④$$

From equation ④:

$$P = 6.585 \text{ kmol}$$

Substituting equations ① and ④ in ② and ④ and solving,

$$M = 46.708 \text{ kmol}; \quad x = 0.774 \text{ and } y = 0.185.$$

Therefore, the kmol of H_2, N_2 and argon leaving in the purge stream for every 100 kmol of fresh feed are:

$$H_2 = 5.097 \text{ kmol}, \quad N_2 = 1.218 \text{ kmol} \quad \text{and} \quad Ar = 0.27 \text{ kmol}$$
$$(= 6.585 \text{ kmol in all})$$

The recycle rate may be calculated by taking a N_2 balance at Q to obtain the reactor feed rate as a function of R and hence, by considering the N_2 conversion in the reactor, to obtain ammonia generation as a function of R. (This is similar to the procedure followed in Example 2.5.)

Basis, as before, is 100 kmol of fresh feed
N_2 feed to reactor $= 24.57 + R \times 0.185$ (from N_2 balance at Q)
N_2 converted $= 0.17(24.57 + R \times 0.185)$ kmol

NH_3 produced $= 2 \times 0.17(24.57 + R \times 0.185) = 8.354 + 0.0629R$ kmol
$$= 46.708 \text{ kmol (previously calculated)}$$

Hence,

$$R = 609.7 \text{ kmol}$$

To summarise:

Fresh feed $=100$ *kmol*; *recycle stream* $= 609.7$ kmol and
purge stream $= 6.315$ kmol

Hence, the recycle and purge ratios based on total flows are:

Recycle ratio = recycle/fresh feed = 609.7/100 = 6.097

Purge ratio = purge/recycle = 6.585/609.7 = 0.0108

The ratios expressed on an inert-free basis are:

Recycle ratio = [(N_2+H_2) in recycle/(N_2+H_2) in fresh feed] = 584.7/99.73
= 5.863

Purge ratio = [(N_2+H_2) in purge/(N_2+H_2) in recycle] = 6.315/584.7
= 0.0108

2.3   Thermal calculations on reactor systems

2.3.1   Basic concepts

When carrying out 'heat' (or, more correctly, energy) balances, it is best to work in terms of enthalpies. Enthalpy [5–8] is a **function of state** given by

$$H = U + PV \qquad (2.29)$$

where H, U, P and V are, respectively, the enthalpy, internal energy, pressure and volume of the system of interest.

The definition of the enthalpy function is such that it has convenient properties both when applied to closed and to continuous flow systems.

2.3.1.1   Enthalpy changes in closed systems (e.g. batch reactors) related to input of heat and work. The increase in enthalpy ΔH accompanying an isobaric change in a closed system during which an amount of heat ΔQ enters the system and an amount of 'shaft work' ΔW_s is done on the system is given by

$$(\Delta H = \Delta Q + \Delta W_s)_p \qquad (2.30)$$

(cf. reference [7, p. 65]).

(The 'shaft work' is external work done on the system in addition to the work $-P\Delta V$ done on it by the environment during a change ΔV in the system volume.)

Equation (2.30) is directly applicable to a batch reactor operated at constant pressure. Such reactors require a stirrer or recirculating pump, the energy input to which is shown as ΔW_s in Figure 2.11.

However, in practice it is usually found that, except when the contents of the reactor are very viscous, the value of ΔW_s is negligible compared with the energy released or taken-up by the reaction over the period considered. As a good approximation therefore:

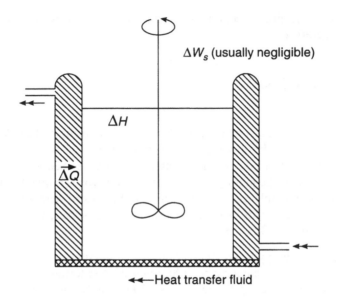

ΔW_s (usually negligible)

ΔH

$\overrightarrow{\Delta Q}$

◄◄—Heat transfer fluid

Figure 2.11 Increase ΔH in enthalpy of contents of batch reactor equated to the sum of heat (ΔQ) entering and shaft work ΔW_s done on it. (The last-named term is usually negligible, see equation (2.30) and text.)

$$(\Delta H = \Delta Q)_p \qquad (2.31)$$

and, under adiabatic conditions,

$$(\Delta H = 0)_p \qquad (2.32)$$

2.3.1.2 Difference between inlet and outlet enthalpies of flow systems (e.g. continuous reactors) when at steady state, related to inputs of heat and work. In a **steady flow** system in which changes in potential energy and kinetic energy between the inlet and outlet are negligible (which is normally true in reactor systems), the excess of the enthalpy of a given mass ($\underline{m}$) of material leaving the system ($\underline{H}_{out}$) over that of an equal mass of material entering the system ($\underline{H}_{in}$) is given (see, e.g., reference [7, p. 65]) by:

$$(\underline{H}_{out} - \underline{H}_{in}) = \Delta\underline{Q} + \Delta\underline{W} \qquad (2.33)$$

where $\Delta\underline{Q}$ is the heat entering the system as mass $\underline{m}$ of feed enters (and an equal mass of the product stream leaves) and $\Delta\underline{W}$ is the shaft work done on the system. As discussed below, the magnitude of $\underline{m}$ is normally implicit in the **basis** chosen for the calculation.

When equation (2.33) is applied to steady flow reactor systems (for example, tubular reactors or even continuous stirred tank reactors operating at steady-state conditions) the term $\Delta\underline{W}$ is normally either zero or so small as to have a negligible effect on the thermal calculations. As a good approximation, therefore:

$$(\underline{H}_{out} - \underline{H}_{in}) = \Delta\underline{Q} \qquad (2.34)$$

Under adiabatic conditions,

$$(\underline{H}_{out} - \underline{H}_{in}) = 0 \qquad (2.35)$$

Equation (2.34) is the starting point for the calculations of heat loads and temperatures in steady flow reactors given in Section 2.3.4.

In these calculations a convenient **basis** is chosen (see Section 2.2) and the numerical value of the mass $\underline{m}$ flowing into the reactor is implicit in the basis chosen. If, for example, the basis chosen is 100 moles of reactant A entering the system, the mass $\underline{m}$ of fluid entering is the mass associated with 100 moles of A. In equation (2.34), therefore:

$\underline{H}_{out}$ is the enthalpy leaving per 100 moles of A entering
$\underline{H}_{in}$ is the enthalpy entering per 100 moles of A entering
$\Delta\underline{Q}$ is the heat entering the reactor per 100 moles of A entering.

An alternative basis might be an operational period of one second, in which case $\underline{H}_{out}$ would be the enthalpy of the fluid leaving the reactor in one second, and so on.

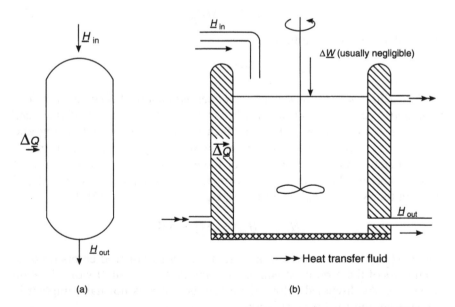

(a)   (b)

Figure 2.12 Excess of enthalpy of material leaving flow reactor at steady state ($\underline{H}_{out}$) over that of material entering ($\underline{H}_{in}$) equated to heat entering ($\Delta\underline{Q}$). (Shaft work and also kinetic and potential energy differences between inlet and outlet streams are taken to be negligible, see equation (2.34) and text.)

It is important to recognise at this stage that the validity of equations (2.33) to (2.35) does not depend on the flow conditions with the reactor systems. These equations are equally applicable to plug flow, perfectly mixed and partially mixed situations.

The reason for expressing heat flows in terms of enthalpy differences (as in equation (2.31) to (2.34) is that, unlike ΔQ, enthalpy differences can be calculated in a conceptually satisfactory manner from the inlet and outlet conversions and temperatures (in the case of flow reactors) or the initial and final conversions and temperatures (in the case of batch reactors). This is because, unlike Q, H is a function of state, i.e. it is a property which can be uniquely specified in terms of appropriate clearly defined variables.

2.3.2   Enthalpies of reactor fluids as functions of temperature and conversion

The variables which determine the enthalpies of the batches and samples of reactor fluid in which we shall be interested are mass, temperature and, to a smaller extent, pressure. The pressure effect in many important situations is sufficiently small to be neglected altogether. The enthalpy of a perfect gas, for example, is independent of pressure [7] and, as has been seen in Section 2.1.4.2, very many of the gas phase reactions to be considered take place under conditions where the perfect gas laws apply as a good approximation. Changes in the enthalpies of liquids (and solids) with pressure are also sufficiently small to be neglected for most practical purposes unless pressure ranges in excess of about 100 bar are considered. In the following sections, enthalpy changes with pressure will not be considered and the enthalpy of the reactor fluid of interest will be taken to be a function only of the mass of the fluid and its temperature and composition.

In batch and steady flow reactor systems for which the feed conditions and reaction stoichiometry are known, the moles of each component present in batch or leaving the flow reactor in a given period is determined by the extent to which the reaction has proceeded, i.e. by the conversion X_A of a specified reactant A. (The appropriate equations are equation (2.21) to (2.26) for the batch reactor and equation (2.21a) to (2.26a) for flow reactors under steady-state conditions.)

In the above situations the enthalpy of a given element of fluid within the reactor is a function only of the mass of the fluid element, its temperature and the conversion X_A of the limiting reactant within it.

Changes of enthalpy with temperature can be calculated from the heat capacity of the reactor fluid while, as described in Sections 2.3.3 and 2.3.4, changes with conversion can be calculated from the enthalpy of reaction.

2.3.2.1   The enthalpy of reaction.   The largest contribution to the enthalpy changes which occur during a reaction under isothermal conditions arises

from the making and breaking of chemical bonds which occurs when molecular species are converted from one form to another. There are secondary effects, arising from physical interactions between the molecules in the reactor fluid, but these are normally much smaller in magnitude (for example, the reactants must mix before they react, leading to a 'heat of mixing' term and there will be physical interactions also between the product molecules and the surrounding reactor fluid).

A measure of the increase in the enthalpy of the reactant/product mixture which takes place as the reaction proceeds is provided by the enthalpy of reaction (sometimes called the heat of reaction). This is positive for endothermic reactions and negative for exothermic ones.

(If heat is absorbed (i.e. if the system is endothermic) ΔQ in equation (2.30) and $\Delta \underline{Q}$ in equation (2.33) are positive resulting in an increase in system enthalpy in the first case and in $(\underline{H}_{out} - \underline{H}_{in})$ in the second, i.e. to positive values for the enthalpy of reaction. For exothermic reactions the reverse is true).

The enthalpy of reaction is a function of temperature and, to a much smaller degree, of pressure also. For reasons already discussed, the pressure dependence will be taken to be negligible in the following sections.

It is convenient to specify the magnitude of the enthalpy of reaction at given temperature in terms of the enthalpy increase corresponding to the consumption of one mole of a named reactant (preferably the limiting reactant A) at that temperature. The enthalpy of reaction per mole of reactant A consumed at temperature T will, throughout this book be designated $\Delta H_{R,A}(T)$.

$$\Delta H_{R,A}(T) = \text{increase in enthalpy of reactor fluid per mole of reactant } A \text{ consumed at temperature } T. \qquad (2.36)$$

As defined above, the enthalpy of reaction can, due to heat of mixing effects, be a function of composition. However, as discussed in Section 2.3.5, such effects are normally negligible. In the following sections $\Delta H_{R,A}(T)$ is regarded as a function of temperature only. If the stoichiometric coefficient for reactant A is taken as (-1), $\Delta H_{R,A}(T)$ then becomes identical with the standard enthalpy change for the reaction, as discussed in Section 2.3.5.

2.3.3   Thermal calculations in batch reactors: heat capacities of reactor fluids

In the case of a **batch reactor**, the fluid of interest is the entire contents of the reactor. The difference is required between the enthalpy H_0 of this fluid in an initial state (immediately after charging) and the enthalpy H_F at a later time when the conversion of specified reactant A has increased from zero to X_{AF}.

Suppose that the temperatures in the initial and final states are T_0 and T_F and that the numbers of moles of all components $(n_{A0}, n_{B0}, \ldots, n_{i0})$ initially present in the reactor are known, together with the stoichiometry of the reaction which takes place. Under these conditions, as noted in Section 2.3.2 the fluid enthalpy H is a function of conversion and temperature only. Consequently changes dH in H corresponding to changes dT in temperature and dX_A in the conversion of A are given by:

$$dH = \left(\frac{\partial H}{\partial T}\right)_{X_A} dT + \left(\frac{\partial H}{\partial X_A}\right)_T dX_A \qquad (2.37)$$

where

$$\left(\frac{\partial H}{\partial T}\right)_{X_A} = C_p(X_A, T) \qquad (2.38)$$

in which $C_p(X_A, T)$ is the total heat capacity of the reactor fluid when the conversion is X_A and the temperature is T. (Heat capacities are normally temperature dependent, though over the temperature ranges encountered in adiabatic reactors it is often adequate to handle this dependence in an approximate way, as in Examples 2.8 and 2.9.)

From equation (2.19), moles of A used up in an infinitesimal change in which the conversion of A increases by dX_A is given by $(-dn_A) = n_{A0}dX_A$. From the defining equation for $\Delta H_{R,A}(T)$ therefore:

$$\left(\frac{\partial H}{\partial X_A}\right)_T = -\left(\frac{\partial H}{\partial n_A}\right)_T n_{A0} = n_{A0}\Delta H_{R,A}(T) \qquad (2.39)$$

where $\Delta H_{R,A}(T)$ is the enthalpy of reaction per mole of reactant A consumed at the temperature T (see equation (2.36))

Substituting equations (2.38) and (2.39) into (2.37),

$$dH = C_p(X_A, T)dT + n_{A0}\Delta H_{R,A}(T)dX_A \qquad (2.40)$$

The difference $H_F - H_0$ between the enthalpies of the batch reactor fluid in the final and initial states is obtained by integrating (2.40) between these states. Since the enthalpy is a function of state, the value obtained should be independent of the integration track used and any convenient path may be followed.

In practice, the most convenient route is usually to integrate with respect to conversion from zero to X_{AF} while holding the temperature constant at T_0 and then to integrate with respect to temperature from T_0 to T_F while holding the conversion constant at its final value X_{AF} so that, on incorporating (2.30),

$$\Delta Q = (H_F - H_0) = (n_{A0})\,\Delta H_{R,A}(T_0)[X_{AF}] + \int_{T_0}^{T_F} C_p(X_{AF}, T)\mathrm{d}T$$

Batch reactor   (2.41)

where ΔQ is the heat entering the batch reactor when the conversion increases from zero to X_{AF}.

For adiabatic operation:

$$\int_{T_0}^{T_F} C_p(X_{AF}, T)\mathrm{d}T = -n_{A0}\Delta H_{R,A}(T_0)X_{AF} \qquad (2.42)$$

The integration track leading to equation (2.41) is shown schematically by lines AB and BH on Figure 2.13 (An alternative track is shown by the broken lines AQ and QH.)

2.3.3.1 Additivity rule for calculating total heat capacity.

Before equation (2.41) can be used, values are required for the enthalpy of reaction $(\Delta H_{R,A}(T_0)$ at the reactor feed temperature and also for the total heat capacity $(C_p(X_{AF}, T))$ of the final reactor mixture over the temperature range T_0 to T_F.

For near ideal mixtures of gases or of liquids and also for many non-ideal mixtures for which the heat of mixing is small or can be considered to be independent of temperature, the total heat capacity C_p of the mixture can be

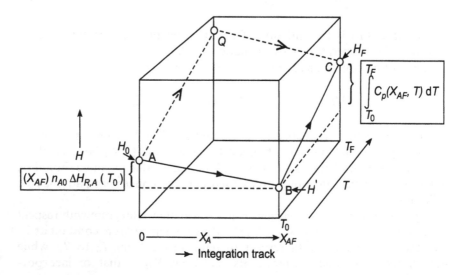

Figure 2.13 Integration track leading to equation (2.41) for $H_F - H_0$.

obtained as a sufficiently close approximation by adding together the total heat capacities of the pure components from which the mixture was constituted.

For a homogeneous binary mixture containing n_A moles of component A and n_B moles of component B, for example.

$$C_p = n_A C_{pmA} + n_B C_{pmB}$$

where C_{pmA} and C_{pmB} are the molar heat capacities of the pure components A and B at the temperature of the mixture. This rule applies well to most mixtures of organic liquids and to the gaseous mixtures considered later.

As an example of the application of the additivity rule to a batch reactor fluid we may suppose that the reactor initially contains n_{A0} moles of a limiting reactant A, n_{B0} moles of reactant B and n_I moles of an inert material.

The reaction between A and B yields products C and D according to the equation

$$(-\nu_A)A + (-\nu_B)B \rightarrow \nu_C C + \nu_D D \tag{2.43}$$

The heat capacity of the reactor fluid is required when the conversion of A is X_A. The number of moles of components A, B, C and D in the reactor (n_A, n_B, n_C and n_D) at this conversion is given by equations (2.21) to (2.24): the number of moles of the inert I remains at n_I. Knowing $n_A \ldots n_I$, the total heat capacity C_p of the mixture at any desired temperature is given by the additivity rule to be

$$C_p = n_A C_{pmA} + n_B C_{pmB} + n_C C_{pmC} + n_D C_{pmD} + n_I C_{pmI} \tag{2.44}$$

where $C_{pmA} \ldots C_{pmI}$ are the molar heat capacities of components $A \ldots I$ in the pure state at the same temperature as the mixture. Pure component heat capacities for a very large number of liquid and gaseous compounds are available in standard data books and data banks [2,3,12] and prediction methods are also available [3].

The values of C_p given by (2.44) over the temperature range of interest are substituted in equation (2.42) and can be used to evaluate the thermal behaviour of the reactor.

The above procedure for calculating C_p may be expressed in a generalised form by writing

$$C_p = \sum_i \left(n_i C_{pmi} \right) \tag{2.45}$$

where n_i is the number of moles of component i in the mixture of interest and C_{pmi} is the molar heat capacity of i in the pure state at the temperature of the mixture. In the summation, i is in turn each component, including inert components, present in the mixture. The number of moles of any component i in a batch reactor is given in terms of the initial number of moles of i (n_{i0}), the stoichiometry of the reaction and the conversion and initial number of moles of limiting reactant A by equation (2.26). Combining (2.45) with (2.26)

$$C_p = \sum_i (n_{i0} C_{pmi}) - n_{A0} X_A \sum_i \left[\left(\frac{\nu_i}{\nu_A} \right) C_{pmi} \right] \qquad (2.46)$$

where n_{i0} is the number of moles of component i (if any) in the feed and ν_i is the stoichiometric coefficient of this component. As in equation (2.6) and subsequently, ν is taken to be positive for products and negative for reactants. It is zero for inert components, as in equation (2.26). An example of the calculation of C_p using the above equations is given in Section 2.3.4.

The simple example below does not involve extensive heat capacity calculations. It is given to illustrate in principle the way in which equation (2.41) can be used to calculate the temperature changes which occur in adiabatic batch reactors or the heat transfer requirements of isothermal ones. More detailed examples are given in Chapter 4.

Example 2.8 *Thermal effects accompanying isomerisation in a batch reactor.* The reaction $A \rightarrow B$ is a first-order isomerisation reaction. Both A and B are liquids of very low volatility with a 'relative molecular mass' (or molecular weight) of 250. Both liquids have average molar heat capacities C_p of $525 \, \text{J mol}^{-1} \text{K}^{-1}$ in the temperature range of interest. The reaction is to be carried out in the liquid phase in a batch reactor and a final conversion of 0.97 is specified. The reactor size is such that 500 kg of A can be processed per batch. The enthalpy of reaction per mole of A consumed at 436 K is given by:

$$\Delta H_{R,A}(436) = -87.15 \, \text{kJ mol}^{-1}$$

(a) If the reactor is to be operated isothermally at full capacity at 436 K, calculate the heat which must be removed from the reactor shell over the entire period of reaction.

(b) If the reactor is operated adiabatically with a starting temperature of 436 K, calculate the reactor temperature when the conversion reaches 0.97.

Solution
In both parts (a) and (b), moles of A fed to reactor $= n_{A0} = 5 \times 10^5/250 = 2000$.

Since one mole of product is formed for each mole of reactant consumed and the average molar heat capacities of the reactant and product are equal, the total heat capacity C_p in equation (2.44) is independent of conversion, the average value over the temperature range of interest being equal to the average heat capacity of the feed over that range.

$$C_p(\text{average}) = n_{A0} C_{pmA} = 0.525 n_{A0} \, \text{kJ K}^{-1}$$

Because of the nature of the reaction involved and the insensitivity of the component heat capacities to temperature change, C_p in this case can be taken to be independent of temperature and conversion. Equation (2.41) then gives

$$\Delta Q = -(0.97n_{A0})87.15 + 0.525n_{A0}(T - 436) \text{ kJ}$$

For isothermal operation, $T = 436$ K and

$$\Delta Q = (-0.97) \times (2000) \times (87.15) \text{ kJ}$$
$$= -169 \text{ MJ}$$

So it is necessary to remove 169 MJ of heat from the reactor shell for each batch processed.

In an adiabatic operation, $\Delta Q = 0$ and

$$T = 436 + (0.97) \times (87.15/0.525) = 597 \text{ K}$$
$$= \text{Reactor temperature at the end of the reaction.}$$

2.3.4 Thermal calculations for flow reactors at steady state (see also Figure 2.14)

Before equation (2.33) can be used to calculate the thermal characteristics of flow reactors at steady state (for example, heat loads on isothermal reactors or temperature changes in adiabatic ones) an expression is required relating the enthalpy change ($H_{\text{out}} - H_{\text{in}}$) to the composition of the inlet and outlet streams. In deriving such an expression we will consider a continuous reactor with an inlet stream consisting of known proportions of reactants A and B and also containing inert material (I). The inlet and outlet stream temperatures are T_0 and T_F respectively and the conversion of (limiting) reactant A in the outlet stream is X_{AF}. We will take as basis $\underline{n}_{A0}$ moles of A entering. This basis specifies the quantity of fluid to be considered in the calculation. On this basis $\underline{n}_{I0}$ and $\underline{n}_{B0}$ moles of other components (I and B) also enter the reactor. The enthalpy of this quantity of fluid will be designated $\underline{H}_{\text{in}}$. Also, for every $\underline{n}_{A0}$ moles of A entering, $\underline{n}_I$ moles of I, together with $\underline{n}_A$ and $\underline{n}_B$ moles of unconsumed reactants and $\underline{n}_C$ and $\underline{n}_D$ moles of products will leave the reactor at steady state, where

$$\underline{n}_I = \underline{n}_{I0}$$
$$\underline{n}_A = \underline{n}_{A0}(1 - X_{AF})$$
$$\underline{n}_B = \underline{n}_{B0} - \left(\nu_B/\nu_A\right)\underline{n}_{A0}X_{AF}$$
$$\underline{n}_C = \left(\frac{-\nu_C}{\nu_A}\right)\underline{n}_{A0}X_{AF}$$
$$\underline{n}_D = \left(\frac{-\nu_D}{\nu_A}\right)\underline{n}_{A0}X_{AF}$$

(see equations (2.21a) to (2.25a)).

The mass of the outlet fluid is the same as that of the inlet on the above basis and its enthalpy will be designated $\underline{H}_{out}$. (In the steady state the mass of the fluid leaving will be the same as that of the fluid entering.)

As pointed out in Section 2.3.2, the enthalpy per unit mass of the reactor fluid in the above situation is a function only of the temperature T and conversion X_A. The difference $(\underline{H}_{out} - \underline{H}_{in})$ between the enthalpies of equal masses of the inlet and outlet streams on the given basis can accordingly be calculated from the inlet and outlet temperatures and the outlet conversion using standard thermodynamic relationships. (The inlet conversion is zero.) Since enthalpy is a function of state any convenient sequence of changes in conversion and temperature by which the specified quantity of fluid with the temperature and composition of the inlet to the reactor is converted to the same mass of fluid at the outlet conditions may be used for this evaluation provided the enthalpy increment at each stage of the route can be calculated. We use a purely hypothetical route in which a batch of the inlet liquid containing $\underline{n}_{A0}$ moles of A with corresponding numbers of moles of the other components is allowed to react as a closed system until the conversion of A reaches the value (X_{AF}) equal to that in the outlet stream from the continuous reactor. The batch is then brought to the reactor outlet temperature T_F while holding the conversion of A constant at its final value X_{AF} as in Figure 2.13. The conditions of temperature and conversion in the batch are then identical with that in the stream leaving the reactor. The mass of the batch is also identical with that on which the enthalpy calculation was based and its enthalpy is accordingly equal to that of the outlet stream on this basis. The increase in enthalpy of our batch of fluid is thus equal to the difference $\underline{H}_{out} - \underline{H}_{in}$ between the steady-state inlet and outlet stream enthalpies. Expressed in symbols,

$$(\underline{H}_{out} - \underline{H}_{in}) = \left(H_{final}^{batch} - H_{initial}^{batch}\right) \qquad (2.47)$$

where

$\underline{H}_{in} =$ enthalpy entering reactor by flow at temperature T_0 over period in which $\underline{n}_{A0}$ moles of reactant A enter

$\underline{H}_{out} =$ enthalpy leaving reactor by flow over same period

$H_{initial}^{batch} =$ initial enthalpy of batch of fluid containing $\underline{n}_{A0}$ moles of reactant A at temperature T_0, the composition of which is identical to that of the feed to the flow reactor

$H_{final}^{batch} =$ enthalpy of same batch of material when brought to the same temperature and degree of conversion as the outlet from the flow reactor

The difference $(H_{batch}^{final} - H_{batch}^{initial})$ between the initial and final enthalpies of a batch reactor in which the above changes in temperature and conversion occur is given by equation (2.41) to be

$$\left(H_{\text{batch}}^{\text{final}} - H_{\text{batch}}^{\text{initial}}\right) = \underline{n}_{A0}X_{AF}\Delta H_{R,A}(T_0) + \int_{T_0}^{T_F} \underline{C}_P(X_{AF}T)dT \qquad (2.48)$$

Combining this result with (2.47) and incorporating also (2.34) (which gives the heat entering a flow reactor), we obtain

$$\Delta\underline{Q}_F = (H_{\text{out}} - H_{\text{in}}) = \underline{n}_{A0}X_{AF}\Delta H_{R,A}(T_0)) + \int_{T_0}^{T_F} \underline{C}_P(X_{AF}, T)dT \quad (2.49)$$

where $\Delta\underline{Q}_F$ is the heat entering the flow reactor for every $\underline{n}_{A0}$ moles of A that flow in.

In the reasoning leading to equation (2.49) no assumption was made about the flow conditions in the actual reactor. The equation is equally applicable to perfectly mixed, plug flow or partially mixed reactors under steady-state conditions.

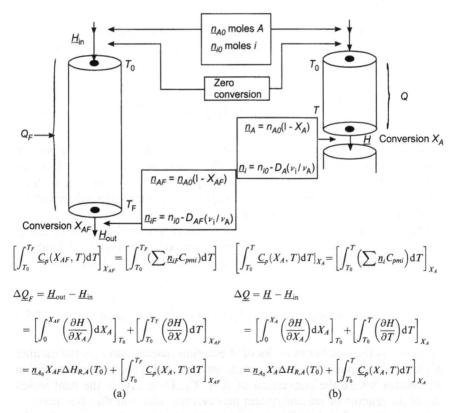

Figure 2.14 Illustrating calculation of (a) enthalpy difference between inlet and outlet streams of flow reactor (or reactor system) at steady state using equation (2.49); (b) difference between inlet stream enthalpy and enthalpy at point within the reactor where conversion of limiting reactant is X_A (equation (2.50)). $D_{AF} = \underline{n}_{A0} X_{AF} =$ number of moles of A consumed in case (a). $D_A = \underline{n}_{A0} X_A =$ number of moles of A consumed in case (b).

$\underline{H}_{out} - \underline{H}_{in}$ in equation (2.49) is the excess of the enthalpy of the stream leaving the reactor over that of the feed to it in the steady state. But the same reasoning which led to this equation would apply equally to the calculation of the excess of the enthalpy $\underline{H}$ of the stream crossing any arbitrary boundary within the reactor over that of the feed to the reactor, the only difference being that the conversion and temperature at the boundary replace those at the reactor outlet, as in equation (2.50) below. In arriving at this expression for $\underline{H} - \underline{H}_{in}$ we need to consider a section of the reactor extending from the feed point to the arbitrary boundary rather than from the feed point to the reactor outlet as previously (see Figure 2.14(a)). If the conversion of A in the stream leaving this section is X_A and the temperature is T, the relevant equation (analogous to (2.49)) is

$$\Delta \underline{Q} = (\underline{H} - \underline{H}_{in}) = \underline{n}_{A0} X_A \Delta H_{R,A}(T_0) + \int_{T0}^{T} \underline{C}_p(X_A, T) dT \qquad (2.50)$$

Equation (2.50) is clearly more general than (2.49) in that it may equally be applied over can entire reactor or reactor system or over a part of it.

For adiabatic operation equation (2.50) rearranges to give

$$X_A = \left(\frac{-1}{\underline{n}_{A0} \Delta H_{R,A}(T_0)} \right) \left(\int_{T_0}^{T} \underline{C}_p(X_A, T) dT \right) \qquad (2.51)$$

Equation (2.51) gives the temperature at any point within an adiabatic reactor at steady state as a function of the conversion of reactant A at that point. If changes in the total heat capacity $\underline{C}_p(X_A, T)$ of the reactor fluid with conversion and temperature are negligible, a linear dependence of X_A on T is predicted (see Example 2.19).

The heat capacity $\underline{C}_p(X_{AF}, T)$ in equations (2.49) to (2.57) can be calculated from the additivity rule (eq. (2.45)) described in Section 2.3.3.1 subject to the provisos there listed. Combining equation (2.45) with (2.50), for example

$$\Delta \underline{Q} = \left(n_{A0} X_A \Delta H_{R,A}(T_0) \right) + \int_{T_0}^{T} \left(\sum_i n_i C_{pmi} \right) dT \qquad (2.52)$$

where n_{A0} is the number of moles of A entering reactor and n_i is the number of moles of component i which over the same period cross a boundary within the reactor where the conversion of A is X_A. Here $\Delta \underline{Q}$ is the heat which enters the reactor or reactor system between the inlet and that boundary.

The number of moles of each component which are then present can be calculated from the conversion and the reaction stoichiometry using equations (2.21a) to (2.25a) as in Example 2.9 below. Alternatively, the generalised expression

$$\sum_i \left(\underline{n}_i C_{pmi} \right) = \sum_i \left(\underline{n}_{i0} C_{pmi} \right) - \underline{n}_{A0} X_A \sum_i \left(\frac{\nu_i}{\nu_A} C_{pmi} \right) \qquad (2.46a)$$

may be used. This equation is analogous to (2.46) for the batch reactor case, though in (2.46a) $\underline{n}_{i0}$ is the number of moles of component i entering the reactor in the period during which $\underline{n}_{A0}$ moles of A enter. $\underline{n}_i$ is the number of moles of i which simultaneously cross a boundary within the reactor where the conversion of a is X_A. The stoichiometric coefficients ν follow the usual sign convention and are 0 for inert components.

Equation (2.52) is basically an energy balance which states that the heat evolved by the reaction plus the heat entering the reactor by conduction is used to raise the stream temperature from T_0 to T.

The energy balance may be expressed in terms of molar flow rates by applying equation (2.52) over unit time so that

$\underline{n}_{A0}$ = moles of A entering reactor in unit time
= F_{A0} = molar flow rate of A at entry
$\underline{n}_i$ = moles of i which in unit time cross the boundary where conversion of A is X_A
= molar flow rate of i at this boundary
= F_i
$\underline{Q}$ = heat entering reactor or reactor system in unit time
= rate of heat input
= $\dot{Q}$

Making these substitutions in (2.52),

$$\dot{Q} = \left(F_{A0} X_A \Delta H_{R,A}(T_0) \right) + \int_{T_0}^{T} \sum_i F_i C_{pmi}\, dT \qquad (2.53)$$

If, for example, components A and B are present when the conversion of A is X_A,

$$\dot{Q} = \left(F_{A0} X_A \Delta H_{R,A}(T_0) \right) + \int_{T_0}^{T} F_A C_{pmA}\,dT + \int_{T_0}^{T} F_B C_{pmB}\, dT \qquad (2.53a)$$

and so on.

Equations (2.52), (2.53) and (2.53a) may be applied between any two points in a continuous unbranched flow reactor system, though it has hitherto been convenient to regard the first point (indicated by subscript 0) as the feed point to a reactor. In general, if applied between points 1 and 2 in an unbranched flow system, equation (2.53) becomes

$$\dot{Q} = [F_{A0}(X_{A2} - X_{A1})\Delta H_{R,A}(T_1)] + \int_{T_1}^{T_2} \sum_i F_i C_{pmi}\, dT \qquad (2.53b)$$

where X_{A2} and T_2 are the temperature and conversion of A at the downstream point, and X_{A1} and T_1 are the conversion and temperature at the upstream point. $\dot{Q}$ is the rate at which heat is entering the section of reactor considered. F_{A0} is the molar flow rate of A at zero conversion.

Example 2.9 *Temperature rise accompanying adiabatic oxidation of* SO_2 *in a tubular reactor operating at steady state.*
The feed to the reactor has the following composition in mole percentages: $SO_2 = 9.0, O_2 = 12.0$ and $N_2 = 79.0$. The feed temperature is 750 K and the reactor pressure is close to 1 atmosphere. If the outlet conversion of SO_2 (moles of SO_2 reacted per mole of SO_2 in the feed) is 0.60, what is the temperature of the product stream?

Data: The oxidation reaction will be written in the form

$$SO_2 + (1/2)O_2 \rightarrow SO_3$$

the enthalpy of reaction at 750 K being

$$\Delta H_R = -89\,300 \text{ J per mole of } SO_2 \text{ consumed.}$$

The mean molar specific heats (in $J\,mol^{-1}\,K^{-1}$) over the temperature range of interest are taken to be as follows:

$$SO_2 = 52.0, \quad O_2 = 33.0, \quad N_2 = 31.0 \quad \text{and} \quad SO_3 = 78.0$$

(The above values may be subject to adjustment when the actual temperature range is known more closely. In this event the calculation would be repeated with the improved average heat capacities.)

Solution
Let the temperature of the product stream be T_F and take 1 mole SO_2 entering as basis (SO_2 is the limiting reactant). On this basis
  Moles O_2 entering $= 12.0/9.0 = 1.33$
  Moles N_2 entering $= 79.0/9.0 = 8.78$
  Moles SO_2 leaving $= (1 - 0.60) = 0.40$
  Moles O_2 leaving $= 1.33 - (0.60/2) = 1.03$
  Moles SO_3 leaving $= 0.60$
  Moles N_2 leaving $= 8.78$
The material leaving the reactor for every mole of SO_2 entering has the following total heat capacity $\underline{C}_p(X_{A1} = 0.60)$

$$\underline{C}_p(X_{AF} = 0.60) = (0.40) \times (52.0) + (1.03) \times (33.0) + (0.60) \times (78.0)$$
$$+ (8.78) \times (31.0)$$
$$= 21 + 34 + 272 + 47 = 374 \, J\,K^{-1}$$

Alternatively, the generalised expression for $\underline{C}_p$ given by equation (2.46a) may be used giving

$$C_p(X_{AF}=0.60)=(52+1.33 \times 33 + 8.78 \times 31) - 0.6(52+1/2 \times 33 - 1 \times 78)$$
$$= 374 \, \text{J K}^{-1}(\text{as above})$$

Since the reactor operates adiabatically, ΔQ_F in equation (2.49) is zero. Making this substitution and setting $(\underline{C}_p(X_{AF}, T) = 374 \, \text{J K}^{-1}$ as an average figure over the temperature range 750 to T_F then gives:

$$0.6 \, (-89\,300) + 374 \, (T_F - 750) = 0$$

that is:

$$T_F = 750 + (89\,300 \times 0.6)/374 = 893 \, \text{K}$$
$$= \text{temperature of product stream.}$$

2.3.5   The standard enthalpy change $(\Delta_r H^\theta(T))$ for a reaction and its use for thermal calculations on reactors.

The 'enthalpy of reaction' per mole of a specified reactant consumed at temperature T (as defined by equation (2.36)) should be distinguished, conceptually at least, from the so-called 'standard enthalpy change' for the reaction at the temperature T, which is given the symbol $\Delta_r H^\theta(T)$.

$\Delta_r H^\theta(T)$ = increase in enthalpy which occurs when stoichiometric amounts of the products of a reaction **in their pure states** are constituted from stoichiometric amounts of the pure reac- (2.54) tants at temperature T. Both the initial reactants and the final products are in their standard states at the temperature T and the reaction is at a standard pressure of 1 bar.

$\Delta_r H^\theta(T)$ can readily be calculated from the standard heats of formation and heat capacities of the pure component reactants and products: under normal conditions the enthalpy of reaction, as defined by (2.36), can then be obtained from it as a very close approximation.

In tabulations of the standard properties of substances, the phase of the substance when in its standard state is conventionally given in brackets following the name of the substance, the symbols used to denote solid, liquid or ideal gas state being (s), (l) and (g) respectively (see Table 2.2 for example)

The properties shown for solids (s) or liquids (l) at temperature T and 1 bar pressure are taken to be the equilibrium properties of the actual pure substance at this temperature and pressure. The properties shown for the ideal gas state (g) are taken to be those which the substance would have at temperature T and 1 bar pressure if it behaved as an ideal gas under **these conditions**.

The enthalpies of ideal gases are independent of pressure, and the enthalpies (and heat capacities) listed for the hypothetical ideal gas state at 1 bar are in this case those of the vapour of the substance at pressure sufficiently low for it to behave ideally.

In practice, deviations of vapour behaviour from ideality at pressures below 1 bar are normally very small, and ideal gas state enthalpies can be obtained by adding the enthalpy of vaporisation at temperature T to the corresponding liquid phase enthalpy.

2.3.5.1 Relation between the enthalpy of reaction per mole of specified reactant and the standard enthalpy change for the reaction.

As pointed out by P.W. Atkins [6] the standard enthalpy change refers to an overall process with three stages.

(1) The reactants (initially in their standard states) are mixed.
(2) They react.
(3) The products are separated and brought to their standard states.

Considering, for example, the reaction shown in equation (2.43):
Increase in enthalpy during the three stages $= \Delta_r H^\theta(T) =$ (enthalpy increase on mixing $(-\nu_A)$) moles of reactant A with $(-\nu_B)$ moles of B)   (stage 1)
$+\nu_A \Delta H_{R,A}(T)$ (stage2)
$+$ (enthalpy increase associated with the separation of ν_C moles of C and ν_D moles of D from the reactor fluid)   (stage 3).

In practice, stage 3 does not occur within reactors and this is often the case also with stage 1. Fortunately, however, enthalpy changes associated with the separation and mixing processes in stages 1 and 3 are normally several orders of magnitude smaller than the heat of reaction $\Delta H_{R,A}(T)$.

Hence, as a good approximation,

$$\Delta H_{R,A}(T) = \left(\frac{\Delta_r H^\theta_{(T)}}{-\nu_A}\right) \tag{2.55}$$

(For homogeneous fluid systems, the enthalpy changes in stages 1 and 3 are due solely to heats of mixing. For organic liquids, heats of mixing are typically not greater than about one-tenth of the latent heat of the components being mixed. For ideal gas systems, to which many of the systems encountered in gas phase reactions closely approximate, the heat of mixing is zero. This is also true for ideal solutions of liquids.)

Although equation (2.55) is applicable under most normal conditions exceptions can arise particularly at very elevated pressure where isobaric and isothermal mixing operations in some homogeneous systems are associated with a transition from gas-like behaviour at one end of the composition range to liquid-like behaviour at the other. In these cases the mixing operation can result in very substantial density and thermal changes.

2.3.5.2 Calculation of the standard enthalpy change for a reaction at 298 K from the enthalpies of formation of the reactants and products at this temperature. It is convenient initially to continue to consider the reaction

$$(-\nu_A)A + (-\nu_B)B \rightarrow (\nu_c)C + (\nu_D)D \qquad (2.43)$$

though a more generalised approach will develop from this.

From the definition of $\Delta_r H^\theta(T)$ given by equation (2.54), the standard enthalpy change for the above reaction at temperature T is given by

$\Delta_r H^\theta(T)$ = (Excess of product enthalpies (after separation) over pure
$\qquad$ reactant enthalpies)

$$= \nu_C H_C^\theta(T) + \nu_D H_D^\theta(T) - [(\nu_A)H_A^\theta(T) + (-\nu_B)H_B^\theta(T)]$$

$$= \nu_A H_A^\theta(T) + \nu_B H_B^\theta + \nu_C H_C^\theta(T) + \nu_D H_D^\theta(T) \qquad (2.56)$$

where $H_A^\theta(T), H_B^\theta(T), H_C^\theta(T)$ and $H_D^\theta(T)$ are the molar enthalpies of the pure substances A, B, C and D in their standard states at temperature T.

In general, for any reaction, we can write:

$$\Delta_r H^\theta(T) = \sum_i \nu_i H_i^\theta(T) \qquad (2.57)$$

where $H_i^\theta(T)$ is the molar enthalpy of pure substance i in its standard state at temperature T.

The summation is taken over all reactants and products **participating in the reaction**. For reaction (2.43), for example, i is in turn equal to A, B, C, and D and (2.57) expands to give the summation (2.56).

If the enthalpies of the constituent elements in their standard states are set equal to zero at the temperature of interest (an arbitary choice which is convenient for isothermal calculations):

$$H_i^\theta(T) = \Delta_f H_i^\theta(T) \qquad (2.58)$$

where $\Delta_f H_i^\theta(T)$ = standard enthalpy of formation of component i at temperature T.
$\qquad\qquad\qquad$ = increase in enthalpy which occurs when one mole of component i is formed from its elements at the specified temperature.

Standard enthalpies of formation are listed in may texts. Usually, though not invariably, the data are given for 298.15 K only. A few typical values at

298.15 K are listed in Table 2.2 (where the abbreviation 298 K is used, 298.15 K is implied). Because of the ready availability of standard enthalpy of formation data at 298.15 K, and for computational convenience, this section is restricted to the calculation of the standard enthalpy change at this temperature. The calculation of $\Delta_r H^\theta$ at temperatures other than 298.15 K is described in Section 2.3.5.3. Substituting (2.58) into (2.57) with $T = 298.15K$

$$\Delta_r H^\theta(298) = \sum_i \nu_i H_i^\theta(298) = \sum_i (\nu_i \, \Delta_f \, H_i^\theta(298)) \qquad (2.59)$$

Example 2.10
Calculate the standard enthalpy change for the SO_2 oxidation reaction at 298.15 K written in the form

$$SO_2(g) + (1/2)O_2(g) \rightarrow SO_3(g)$$

From Table 2.2

$$\Delta_f H^\theta(298) \text{ for } SO_2(g) \text{ is } - 296.83 \, kJ \, mol^{-1}$$

$$\Delta_f H^\theta(298) \text{ for } SO_3(g) \text{ is } - 395.72 \, kJ \, mol^{-1}$$

$$(\Delta_f H^\theta(298) \text{ for } O_2(g) \text{ is zero since } O_2 \text{ is an element})$$

Table 2.2 Standard enthalpies of formation of some selected compounds at 298.15 K

Substance	$\Delta_f H_{298}^\theta (kJ \, mol^{-1})$	Substance	$\Delta_f H_{298}^\theta (kJ \, mol^{-1})$
CO (g)	−110.53	NH_3 (g)	−46.11
CO_2 (g)	−393.51	S_2 (g)	+128.37
H_2 (g)	0	SO_2 (g)	−296.83
H_2O (l)	−283.83	SO_3 (g)	−395.72
H_2O (g)	−241.82	H_2SO_4 (l)	813.99
N_2 (g)	0		
NO_2 (g)	+33.18	N_2O_4 (g)	+9.16

Substituting the above data in equation (2.59) with $\nu_{SO_2} = -1$, $\nu_{O_2} = -\frac{1}{2}$ and $\nu_{SO_3} = +1$ gives:

$$\Delta_r H^\theta(298) = -395.72 - (-296.83) \, kJ = -98.89 \, kJ$$

Also, from equation (2.55):
$\Delta H_{R,SO_2}(298) = $ Enthalpy of reaction per mole of SO_2 consumed at 298.15 K
$$= -98.89 \, kJ \, mol^{-1}.$$

2.3.5.3 Dependence of the standard enthalpy change for a reaction on temperature: calculation of $\Delta_r H^\theta$ at other temperatures from value at 298.15 K. Differentiating (2.57) with respect to temperature gives

$$\frac{d\Delta_r H^\theta(T)}{dT} = \sum_i \nu_i C_{pmi}^\theta(T) \qquad (2.60)$$

where $C_{pmi}^\theta(T)$ is the molar heat capacity of pure component i in its standard state at temperature T. (The corresponding expression for the temperature dependence of the enthalpy of reaction (or heat of reaction) per mole of A consumed, obtained by combining equation (2.60) with (2.55) is

$$\frac{d\Delta H_{R,A}(T)}{dT} = \sum_i \left(\frac{-\nu_i}{\nu_A} C_{pmi}(T) \right) \qquad (2.60a))$$

Applying equation (2.57) at temperatures T_1 and T_2:

$$\Delta_r H^\theta(T_2) = \Delta_r H^\theta(T_1) + \sum_i \nu_i \left(H_i^\theta(T_2) - H_i^\theta(T_1) \right)$$

Since $\dfrac{dH_i^\theta}{dT} = C_{pmi}^\theta$,

$$\Delta_r H^\theta(T_2) = \Delta_r H^\theta(T_1) + \sum_i \left[\nu_i \int_{T_1}^{T_2} C_{pmi}^\theta \, dT \right] \qquad (2.60b)$$

Example 2.11
The enthalpy of reaction per mole of SO_2 consumed in the SO_2 oxidation reaction at 298 K is -98.89 kJ mol^{-1} (see Example 2.10). Calculate the enthalpy of reaction at 750 K and at 905 K.

Data: The molar heat capacities at constant pressure in J mol^{-1} K^{-1} for SO_2, O_2 and SO_3 may be represented as functions of absolute temperature (K) as follows in the range 298 to 900 K:

SO_2 : $C_{pm} = 23.85 + 0.0670T - 4.96 \times 10^{-5}T^2 + 1.328 \times 10^{-8}T^3$
O_2 : $C_{pm} = 28.11 - 3.68 \times 10^{-6}T + 1.746 \times 10^{-5}T^2 - 1.065 \times 10^{-8}T^3$
SO_3 : $C_{pm} = 16.37 + 0.146T - 11.20 \times 10^{-5}T^2 + 3.242 \times 10^{-8}T^3$

Solution
The reaction will be written in the same form as in Example 2.10 so that

$$\nu_{SO_2} = -1, \quad \nu_{O_2} = -\tfrac{1}{2}, \quad \nu_{SO_3} = +1$$

Since $\nu_{SO_2} = -1$, the enthalpy of reaction per mole of SO_2 consumed at temperature T (ΔH_R) will be identified with the standard enthalpy change for the reaction $\Delta_r H^\theta(T)$ (see equation (2.55)).

A general expression for $\Delta_r H^\theta$ as a function of T for this system will first be obtained. The required values for T will then be substituted in this expression.

From equation (2.60b),

$$\Delta_r H^\theta(T) = \Delta_r H^\theta(298) + \sum_i \left\{ \nu_i \int_{298}^{T} C_{pmi} \, dT \right\}$$

If C_{pmi} is expressed as a power series in T (as in the heat capacity data given)

$$C_{pmi} = a_i + b_i T + c_i T^2 + d_i T^3 \qquad (2.60c)$$

and

$$\Delta_r H^\theta(T) = \Delta_r H^\theta(298) + \int_{298}^{T}(\underline{a} + \underline{b}T + \underline{c}T^2 + \underline{d}T^3)dT$$

where

$$\underline{a} = \sum_i \nu_i a_i$$

$$\underline{b} = \sum_i \nu_i b_i$$

$$\underline{c} = \sum_i \nu_i c_i \qquad (2.61)$$

$$\underline{d} = \sum_i \nu_i d_i$$

Carrying out the integration,

$$\Delta_r H^\theta(T) = Z + \underline{a}T + \underline{b}T^2/2 + \underline{c}T^3/3 + \underline{d}T^4/4 \qquad (2.62)$$

where

$$Z = \Delta_r H^\theta(298) - 298\underline{a} - (298^2/2)\underline{b} - (298^3/3)\underline{c} - (298^4/4)\underline{d} \qquad (2.63)$$

In the present example,

$$\underline{a} = (16.37 - 28.11/2 - 23.85) = -21.53 \, \text{J mol}^{-1}\text{K}^{-1}$$
$$\underline{b} = (0.1460 + 0.000 - 0.0670) = 0.0789 \, \text{J mol}^{-1}\text{K}^{-2}$$
$$\underline{c} = (-11.20 - 1.746/2 + 4.96) \times 10^{-5} = -7.11 \times 10^{-5} \, \text{J mol}^{-1} \, \text{K}^{-3}$$
$$\underline{d} = (3.24 - 1.06/2 - 1.33) \times 10^{-8} = 1.39 \times 10^{-8} \, \text{J mol}^{-1} \, \text{K}^{-4}$$

From Example 2.10,

$$\Delta_r H^\theta(298) = -98.89 \times 10^3 \, \text{J}$$

So, from (2.63),

$$Z = -95.38 \times 10^3 \, \text{J}$$

Setting $T = 750 \, \text{K}$ in equation (2.62) then gives

$$\Delta_r H^\theta(750) = -98.24 \, \text{kJ mol}^{-1}$$

Setting $T = 905 \, \text{K}$ gives

$$\Delta_r H^\theta(905) = -97.81 \, \text{kJ mol}^{-1}$$

From equation (2.55), since $\nu_{SO_2} = -1$, the enthalpies of reaction per mole of SO_2 consumed at 750 and 905 K are equal to the standard enthalpy changes, i.e.

$$\Delta H_{R,SO_2}(750) = -98.24 \text{ kJ mol}^{-1}$$
$$\Delta H_{R,SO_2}(905) = -97.81 \text{ kJ mol}^{-1}$$

2.4 Chemical equilibrium calculations: equilibrium conversions in reactor systems

2.4.1 Introduction: Le Chatelier's principle and other basic concepts

As pointed out in Section 2.1, many reactions, known as reversible reactions, do not proceed until the limiting reactant has been completely exhausted. Instead they proceed until the rate of the forward reaction is equal to the rate of the reverse reaction. (The reverse reaction rate increases as the concentrations of the products increase while the forward reaction rate decreases as the concentrations of the reactants decrease.) When the two rates are equal no further changes in composition occur and the system is at equilibrium. The concentrations of the reactants and products at equilibrium depend on the temperature, pressure and the initial concentrations of the substances involved. The way in which the position of the equilibrium (and hence the attainable conversion) varies with pressure and temperature can be predicted qualitatively using **Le Chatelier's principle**. According to this principle, if a restraint, such as lowering the temperature is imposed on a reacting system which is initially at equilibrium, the system will react in a manner which tends to oppose the restraint; in the above case it will produce heat.

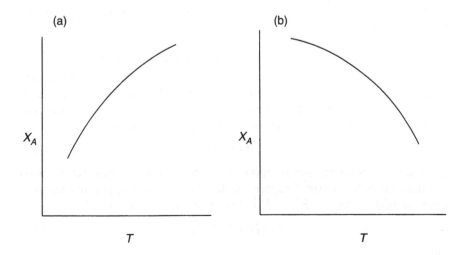

Figure 2.15 Equilibrium conversions (X_A) as functions of temperature in (a) an endothermic system and (b) an exothermic system.

If the forward reaction is exothermic (and the back reaction therefore endothermic) production of heat can only be achieved by an improvement in the equilibrium conversion. For an exothermic reaction, therefore, Le Chatelier's principle predicts an increase in equilibrium conversion with a lowering of temperature. For an endothermic reaction the reverse is true (Figure 2.15). According to Le Chatelier's principle also, if the pressure of a gaseous reacting system (initially at equilibrium) is raised, the system will oppose the restraint by reducing the number of molecules (if the reverse were true the system would of course be unstable). If the forward reaction results in a reduction in the number of molecules, the equilibrium conversion will therefore be enhanced by increase in pressure; if the forward reaction results in an increase in the number of molecules, the equilibrium conversion will fall.

As examples of the above effects of temperature and pressure changes, the reactions

$$SO_2 + \frac{1}{2}O_2 \rightleftharpoons SO_3 \quad \text{and} \quad CO + 2H_2 \rightleftharpoons CH_3OH$$

may be considered. In each case the forward reaction results in the evolution of heat ($\Delta_r H^\theta$ is negative) and a decrease in the number of molecules. In both cases Le Chatelier's principle is followed in that the equilibrium conversion for given feed composition falls as the temperature rises and can be improved by raising the pressure. (Figure 2.16 shows the effect of temperature on conversion in the SO_2 oxidation reaction.)

Quantitative information about the relative amounts of the reactant and product components at the equilibrium point, and the way in which the composition at this point depends on the temperature and pressure for a given feed, is given by an equilibrium constant for the reaction and its temperature dependence.

2.4.1.1 Equilibrium constants: the kinetic approach.
At the equilibrium point the rate of the forward reaction is balanced by the rate of the backward reaction. If the kinetics of the reaction are simple, this fact can lead to useful expressions for the relative concentrations of the reactants and products at the equilibrium point. Considering, for example, the reaction

$$A + B \rightarrow C + D \tag{2.64}$$

and the corresponding back reaction. If the forward and backward reactions are both elementary (see Chapter 3), the forward and backward reaction rates per unit volume $(-\overrightarrow{r_A})$ and $(\overleftarrow{r_A})$ respectively will be given by

$$(-\overrightarrow{r_A}) = k_1 C_A C_B \tag{2.65}$$

and

$$(\overleftarrow{r_A}) = k_2 C_C C_D \tag{2.66}$$

where C_A, C_B, C_C and C_D are the concentrations of components A, B, C and D and k_1 and k_2 are the rate constants for the forward and backward reactions respectively.

$(-\overrightarrow{r_A})$ is the rate of consumption of reactant A by the forward reaction and $(\overleftarrow{r_A})$ is the rate of formation of A by the back reaction. At equilibrium the forward and backward reaction rates are equal and

$$\frac{k_1}{k_2} = K_c = \frac{C_{Ce} C_{De}}{C_{Ae} C_{Be}} \tag{2.67}$$

where C_{Ae}, C_{Be}, C_{Ce} and C_{De} are the concentrations of components A, B, C and D when the reaction has come to equilibrium. K_c is the 'equilibrium constant' for the simple reaction (2.64) expressed in terms of concentrations. In the kinetic treatment it is given as the ratio of the forward to the backward reaction rate constants (k_1 and k_2) and is independent of composition at given temperature. Since the rate constants are normally strongly temperature dependent (Chapter 3), K_c is temperature dependent.

Equation (2.67) refers to the simple reaction (2.64). For a reaction of the more general form

$$(-\nu_A)A + (-\nu_B)B \rightleftharpoons \nu_C C + \nu_D D \tag{2.68}$$

the forward and backward reaction rates are given by (see Chapter 3)

$$(-\overrightarrow{r_A}) = k_1 C_A^{-\nu_A} C_B^{-\nu_B} \quad \text{and} \quad (\overleftarrow{r_A}) = k_2 C_C^{\nu_C} C_D^{\nu_D}$$

giving

$$K_c = \frac{k_1}{k_2} = \frac{C_{Ce}^{\nu_C} C_{De}^{\nu_D}}{C_{Ae}^{-\nu_A} C_{Be}^{-\nu_B}} \tag{2.69}$$

In general,

$$K_c = \prod_i (C_{ie})^{\nu_i} \tag{2.69a}$$

where the symbol $\prod_i$ denotes repeated product for all components i and C_{ie} is the equilibrium concentration of component i.

Provided the expressions for the backward and forward reaction rates as functions of concentration are valid, K_c in the above equations should be constant at given temperature, independent of the magnitudes of the concentrations C_{Ae}, C_{Be}, C_{Ce} and C_{De}. It is this property which makes the equilibrium constant useful in calculations. When, for example, the equilibrium constant is known, it is possible to calculate the equilibrium conversion attainable for given feed conditions (see Example 2.12).

Up to this point we have considered the equilibrium constant for a single elementary reaction such as (2.64) or (2.68). However, it may readily be shown that the overall equilibrium constants for reactions which proceed by a sequence of elementary steps will also be independent of composition being given by the product of the ratios of the backward to the forward rate constants for each step.

Suppose, for example, that the reaction

$$A + B \rightarrow D + E + F$$

proceeds via the elementary steps

$$A + B \xrightarrow{k_1} C + D$$

$$C \xrightarrow{k_3} E + F$$

with the corresponding backward reactions

$$C + D \xrightarrow{k_2} A + B$$

$$E + F \xrightarrow{k_4} C$$

Since the steps are elementary and at equilibrium the rates of the forward and backward reactions are equal,

$$k_1 C_{Ae} C_{Be} = k_2 C_{De} C_{Ce} \quad \text{and} \quad k_3 C_{Ce} = k_4 C_{Ee} C_{Fe}$$

giving

$$(k_1/k_2)(k_3/k_4) = \text{Constant at given temperature} = \frac{C_{De} \times C_{Ee} \times C_{Fe}}{C_{Ae} \times C_{Be}}$$

But $(C_{De} \times C_{Ee} \times C_{Fe})/(C_{Ae} \times C_{Be})$ is the overall equilibrium constant K_c for this reaction and is thus shown to be independent of composition.

$$K_c = \frac{C_{De} \times C_{Ee} \times C_{Fe}}{C_{Ae} \times C_{Be}}$$

Example 2.12

Components A and B in aqueous solution react according to equation (2.64), the equilibrium constant at 25°C being 30. If an M/10 solution of A is mixed with an equal volume of an M/5 solution of B in a batch reactor maintained at 25°C, what will be the equilibrium conversion (X_{Ae}) of component A?

Solution

Concentration of A immediately after mixing but before reaction
$= M/20 = C_{A0}$
Concentration of B immediately after mixing but before reaction
$= M/10 = C_{B0}$

If volume changes during the reaction are negligible,

C_{Ae} = concentration of A when equilibrium has been reached

$$= C_{A0}(1 - X_{Ae}) \tag{i}$$

C_{Be} = concentration of B when equilibrium has been reached

$$= C_{B0} - C_{A0}X_{Ae} \tag{ii}$$

C_{Ce} = concentration of C when equilibrium has been reached

$$= C_{A0}X_{Ae} \tag{iii}$$

C_{De} = concentration of D when equilibrium has been reached

$$= C_{A0}X_{Ae} \tag{iv}$$

Since components A and B react according to the equation

$$A + B \rightarrow C + D$$

the stoichiometric coefficients of A, B, C and D are $-1, -1, +1$ and $+1$ respectively and from (2.69) (or more directly from (2.67))

$$K_c = (C_{Ce}C_{De})/(C_{Ae}C_{Be}) = 30.0 \tag{v}$$

Setting $(C_{B0}/C_{A0}) = 2$ and combining results (i) to (v):

$$30(1 - X_{Ae})(2 - X_{Ae}) = X_{Ae}^2$$

that is,

$$X_{Ae} = (90 \pm 33.8)/58$$

From its definition X_{Ae} must lie between zero and unity. So

$$X_{Ae} = (90.0 - 33.8)/58 = 0.9696$$

Equilibrium conversion of A = 97%

In addition to K_c, a variety of other expressions for the equilibrium constant exist. For reactions in an ideal gas phase, for example, $C_i = (p_i/RT)$, where p_i is the partial pressure of component i, and it follows from (2.70) that

$$\left(\prod_i (p_{ie})^{\nu_i} \right)(RT)^{-\Sigma\nu} = \text{Constant} = K_c$$

where $\Sigma\nu$ is the increase in the number of moles during the reaction. The repeated product $\prod_i (p_{ie})^{\nu_i}$ is here designated K_p' giving

$$K_p' = \prod_i (p_{ie})^{\nu_i} = K_c(RT)^{-\Sigma\nu}$$

$$= \left(\prod_i (y_{ie})^{\nu_i} \right)(P)^{\Sigma\nu}$$

$$= K_c(RT)^{\Sigma\nu} \tag{2.70}$$

It is evident that K_p' defined as above will have dimensions if $\Sigma\nu \neq 0$.

In this regard it should be distinguished from the dimensionless equilibrium constant $\underline{K}_p$ obtained from the thermodynamic treatment described in the next section. However, if pressures are measured in bar, $\underline{K}_p$ is numerically equal to K_p'.

2.4.2 The thermodynamic treatment of chemical equilibrium in gas phase reactions

The kinetic approach described briefly above provides a valuable 'lead in' to the concept of an equilibrium constant and to the use of equilibrium constants for calculating equilibrium conversions. However, it can present pitfalls if used to infer kinetic information [5]. The thermodynamic approach described in the next section is not based on details of the kinetics of the reaction but is founded on the tendency of systems to minimise their Gibbs free energy at given temperature and pressure so that at equilibrium the Gibbs free energy is at its lowest value. The treatment is developed in terms of the standard Gibbs energy change for the reaction and the 'thermodynamic equilibrium constant' which is directly related to it. When applied to gas phase reactions under conditions where the ideal gas equations apply as a good approximation, this approach is more powerful than the kinetic one in that it enables equilibrium conversions to be calculated from pure component data only.

Before turning to the ideal gas approximation in detail, a brief summary will be given in Sections 2.4.2.1 to 2.4.2.3 of the principal concepts and quantities used in the thermodynamic approach in general and the way in which these quantities can be evaluated.

2.4.2.1 The standard Gibbs energy change for the reaction.
This quantity is defined in an analogous manner to the standard enthalpy change for the reaction (Section 2.3.5) and will be written $\Delta_r G^\theta(T)$ or, in some situations, $\Delta_r G_T^\theta$.

In words, $\Delta_r G^\theta(T)$ = increase in Gibbs function which occurs when stoichiometric amounts of the products of a reaction in their pure states are constituted from stoichiometric amounts of the pure reactants at temperature T. Both the initial reactants and the final products are in their **standard states** at temperature T (cf. Section 2.3.5). In symbols,

$$\Delta_r G^\theta(T) = \sum_i \nu_i G_i^\theta(T) \qquad (2.71)$$

where $G_i^\theta(T)$ is the molar Gibbs energy of the pure component i when in its standard state at temperature T (see also equation (A2.7)).

If the standard Gibbs energy change for a reaction is positive, the conversion which can be achieved from a stoichiometric feed is low, while if it is strongly negative the equilibrium conversion is high.

For homogeneous gas phase reactions the appropriate standard state for the reactants and products is the hypothetical ideal gas state at 1 bar pressure and the specified temperature. This is the standard state with which we will be concerned in the examples in this chapter. (For homogeneous liquid phase reactions the appropriate standard state for each reactant and product would normally be the liquid state at 1 bar pressure. For heterogeneous reactions, the standard states used are appropriate to the phase in which each reactant or product is present.)

The numerical value of the standard Gibbs energy change for a reaction clearly depends on the standard state selected. As a reminder of this, the symbol $\Delta_r G^{\theta,g}$ will be used where appropriate to denote standard Gibbs energy changes based on the ideal gas standard state at 1 bar.

2.4.2.2   The standard Gibbs energy of formation $(\Delta_f G_i^{\theta}(T)$ for each reactant and product i   By definition:

$\Delta_f G_i^{\theta}(T) =$ Increase in Gibbs energy associated with formation of one mole of compound i in its standard state from its elements at temperature T.   (2.72)

Standard Gibbs energies of formation for a wide range of compounds have been evaluated and are available in many texts and data banks, particularly at 298.15 K. (Values at other temperatures are available in some more specialised texts.) A few typical values at '298 K' are listed in Table 2.3. (As elsewhere, 298 is here used as an abbreviation for 298.15 K.) Having

Table 2.3 Standard Gibbs energies of formation of some selected compounds at 298.15 K

Substance	$\Delta_f G_{298}^{\theta}(\text{kJ mol}^{-1})$	Substance	$\Delta_f G_{298}^{\theta}(\text{kJ mol}^{-1})$
CO (g)	-137.17	N_2O_4 (g)	97.89
CO_2 (g)	-394.36	NH_3 (g)	-16.45
H_2 (g)	0	S_2 (g)	79.30
H_2O (l)	-237.13	SO_2 (g)	-300.19
H_2O (g)	-228.57	SO_3 (g)	-371.06
N_2 (g)	0	H_2SO_4 (l)	-690.00
NO_2 (g)	51.31		

obtained the standard free energies of formation for the reactants and products of a reaction at a temperature T_0 at which these data are available, the standard Gibbs energy change for the reaction at this temperature can be calculated from the equation

$$\Delta_r G^\theta(T_0) = \sum_i (\nu_i \Delta_f G_i^\theta(T_0)) \tag{2.73}$$

(see Example 2.13).

Equation (2.73) is analogous to equation (2.59) used to calculate the standard enthalpy change for a reaction from the standard enthalpies of formation, and the arguments leading to it are similar.

Having calculated the standard Gibbs energy change for a reaction at a temperature T_0 from equation (2.73), the corresponding thermodynamic equilibrium constant $\underline{K}$, discussed below, may be calculated from it. This quantity at other temperatures may then be calculated as described in Section 2.4.2.2.

The thermodynamic equilibrium constant is logarithmically related to the standard Gibbs energy change for the reaction and is linked in a simple way to pressure and composition in near-ideal gaseous systems.

Example 2.13

Use the standard Gibbs energy of formation data listed in Table 2.3 to calculate the standard Gibbs energy change $\Delta_r G^{\theta,g}$ for the reaction

$$2NO_2(g) \rightarrow N_2O_4(g) \text{ at } 25°C$$

Solution
At 25°C:

$$\Delta_f G_{N_2O_4}^{\theta,g} = 97.89 \text{ kJ mol}^{-1} (\text{Table 2.3}) \text{ also } \nu_{N_2O_4} = 1$$

$$\Delta_f G_{NO_2}^{\theta,g}(g) = 51.31 \text{ kJ mol}^{-1} (\text{Table 2.3}) \text{ also } \nu_{NO_2} = -2$$

From equation (2.73),

$$\Delta_r G_{298}^\theta = 97.89 - 2 \times 51.31 = -4.73 \text{ kJ mol}^{-1}$$

2.4.2.3   The thermodynamic equilibrium constant K.   This is related to the standard Gibbs energy change for the reaction (calculated with respect of specified component standard states) as follows:

$$RT \ln K = -\Delta_r G^\theta(T) \tag{2.74}$$

The numerical magnitude of K depends on the component standard states used in the evaluation of $\Delta_r G^\theta(T)$.

Because of the manner in which K and $\Delta_r G^\theta$ are defined (equations (2.71) and (2.74)), K is in every case a function of temperature but not of pressure or composition. It is positive and dimensionless and is a useful instrument for equilibrium calculations.

2.4.3 The thermodynamic equilibrium constant based on ideal gas standard state at 1 bar

We shall be concerned in this chapter with standard Gibbs energy changes $\Delta_r G^{\theta,g}$ which have been evaluated taking the ideal gas state at 1 bar and the specified temperature as standard state for all components participating in the reaction. We shall designate the corresponding equilibrium constant as $\underline{K}$ so that

$$RT \ln \underline{K} = -\Delta_r G^{\theta,g}(T) = -\sum_c \nu_i G_i^{\theta,g} \tag{2.74a}$$

where $G_i^{\theta,g}$ is the molar Gibbs energy of pure component i normalised to the ideal gas state at 1 bar pressure.

As outlined in Appendix 2.1, $\underline{K}$ can be related to the fugacities of the components present in the equilibrium mixture formed when the reaction has ceased by the equation

$$\underline{K} = \prod_i (f_{ie}/p^\theta)^{\nu_i} \tag{2.75}$$

where ν_i is the stoichiometric coefficient for each component i (taken as positive for products and negative for reactants) f_{ie} is the fugacity of each component i when equilibrium has been reached, p^θ is the standard state pressure (1 bar) and the symbol $\prod_i$ indicates that a repeated product of the ratio $(f_{ie}/p^\theta)^{\nu_i}$ is to be taken for all components present (using the same terminology $\prod_{i=1}^3 x_i = x_1 x_2 x_3$). For the NO_2 dimerisation reaction, for example, written as in Example 2.13

$$\underline{K} = \left(f_{N_2O_4}/(f_{NO_2})^2\right)p^\theta$$

where $p^\theta = 1$ bar.

2.4.3.1 Thermodynamic equilibrium constant under ideal gas conditions expressed in terms of partial pressures or mole fractions. **If the gas phase is near-ideal**, the component fugacities f_i approximate closely to the corresponding partial pressures p_i, and so

$$\left[\underline{K} = \prod_i (p_{ie}/p^\theta)^{\nu_i}\right]_{\text{ideal gas}} \tag{2.76}$$

in an ideal gas reaction mixture. p_{ie} in equation (2.76) is the partial pressure of component i in the equilibrium mixture and $p^\theta = 1$ bar.

The repeated product is taken over all components participating in the reaction. It is convenient to use the symbol $\underline{K}_p$ to describe this repeated product.

By definition, therefore,

$$\underline{K}_p = \prod_i (p_{ie}/p^\theta)^{\nu_i} = \left(\prod_i (p_{ie})^{\nu_i} \right) (p^\theta)^{-\Sigma \nu_i} \qquad (2.77)$$

Under ideal gas conditions, $\underline{K}_p = K = \exp(-\Delta_r G^{\theta,g}/RT)$. $\underline{K}_p$, defined by equation (2.77), can of course be calculated from equilibrium data under any conditions, but it is only under ideal gas conditions that it is equal to the thermodynamic equilibrium constant and can be predicted from standard Gibbs energy change of the reaction and, hence, from standard Gibbs energy of formation data for the pure components.

Since the partial pressure of i is given by

$$p_i = P y_i$$

where y_i is the mole fraction of component i in the gas mixture and P is the total pressure, $\underline{K}_p$ can be written in terms of mole fractions.

$$\underline{K}_p = \prod_i (y_{ie} P/p^\theta)^{\nu_i} = \left(\prod_i (y_{ie})^{\nu_i} \right) (P/p^\theta)^{\Sigma \nu_i} \qquad (2.78)$$

For the dimerisation reaction

$$2NO_2 \rightarrow N_2O_4,$$

for example, $\nu_{NO_2} = -2, \nu_{N_2O_4} = 1, \Sigma \nu = -1$. So

$$\underline{K}_p = \left(y_{N_2O_4}/y_{NO_2}^2 \right) (p^\theta/P).$$

where $y_{N_2O_4}$ and y_{NO_2} are the equilibrium mole fractions of N_2O_4 and NO_2.

Equation (2.78) is very useful since, as we shall see, it enables gas compositions at the equilibrium point to be calculated from the equilibrium constant and feed composition data.

$\underline{K}_p$ defined equation (2.77) is of course always dimensionless though it obviously bears a strong resemblance to the quantity K_p' given by the kinetic approach (equation (2.70)).

If partial pressures are measured in bar, K_p' and $\underline{K}_p$ are numerically equal, though they differ in that K_p' has dimensions if the reaction results in an increase or decrease in the number of molecules. To maintain dimensional consistency it is best to use the dimensionless form $\underline{K}_p$ in thermodynamic calculations.

Example 2.14

(1) Calculate the thermodynamic equilibrium constant $\underline{K}$ for the association of NO_2 at 25°C from the standard Gibbs energy change for the reaction.
(2) Assuming ideal gas behaviour, calculate the molar ratio of NO_2 to N_2O_4 in the equilibrium mixture at 25°C and 1 bar pressure.

Solution

(1) From Example 2.13, the standard Gibbs energy change $\Delta_r G_{298}^{\theta,g}$ for the reaction is -4.73×10^3 J mol^{-1}. So

$$\ln K = 4730/(8.214 \times 298.15)$$
$$= 1.908$$
$$\underline{K} = 6.74$$

Under ideal gas conditions $\underline{K} = \underline{K_p}$, where, from equation (2.78) and following text,

$$\underline{K_p} = \left(y_{N_2O_4}/y_{NO_2}^2 \right) (p^\theta/P)$$

p^θ is 1 bar and P the system pressure (1 bar in the present example).
   At equilibrium, therefore,

$$(y_{N_2O_4})/(y_{NO_2})^2 = (1 - y_{NO_2})/(y_{NO_2})^2 = 6.74$$

that is:

$$y_{NO_2} = 0.32 \quad \text{and} \quad y_{N_2O_4} = (1 - y_{NO_2}) = 0.68$$

Ratio of NO_2 to N_2O_4 in the equilibrium mixture $= 0.32/0.68 = 0.47$

2.4.3.2   Dependence of the standard Gibbs energy change for a reaction and also of the thermodynamic equilibrium constant on temperature: Calculation of equilibrium constants at other temperatures from value of 298.15 K.   The standard Gibbs energy change $\Delta_r G^\theta$ varies with temperature according to the equation

$$\left(\frac{d(\Delta_r G^\theta/T)}{dT} \right) = -\left(\frac{\Delta_r H^\theta}{T^2} \right) \tag{2.79}$$

where $\Delta_r H^\theta$ is the standard enthalpy change for the reaction, which is itself a function of temperature. (Equation (2.79) is obtained by applying the Gibbs – Helmholtz equation

$$H = -T^2 \left(\frac{\partial}{\partial T} \left(\frac{G}{T} \right) \right)_p$$

to the reactants and products of the reaction in their standard states as described in reference [7].)

The dependence of the thermodynamic equilibrium constant K on temperature is obtained by combining equation (2.79) with the defining equation for K (equation (2.74)),

$$\left(\frac{d \ln K}{dT}\right) = \left(\frac{\Delta_r H^{\theta}}{RT^2}\right) \tag{2.80}$$

or

$$\left(\frac{d \ln K}{d(1/T)}\right) = -\left(\frac{\Delta_r H^{\theta}}{R}\right) \tag{2.81}$$

where $\Delta_r H^{\theta}$ is the standard enthalpy change for the reaction.

Having calculated K at, for example, $298.15\,K$ from standard Gibbs energy of formation data, the value at other temperatures may be found by integrating equation (2.80) or (2.81) over the appropriate temperature range. Over limited temperature ranges the enthalpy change for the reaction $(\Delta_r H^{\theta})$ may be regarded as constant so that, from equation (2.81)

$$\ln(K) = A + B/T \tag{2.82}$$

where $B = (-\Delta_r H^{\theta}/R)$. This form of equation is often used to represent equilibrium constants as functions of T.

2.4.3.3 Application of the thermodynamic equations to the calculation of chemical equilibrium in near-ideal gaseous systems. As pointed out in Section 2.1.4.2, many (and probably most) industrial gas phase reactions take place under conditions where the gas is near-ideal (this is not true in every case, of course, the ammonia synthesis reaction is a well-known example of a reaction, which, at the pressures currently employed, takes place under distinctly non-ideal conditions [10]). Under near-ideal conditions the thermodynamic equations become very easy to use and enable unambiguous predictions to be made based on tabulated pure component data only. The remainder of this chapter is devoted to examples of equilibrium and thermal calculations in homogeneous gaseous systems of this type. The relevant standard state condition in this situation is the ideal gas state at 1 bar for all participating components.

The equilibrium equations which will be used are summarised below. In these equations the superscript θ, g indicates that the ideal gas standard state at 1 bar applies and p^{θ} = standard state pressure = 1 bar.

$$\Delta_r G^{\theta,g} = \sum_i (\nu_i \Delta_f G_i^{\theta,g}) \tag{2.83}$$

$$RT \ln \underline{K}_p = -\Delta_r G^{\theta,g} \tag{2.84}$$

$$K_p = \prod_i (p_i/p^\theta) \tag{2.85}$$

$$\frac{\mathrm{d}\ln \underline{K_p}}{\mathrm{d}T} = \frac{\Delta_r H^{\theta,g}}{RT^2} \tag{2.86}$$

In the above equations, the summation $\sum$ and the repeated product $\prod$ are taken over all components i participating in the reaction.

To illustrate the use of the above equations, examples will centre on the oxidation of SO_2

$$SO_2 + \frac{1}{2}O_2 \rightarrow SO_3 \tag{2.87}$$

a reaction which is currently carried out at temperatures between 600 and 900 K and pressures which are usually between 1.2 and 1.5 bar, though pressures up to 8 bar are used in a few cases [10]. In the following examples:

1. $\underline{K_p}$ for the above reaction will be obtained at 25°C.
2. Based on this value, $\underline{K_p}$ will be calculated at the higher temperatures required for plant operations, equation (2.86) being integrated for this purpose.
3. The equilibrium constants at the upper end of the temperature range considered will be fitted to an equation of the form of (2.82).
4. Equilibrium conversions for a given feed composition will be calculated from the equilibrium constant data.
5. The conversion achievable in an idealised adiabatic reactor in which the bed outlet conversion is equal to the equilibrium value will be calculated.

Example 2.15 *Calculation of $\underline{K_p}$ for the SO_2 oxidation reaction (equation (2.87)) at 298.15 K, assuming ideal gas conditions so that $K_p = \underline{K}$.*
From Table 2.3, the standard Gibbs energies of formation of SO_3 and SO_2 in the ideal gas state at 1 bar are:

$$\Delta_f G^{\theta,g}_{SO_3} = -371\,060 \text{ J mol}^{-1}$$

$$\Delta_f G^{\theta,g}_{SO_2} = -300\,190 \text{ J mol}^{-1}$$

Since O_2 is an element

$$\Delta_f G^{\theta,g}_{O_2} = 0$$

From the reaction stoichiometry equation (2.87),

$$\nu_{SO_3} = 1, \nu_{SO_2} = -1 \quad \text{and} \quad \nu_{O_2} = \left(-\frac{1}{2}\right)$$

Hence (from equations (2.83) and (2.84))

$$-\ln \underline{K}_p = (-371\,060 + 300\,190 - 0)/(8.314 \times 298.15) = -28.59$$

that is,

$$\underline{K}_p = 2.61 \times 10^{12} \text{ at } 298.15\,\text{K}$$

Provided that SO_2 is the limiting reactant the above very large equilibrium constant would correspond to virtually complete conversion of SO_2 to SO_3. To obtain a measurably rapid reaction rate, however, it is necessary to use catalysts and to raise the temperature substantially above this reference value. (Catalysts do not alter the reaction equilibrium but do improve reaction rates and/or selectivity.)

A method for calculating equilibrium constants at higher temperatures from the value at 25°C is described in the next example. This method is based on equation (2.86) and allows for the temperature dependence of the standard enthalpy change for the reactions. As noted earlier, useful expressions over more limited ranges of temperature may often be obtained by regarding this enthalpy change as independent of temperature.

Example 2.16 *Calculation of $\underline{K}_p$ for the SO_2 oxidation reaction at a series of temperatures between 750 K and 905 K by thermodynamic extrapolation from the value at 298.15 K.*
Using the expression for $\Delta_r H^{\theta,g}$ as a function of T given by equation (2.62) together with the heat capacity equations for SO_2, O_2 and SO_3 given in Example 2.11, calculate thermodynamic equilibrium constants for the reaction

$$SO_2 + \tfrac{1}{2}O_2 \rightarrow SO_3$$

at 750, 800, 850, 900 and 905 K.
Solution
A general expression for $\ln \underline{K}_p$ as a function of T will first be obtained. The required values for T will then be substituted in this expression.
From equations (2.86) and (2.62),

$$\ln \underline{K}_p(\text{T}) = \ln \underline{K}_p(298) + (1/R)$$

$$\times \left[\int_{298}^{T} \frac{Z}{T^2}\mathrm{d}T + \int_{298}^{T} (a/T)\mathrm{d}T + \int_{298}^{T} (b/2)\mathrm{d}T + \int_{298}^{T} (cT/3)\mathrm{d}T \right. \qquad (2.88)$$

$$\left. + \int_{298}^{T} (\underline{d}T^2/4)\mathrm{d}T \right]$$

where

$$Z = \Delta_r H^{\theta}(298) - 298\underline{a} - (298^2/2)\underline{b} - (298^3/3)\underline{c} - (298^4/4)\underline{d} \qquad (2.63)$$

Carrying out the integrations,

$$\ln \underline{K}_p(T) = \ln \underline{K}_p(298) + \frac{1}{R}[A(T) - D] \qquad (2.89)$$

where

$$A(T) = \left[-\left(\frac{Z}{T}\right) + \underline{a} \ln T + (\underline{b}T/2) + (\underline{c}T^2/6) + (\underline{d}T^3/12)\right] \qquad (2.89a)$$

$$D = -\left[-\left(\frac{Z}{298}\right) + \underline{a} \ln(298) + (\underline{b} \times 298/2) + (\underline{c} \times 298^2/6) + (\underline{d} \times 298^3/12)\right] \qquad (2.89b)$$

Inserting into (289b) the values

$$Z = -95\,380 \text{ J mol}^{-1}$$
$$\underline{a} = -21.53 \text{ J mol}^{-1}\text{K}^{-1}$$
$$\underline{b} = 0.0789 \text{ J mol}^{-1}\text{K}^{-2}$$
$$\underline{c} = -7.11 \times 10^{-5} \text{ J mol}^{-1}\text{K}^{-3}$$
$$\underline{d} = 1.39 \times 10^{-8} \text{ J mol}^{-1}\text{K}^{-4}$$

obtained in Example 2.11 gives

$$D = 207.98 \text{ J mol}^{-1}\text{K}^{-1}$$

On inserting this value for D into equation (2.89) together with $A(T)$ from (2.89a) and $\ln \underline{K}_p(298)$ from Example 2.11, the equilibrium constants given in Table 2.4 are obtained at the specified temperatures.

A very sharp fall in the equilibrium constant with increase in temperature for this (exothermic) system is apparent from Table 2.4.

Example 2.17   *Use equation (2.82) to represent* $\ln \underline{K}_p$ *as a function of T over the temperature range 750 to 905 K.*
According to equation (2.82)

$$\ln \underline{K}_p = A + (B/T)$$

where $B = (-\Delta_r H^{\theta,g}/R)$.

In the example below A is calculated from $\ln \underline{K}_p$ at 750 K together with an arithmetic average value for $\Delta_r H^{\theta,g}$ over the range 750 to 905 K, the expression for A being

$$A = \left[\ln \underline{K}_p(T_0) + \Delta_r H^{\theta,g}(AV)/RT_0\right]$$

where $T_0 = 750$ K.

In Example 2.11 it was found that the values of $\Delta_r H^{\theta,g}$ at 750 and 905 K were -98.24 and -97.81 kJ mol^{-1} respectively, the arithmetic mean being -98.03 kJ mol^{-1}, giving $(\Delta_r H^{\theta,g}(\Delta V)/R) = -11\,790$ K.

Inserting $\ln \underline{K}_p = 4.54$ at 750 K from Table 2.4 gives

$$A = 4.54 - (11\,790/750) = -11.18$$

and

$$\ln \underline{K}_p = (11\,790/T) - 11.18 \qquad (2.90)$$

When values of $\underline{K}_p$ calculated from (2.90) in the range between 750 and 905 K are compared with the values shown in the last column of Table 2.4 it is found that agreement is for practical purposes exact.

In conclusion, it is seen that expressions of the form

$$\ln \underline{K}_p = A + B/T$$

can be simple and accurate when applied over a limited temperature range.

Example 2.18 *Calculation of equilibrium conversions (X_{Ae}) assuming ideal gas conditions.*

Generate data for a graph showing equilibrium conversion of SO_2 in the SO_2 oxidation reaction equation (2.87) as a function of temperature at 1 bar pressure for the following feed composition:

SO_2:   9.0 mol%
O_2:   12.0 mol%
N_2:   79.0 mol%

(The above feed may be regarded as fed either to a batch reactor or to a continuous flow reactor (of infinite size!, operating at steady state.) The simplest way of establishing the dependence of X_{Ae} on temperature is in fact to do the reverse and to calculate the temperatures at which X_{Ae} takes a series of preset values. In the present example the equilibrium conversions 0.65, 0.7, 0.8, 0.9 and 0.95 are considered.

Table 2.4 Equilibrium constants for the SO_2 oxidation reaction

T (K)	$A(T)$(J mol^{-1}K^{-1})	$\ln \underline{K}_p$	$\underline{K}_p$
298.15	207.98	28.590	2.6×10^{12}
750	8.053	4.542	93.9
800	-0.126	3.560	35.1
850	-7.328	2.692	14.8
900	-13.727	1.923	6.8
905	-14.325	1.851	6.4

1. A calculation **basis** is chosen (in the present instance 100 moles of feed is convenient, though not mandatory) and on this basis the number of moles of each component i, at each conversion, is calculated.
2. Knowing the number of moles of each component the composition of the mixture is obtained at each conversion and (knowing the total pressure) the partial pressure of each component i (in this case the partial pressure of SO_2, O_2, SO_3 and N_2) is evaluated.
3. The quotient $\prod_i (p_i/p^\theta)^{\nu_i}$ is then generated from these partial pressures for each conversion. (In this case the stoichiometric coefficients ν for SO_3, SO_2 and O_2 are 1, -1 and $-\frac{1}{2}$ respectively and the required quotient is

$$\frac{p_{SO_3}/p^\theta}{(p_{SO_2}/p^\theta)(p_{O_2}/p^\theta)^{1/2}}$$

or

$$\left(\frac{p_{SO_3}}{p_{SO_2}(p_{O_2})^{1/2}}\right)(p^\theta)^{1/2}$$

where $p^\theta = 1$ bar and the partial pressures are expressed in bar.)

   At the equilibrium point this quotient (sometimes called the reaction quotient [6]) is equal to the equilibrium constant $\underline{K}_p$ for the reaction.
4. The temperature at which $\underline{K}_p$ is equal to the above quotient is then determined for each conversion from the known temperature dependence of $\underline{K}_p$, (in this case from equation (2.90):

$$\ln \underline{K}_p = (11\,790/T) - 11.18 \qquad (2.90)$$

Details of the calculation for the SO_2 oxidation reaction at an equilibrium conversion of 0.65 are given below.

A basis of 100 moles of feed is used. On this basis

Moles SO_2 fed $= 9.0$.
Moles SO_2 used up $= 9.0 \times 0.65 = 5.85$
Final moles $SO_2 = 9.0(1 - 0.65) = 3.15$

Moles O_2 fed $= 12.0$
Moles O_2 used up $= (5.85/2) = 2.93$
Final moles $O_2 = 12.0 - 2.93 = 9.07$

Final moles N_2 = moles N_2 fed $= 79$
Final moles SO_3 = moles SO_2 used up $= 5.85$

Total number of moles present when conversion is 0.65 is

$$3.15 + 9.07 + 79.0 + 5.85 = 97.07.$$

Partial pressure of $SO_2 = \left(\dfrac{3.15}{97.1}\right) \times 1.0$ bar $= 0.0324$ bar

Partial pressure of $O_2 = \left(\dfrac{9.07}{97.1}\right) \times 1.0$ bar $= 0.0934$ bar

Partial pressure of $SO_3 = \left(\dfrac{5.85}{97.1}\right) \times 1.0$ bar $= 0.0602$ bar

At the equilibrium point,

$$K_p = \left(p_{SO_3} / \left(p_{SO_2} p_{O_2}^{1/2}\right)\right) (p^{\theta})^{1/2} \qquad = 0.0602/(0.0324 \times 0.0934^{1/2})$$
$$= 6.08$$

($p^{\theta} = 1$ bar). Hence, from equation (2.90)

$$T = \left(\frac{11\,790}{\ln(6.08) + 11.18}\right) = 908 \text{ K}$$

At 908 K, the equilibrium conversion is thus 0.65. Proceeding in the same way, the temperatures required for equilibrium conversions of 0.70, 0.8, 0.9 and 0.95 are found to be 892, 855, 808 and 766 K respectively. These and other values derived as above from equation (2.90) for the given feed composition are shown plotted on Figure 2.16. The sharp fall in equilibrium conversion with increase in temperature is shown. As is apparent in the above example, the equilibrium conversion at given temperature depends on the composition of the feed and in the example would be higher for lower sulphur dioxide concentrations.

2.4.3.4 Calculation of equilibrium conversion from feed temperature and composition in an adiabatic reactor or in a reactor divided into adiabatic sections: representation of the state of the reaction fluid on a temperature/ conversion diagram. The equilibrium conversion is the conversion which would be achieved in an idealised reactor in which the contact time was infinite. It can obviously never be exactly achieved in practice, though quite a close approach to it can be obtained in some situations. It provides an upper limit to reactor conversions and is a useful concept in reactor design for reversible systems.

The calculation of the equilibrium conversion for an isothermal reactor from the feed conditions presents no problems provided equilibrium conversion versus temperature data (calculated as in Example 2.18) are available for the feed composition and pressure of interest. For an isothermal reactor with a feed temperature of 800 K and the feed composition and pressure shown in Figure 2.16, for example, the equilibrium conversion can be read directly from the equilibrium curve as 0.91.

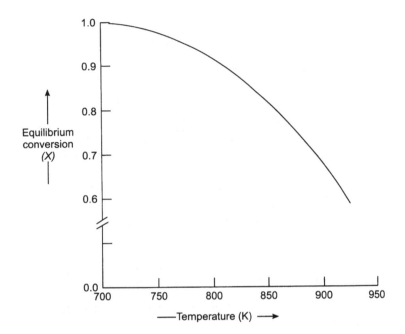

Figure 2.16 Equilibrium conversion diagram for SO_2 in reaction $SO_2 + (1/2)O_2 \rightarrow SO_3$ (reactor feed of 9 mol% SO_2, 12 mol% O_2 and 79 mol% N_2). Pressure: 1 bar.

In the case of an adiabatic reactor, however, the system temperature is not constant but varies with conversion, from which it may be calculated using the enthalpy balance equation (2.51). For exothermic systems the temperature rises as conversion increases while for endothermic systems the reverse is true. In each case the result of the temperature change is to cause the equilibrium conversion to deteriorate as the reaction proceeds.

To overcome the deleterious effects of these temperature changes, some reactors (usually catalytic reactors) are divided into adiabatic sections. Heat transfer takes place between the sections to restore the temperature of the reactor fluid entering each section to a value close to that at the reactor inlet.

In these situations, the reactor as a whole is not adiabatic though the individual sections are.

In illustration of this it is convenient to revert to the SO_2 oxidation reaction (equation (2.87)). In practice, this reaction is usually [10] carried out in a series of about four adiabatic beds of catalytic material with interbed cooling, as shown schematically in Figure 2.17.

Since the oxidation reaction is exothermic, the temperature of the reaction mixture increases on passage through each bed until either the equilibrium conversion at the enhanced temperature is approached as closely as is economically feasible or an upper limit to the catalyst operating range is

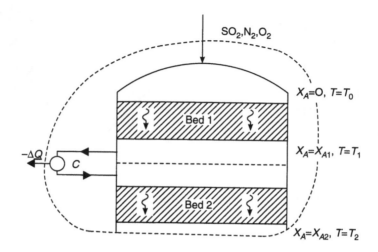

Figure 2.17 Schematic diagram of upper part of a typical reactor for converting SO_2 to SO_3. (The lower part would typically contain about two additional beds of catalyst.)

reached. In either case the reactant/product mixture is then cooled before passage to the next bed of catalyst.

It is convenient to represent the state of the reactant/product mixture as it passes through the sequence of beds and heat exchangers on a conversion/ temperature diagram. Figure 2.18, for example, depicts the variations in temperature and conversion which would occur in the first two beds of an SO_2 converter in an idealised situation in which the mixture leaving each bed has come to equilibrium at the bed outlet temperature.

Both beds operate adiabatically and the only transfer of heat from the surroundings in the sections of the reactor shown is at the interbed cooler (C) in Figure 2.17.

The increase in temperature which occurs as the reaction mixture passes through the first bed is given by the enthalpy balance equation (equation (2.51)) and is shown as line 1–2 on Figure 2.18.

As equilibrium is approached the reaction rate progressively falls and a limit to the conversion achievable in a single adiabatic bed is set by the intersection of line 1–2 with the equilibrium curve which occurs at a temperature of 913 K. On emerging from the first reactor bed the reactant/ product stream enters the cooler and its state then follows the horizontal line 2–3 on Figure 2.18. Along this line the temperature falls from 913 K to 750 K while the conversion remains constant at 0.634. The mixture then enters the second catalyst bed where reaction is resumed and path 3–4 is followed. The construction of lines 3–4 and 1–2 from thermal data for the systems is described in Examples 2.19 and 2.20.

It is noteworthy that the data used to generate the equilibrium curve and also the construction lines 1–2 and 3–4 in Figure 2.18 are (apart from the stoichiometric coefficients) all pure component properties or are calculated from them. Figure 2.18 has in fact been largely generated from pure component data!

Example 2.19
Obtain expressions for the conversion (X_A) of SO_2 as a function of temperature in bed 1 of the reactor shown in Figure 2.17 (i.e. along line 1–2 in Figure 2.18). Feed conditions are as specified beneath Figure 2.16.

Solution
Since the bed is adiabatic, the enthalpy decrease due to (exothermic) reaction at the inlet temperature T_0 may be pictured as equal and opposite to the enthalpy increase when product material is heated from the inlet temperature to the bed temperature T. The conversion X_A of SO_2 is then given as a function of T by equation (2.51)

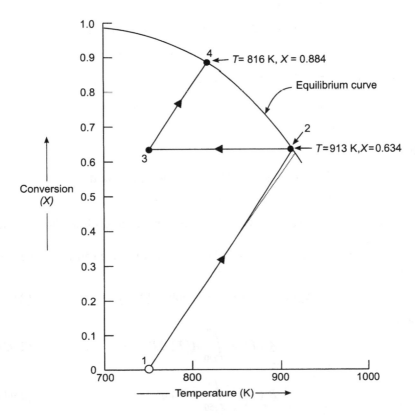

Figure 2.18 Temperature and conversion of reaction mixture passing through the first two beds of reactor for converting SO_2 to SO_3 in hypothetical situation where equilibrium is achieved at each bed outlet.

$$X_A = \left(\frac{-1}{\underline{n}_{A0}\Delta H_{R,A}(T_0)}\right) \int_{T_0}^{T} \underline{C}_p(X_A, T)dT \qquad (2.51)$$

where $\underline{n}_{A0}$ is the number of moles of limiting reactant (in the present case SO_2) entering the bed and $\Delta H_{R,A}(T_0)$ is the enthalpy of reaction per mole of limiting reactant consumed at the inlet temperature T_0, $\underline{C}_p(X_A, T)$ is the heat capacity of the reactor fluid when the conversion of A is X_A and the temperature is T.

Evaluation of terms in equation (2.51)

In the present example T_0 is 750 K and $\Delta H_{R,A}(T_0)$ is $-98.24 \times 10^3 \, \mathrm{J\,mol^{-1}}$. Taking 100 moles of gas entering the bed as basis:

$\underline{n}_{A0}$ = moles SO_2 entering bed = 9.0 and moles of SO_2 when conversion
   is $X_A = 9.0(1 - X_A)$

   Moles O_2 entering bed = 12.0 and moles O_2 when conversion is
   $X_A = (12.0 - 4.5X_A)$

   Moles N_2 entering bed = moles N_2 at conversion X_A

$$\underline{n}_{A0}\Delta H_{R,A}(T_0) = -88.42 \times 10^3 \, \mathrm{J\,mol^{-1}}$$

Using the additivity rule for total heat capacity (Section 2.3.3.1)

$$\begin{aligned}\underline{C}_p &= 9(1 - X_A)C_{pm\,SO_2} + (12 - 4.5X_A)C_{pm\,O_2} + 9X_A C_{pm\,SO_3} + 79C_{pm\,N_2} \\ &= (\underline{A}(T) + [\underline{B}(T)]X_A)\end{aligned} \qquad (2.91)$$

where

$$\underline{A}(T) = 9C_{pm\,SO_2} + 12C_{pm\,O_2} + 79C_{pm\,N_2} \quad \text{and}$$
$$\underline{B}(T) = 9C_{pm\,SO_3} - 9C_{pm\,SO_2} - 4.5C_{pm\,O_2}$$

$C_{pm\,SO_2}$ is the molar heat capacity of SO_2 and the notation is similar for the other components.

Equations (2.51b) and (2.91) may be combined to give (on including the value of $\underline{n}_{A0}\Delta H_{R,A}$)

$$X_A(T) = \underline{A}^1(T)/(884\,200 - \underline{B}^1(T)) \qquad (2.92)$$

where

$$\underline{A}^1(T) = \int_{750}^{T} \underline{A}(T) \, dT \qquad (2.93)$$

$$\underline{B}^1(T) = \int_{750}^{T} \underline{B}(T) \, dT \qquad (2.94)$$

The molar heat capacities (and hence $\underline{A}(T)$ and $\underline{B}(T)$) are temperature dependent, but over the temperature range considered, not strongly so, as

Table 2.5 Temperature dependence of heat capacity given by $C_{pm} = a + bT + cT^2 + dT^3$

Gas	a $(\text{J mol}^{-1}\text{K}^{-1})$	b $(\text{J mol}^{-1}\text{K}^{-2})$	$c \times 10^5$ $(\text{J mol}^{-1}\text{K}^{-3})$	$d \times 10^8$ $(\text{J mol}^{-1}\text{K}^{-4})$	C_{pm} at 750 K $(\text{J mol}^{-1}\text{K}^{-1})$	C_{pm} at 900 K $(\text{J mol}^{-1}\text{K}^{-1})$
$SO_2(1)$	23.85	0.0670	-4.961	1.328	51.80	53.65
$O_2(2)$	28.11	0.0000	$+1.746$	-1.065	33.43	34.49
$SO_3(3)$	16.37	0.1459	-11.200	$+3.242$	76.47	80.59
$N_2(4)$	31.15	-0.0136	2.679	-1.168	31.12	32.13

shown in the last columns of Table 2.5. For this reason, the full calculation of $\underline{A}^1(T)$ and $\underline{B}^1(T)$, allowing for this temperature dependence, gives results very similar to those obtained in approximate calculations in which $\underline{C}_p$ is taken as independent of temperature. Nevertheless the full calculation is outlined briefly below.

Molar heat capacities are given as power series in temperature in several sources and some prefered values of the coefficients in the series

$$C_{pm} = a + bT + cT^2 + dT^3$$

are given in Table 2.5 for the components $SO_2[1], O_2[2], SO_3[3]$ and $N_2[4]$.

Insertion of the power series expressions for $C_{pm\,SO_2}$, $C_{pm\,O_2}$, $C_{pm\,SO_3}$ and $C_{pm\,N_2}$ in (2.91) gives

$$\underline{A}(T) = \underline{a} + \underline{b}T + \underline{c}T^2 + \underline{d}T^3 \quad \text{and} \quad \underline{B}(T) = \bar{a} + \bar{b}T + \bar{c}T^2 + \bar{d}T^3$$

where

$$\underline{a} = 9a_1 + 12a_2 + 79a_4$$
$$\bar{a} = -9a_1 - 4.5a_2 + 9a_3$$
$$\underline{b} = 9b_1 + 12b_2 + 79b_4$$
$$\bar{b} = -9b_1 - 4.5b_2 + 9b_3$$

and so on for the remaining terms.

$\underline{A}^1(T)$ and $\underline{B}^1(T)$ are then given by

$$\underline{A}^1(T) = \left[\underline{a}T + (\underline{b}/2)T^2 + (\underline{c}/3)T^3 + (\underline{d}/4)T^4\right]_{750}^{T}$$

and

$$\underline{B}^1(T) = \left[\bar{a}T + (\bar{b}/2)T^2 + (\bar{c}/3)T^3 + (\bar{d}/4)T^4\right]_{750}^{T}$$

where the coefficients $\underline{a}, \bar{a}$, etc., are as given above.

Values of $\underline{A}^1(T)$ and $\underline{B}^1(T)$ at several temperatures and the values of X_A obtained by inserting them in Equation (2.92) are given in Table 2.6.

These and other values are shown as line 1–2 on Figure 2.18 and the progressive increase in temperature with conversion is apparent. The relationship is nearly linear though line 1–2 is very slightly curved upward at the highest temperatures. Line 1–2 intersects the equilibrium curve at

Table 2.6 Conversions $X_A(T)$ of SO_2 in bed 1 (Figure 2.18) obtained as described in the text

$T(K)$	$\underline{A}^1(T) \times 10^{-2}$ (J mol^{-1}) $\underline{B}^1(T)$ (J mol^{-1})		$X_A(T)$
750	0	0	0
800	16.73	3704	0.190
850	3363	7738	0.384
915	5589	13 280	0.642

a temperature of 913 K and an SO_2 conversion of 0.634. These are the equilibrium temperature and conversion respectively for the single-stage adiabatic processing of the feed shown beneath Figure 2.16.

The broken line marginally below line 1–2 is obtained by taking the molar heat capacities to be independent of temperature and equal to their values at 750 K. $\underline{C}_p$ at 750 K is obtained by substituting molar heat capacities at 750 K from Table 2.5 into Equation (2.91), giving $\underline{C}_p = 3325 + 71.6X$.

Substituting this value into Equation (2.51) gives

$$X_A = 3325\Delta T/(884\,160 - 71.6\,\Delta T) \quad (\Delta T = T - 750 \quad) \qquad (2.95)$$

or

$$T = 750 + \left(\frac{884\,160X_A}{3325 + 71.6X_A}\right)$$

A further degree of approximation can be obtained by regarding the term $71.6\,X_A$ in this equation as negligible compared with 3325 (X_A can never exceed unity), i.e. by regarding $\underline{C}_p$ as independent both of T and X_A. This gives:

$$T = 266X_A + 750$$

To this degree of approximation, the variation of T with X_A is accurately linear. Both approximations give results which are very close to those from the full calculation, giving errors of only 2.0 and 3.3% respectively at the highest temperature.

Example 2.20

Obtain an expression for the conversion of SO_2 as a function of temperature in bed 2 of the unit sketched in Figure 2.17 (i.e. along line 3–4 in Figure 2.18). The feed to bed 1 is as in Example 2.19 and is given beneath Figure 2.18. The conversion of the material leaving bed 1 is assumed to approximate closely to the equilibrium conversion at the bed 1 outlet temperature.

Solution

The basis for the calculation (100 moles feed entering bed 1) will be as in Example 2.19. The system now to be considered extends from the inlet to bed 1, through bed 1 and the cooler and at least partly through bed 2. (The maximum extent is when all bed 2 is included as is shown by the broken envelope curve in Figure 2.17.) Since heat is removed from this system by the cooler, the relevant energy equation is the non-adiabatic equation (2.50):

$$\Delta \underline{Q} = (\underline{H} - \underline{H}_{in}) = \underline{n}_{A0} X_A \Delta H_{R,A}(T_0) + \int_{T_0}^{T} \underline{C}_p(X_A, T) dT \qquad (2.50)$$

which rearranges to give

$$X_A = -\left(\frac{1}{\underline{n}_{A0} \Delta H_{R,A}(T_0)}\right)\left(\int_{T_0}^{T} \underline{C}_p(X_A, T) dT - \Delta \underline{Q}\right) \qquad (2.96)$$

In the present case,

$\underline{n}_{A0} = 9.0 =$ number of moles of SO_2 entering bed 1.

$\quad\quad X_A =$ total conversion achieved by reactor fluid after passing through bed 1 and into bed 2.

$\Delta H_{R,A}(T_0) =$ enthalpy of reaction of SO_2 per mole of SO_2 consumed at the inlet temperature to bed 1 (750 K).

$\quad\quad = -98.4 \times 10^3 \, J \, mol^{-1}$

$\underline{C}_p(X_A, T) =$ total heat capacity of reactor fluid at point in bed 2 where conversion is X_A and temperature is T. The dependence of $\underline{C}_p$ on X_A and T is, as in Example 2.19, given by Equation (2.91).

$(-\Delta \underline{Q}) =$ heat removed by cooler = heat which must be removed to reduce the temperature of the fluid leaving bed 1 from 913 K to 750 K.

$$= \int_{750}^{913} \underline{C}_p(X_{A1}, T) dT \qquad (2.97)$$

where $X_{A1} =$ Conversion achieved in bed 1 = 0.634.

Applying equation (2.51) at the outlet from bed 1 gives (with $T_0 = 750$ K and $T = 913$ K):

$$\int_{750}^{913} \underline{C}_p(X_{A1}, T) dT = \underline{n}_{A0} X_{A1} \Delta H_{R,A}(750) \qquad (2.98)$$

Combining (2.96) with (2.97) and (2.98) and setting $T_0 = 750$ K,

$$X_A = \text{conversion at point in bed 2}$$

$$= X_{A1} - \left(\frac{1}{\underline{n}_{A0} \Delta H_{R,A}(T_0)}\right) \int_{750}^{T} \underline{C}_p(X_A, T) dT \qquad (2.99)$$

As seen from Figure 2.18, the range of temperatures possible in the second bed is substantially smaller than the range experienced in the first. The approximation will therefore be made when evaluating the integral term in

(2.99) of neglecting variations of molar heat capacity with temperature over the range from $T_0 = 750$K to the bed outlet temperature. $\underline{C}_p$ is then given by equation (2.96) so that

$$\int_{750}^{T} \underline{C}_p(X_A, T)dT = 3325\Delta T + 71.6 X_A \Delta T$$

where $\Delta T = (T - 750\,\text{K})$. Inserting this result in Equation (2.99) together with numerical values for $\underline{n}_{A0}, \Delta H_{R,A}$ and X_{A1} gives:

$$X_A = 0.634 + \left(\frac{3325\Delta T}{884\,200 - 71.6\Delta T}\right) \tag{2.100}$$

Line 3–4 on Figure 2.18 shows the variation of conversion with temperature for the reactant/product mixture passing through the second bed as given by equation (2.100). The line intersects the equilibrium curve at $T = 816$ K and $X_A = 0.884$.

The equilibrium conversion given by the two-bed system is thus 0.884 and the corresponding outlet temperature from the second bed is 816 K.

2.4.4   Equilibrium approach temperature

In the previous idealised example the temperature of the gas at the exit from the catalyst bed was equal to the equilibrium value corresponding to the gas composition. This can never be exactly true and the difference between the bed exit temperature and the equilibrium temperature at the given conversion is a useful yardstick both in design and in checking the performance of operating plant. The difference is often called the 'equilibrium approach temperature'. Figure 2.19 shows a situation similar to that in the previous example but with a lower feed temperature and where the equilibrium approach temperature is about 20 K.

2.4.5   Chemical equilibrium in imperfect gas systems

In the examples above, it has been assumed that the reactor fluid can be treated as a perfect gas. Because of the elevated temperatures which are normally involved, this is a good approximation in many industrial reactors, particularly those which operate at pressures only a few bar above ambient. Indeed, if the reduced temperatures of all the reactor components in the pure state exceed about 2, the approximation remains a good one up to substantially higher pressures than this (see Section 2.1.4.2 and Figures 2.1 and 2.2). However, deviations from perfect gas behaviour are appreciable in several important industrial gas phase reactions, examples being the ammonia synthesis reaction

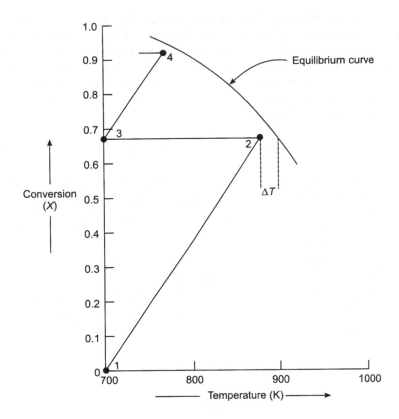

Figure 2.19 Temperature/conversion diagram illustrating typical situation where the equilibrium approach temperature (ΔT) is about 20 K. Lines 1–2 and 2–4 are given by energy balance equations as in Figure 2.18.

$$\tfrac{1}{2}N_2 + \tfrac{3}{2}H_2 \rightarrow NH_3 \qquad (2.101)$$

and the methanol synthesis reactions

$$CO + 2H_2 \rightarrow CH_3OH \qquad (2.102)$$

and

$$CO_2 + 3H_2 \rightarrow CH_3OH + H_2O \qquad (2.103)$$

which were discussed in Section 1.4.2.1.

The ammonia synthesis reaction is today normally carried out at about 400 °C and typical pressures range from 150 to 350 bar, though in the past pressures as high as 750 bar have been used [9]. The methanol synthesis reactions are typically carried out at about 200°C and 200 bar. On inserting the reduced temperatures and pressures for the reactants and products of the above reactions into Figure 2.2 it is found that the compressibility factors Z

for the products NH_3 and CH_3OH are less than 0.95 in each case while the reactant compressibility factors (apart from H_2 which is outside the scale covered in the diagram) are greater than 1.05. A rather similar situation applies to the fugacity/pressure ratio (F/p) for the reactants and products in their pure states. Z tends to be very similar to F/p in many situations where deviations from the gas laws are not large. The two are linked by the thermodynamic relationship

$$\ln F/p = \int_0^p [(Z - 1)/p]\mathrm{d}p \qquad (2.104)$$

For example, the fugacity pressure ratios for NH_3, H_2 and N_2 at 300 bar and 450°C calculated from (2.104) using volumetric data from reference [2] are 0.917, 1.088 and 1.130 respectively while the compressibility factors are 0.929, 1.083 and 1.129 (virtually identical results are given by a corresponding states correlation for (F/p) for gases including hydrogen and helium which was developed over 60 years ago by Newton [13]).

When deviation from perfect gas behaviour is as large as that noted above, it is necessary to work in terms of component fugacities rather than solely in terms of partial pressures when calculating equilibrium compositions. The calculations are then based on (2.74a) and (2.75) rather than on (2.74a) and (2.76).

$$RT \ln \underline{K} = -\Delta_r G^{\theta,g} \qquad (2.74a)$$

$$\underline{K} = \prod_i (f_i e/p^\theta)^{\nu_i} \qquad (2.75)$$

$f_i e$ in equation (2.75) is the fugacity of component i at the equilibrium point. Unfortunately there is no one single simple expression of general validity for component fugacities as functions of composition and pressure, as there was in the case of partial pressures. A rigorous determination of component fugacities in terms of component mole fractions at given temperature and pressure (and consequently the ability to derive equilibrium mole fractions from (2.75)) requires a good equation of state giving the volumetric behaviour of the components and their mixtures in the reactor fluid [8]. Such an equation of state will not necessarily be available. Short of this, some assumptions or approximations are likely to be required. An approximation which has proved very fruitful is given by the Lewis and Randal rule, according to which the fugacity of any component i in the gas mixture is the product of the fugacity of the pure component i at the given temperature and pressure f_i and the mole fraction y_i of i in the gas mixture:

$$(f_i = y_i F_i)_{TP} \qquad (2.105)$$

Combining this rule with (2.75)

$$\underline{K} = \prod_i \left(y_{ie} \frac{F_i}{p^\theta} \right)^{\nu_i}$$

$$= \prod_i \left(\frac{p_{ie}}{p^\theta} \right)^{\nu_i} \prod_i (F_i/p)^{\nu_i} \qquad (2.106)$$

$$= \underline{K}_p \prod_i (F_i/p)^{\nu_i}$$

where $\underline{K}_p$ is given by (2.77). In equation (2.106), p is the system pressure, $p^\theta = 1$ bar, p_{ie} is the partial pressure of component i in the equilibrium mixture and y_{ie} is the mole fraction of component i in the equilibrium mixture. Using equation (2.106), $\underline{K}_p$ may be calculated from $\underline{K}$ which, in turn, is given by the Gibbs energy change for the reaction.

Once $\underline{K}_p$ has been calculated at the temperatures and pressures of interest, equilibrium compositions for a given feed may be obtained as in previous sections.

In order to illustrate the way in which equation (2.75) and hence, in turn, equation (2.106) can be used it is convenient to take ammonia synthesis as an example. For this reaction, written as in equation (2.101), equation (2.75) gives

$$\underline{K} = \left[\frac{f_{NH_3(e)}}{\left(f_{N_2(e)} \right)^{1/2} \left(f_{H_2(e)} \right)^{3/2}} \right] p^\theta \qquad (2.107)$$

where $f_{NH_3(e)}$, $f_{N_2(e)}$ and $f_{H_2(e)}$ are the fugacities of NH_3, N_2 and H_2 in the equilibrium mixture.

On applying the Lewis and Randal rule, (2.106) gives

$$\underline{K}_p = \underline{K} \left(\frac{(F/p)_{N_2}^{1/2} (F/p)_{H_2}^{3/2}}{(F/p)_{NH_3}} \right) \qquad (2.108)$$

This rule is found to work well for the ammonia synthesis reaction at pressures up to about 300 bar, though deviations are apparent at higher pressures [7, 15].

Example 2.20
The Gibbs energy change for the ammonia synthesis reaction written as in (2.101) is 30.3 kJ mol^{-1} at 450°C [10]. Use the Lewis and Randal rule to estimate a value for $\underline{K}_p$ at 300 bar and 450° C.

Solution
As noted above, the fugacity/pressure ratios for nitrogen, hydrogen and ammonia in their pure states at 450 °C and 300 bar are given by

$$(F/p)_{N_2} = 1.130$$
$$(F/p)_{H_2} = 1.088$$
$$(F/p)_{NH_3} = 0.917$$

From data given, with equation (2.74a),

$$\underline{K} = \exp(-30\,300/(8.3144 \times 723.2))$$
$$= 6.5 \times 10^{-3}$$

Substituting these values in equation (2.108),

$$\underline{K}_p = 8.6 \times 10^{-3}$$

This may be compared with a value of 8.8×10^{-3} obtained by interpolation from experimental data in reference [14].

The influence of gas non-ideality on equilibrium in the ammonia synthesis reaction has been discussed in more detail by Gillespie and Beattie [16].

Appropriate fugacity coefficients to correct for non-ideality in the methanol synthesis reactions are given in references [18] and [19] and discussions of non-ideality in these systems are also given in references [10] and [17].

2.4.6   Chemical equilibria in liquid and mixed phase systems

It is beyond the scope of this book to discuss the application of the thermodynamic approach to chemical equilibria in systems other than gaseous ones. It is probably fair to say that this approach is less rewarding when applied to liquid phase reactions than to gaseous ones, particularly if the gas is near-ideal. The situation is well discussed and described in reference [8].

Appendix 2.1

The purpose of this appendix is to assemble the two thermodynamic results (A2.1.4 and A2.8 below) which lead directly to the equilibrium constant equations (2.74a) with (2.75) and (2.84) with (2.85) in the main text and to say a little about their background. This very brief treatment may refresh the memory or lead to further reading as the case may be.

The treatment rests on the behaviour of the total Gibbs energy $\underline{G}$ of a closed reacting system maintained at constant temperature and pressure. As the reaction proceeds (from either direction) the Gibbs energy decreases (Figure A2.1.1) until at equilibrium a minimum is reached (this result is not of course limited to reacting systems: a general criterion for equilibrium in a system maintained at constant temperature and pressure is that $\underline{G}$ has reached its minimum value (reference [7, p. 69]). In illustration of the above, the total Gibbs energy (which is a function of state) may be plotted versus any convenient parameter which changes progressively as the reaction proceeds. In Figure A2.1.1, for example, the parameter chosen is the number of moles (n_A) of the limiting reactant (A) present in the reactant/product mixture.

The right-hand arm of the curve through E in Figure A2.1.1 shows decrease in $\underline{G}$ as equilibrium is approached after first charging the system with reactants at temperature T and pressure P. The left-hand arm shows the decrease in $\underline{G}$ which would occur at temperature T and pressure P as the system composition returned to equilibrium following a disturbance (possibly a temporary change in temperature or pressure) during which n_A fell below n_{Ae}.

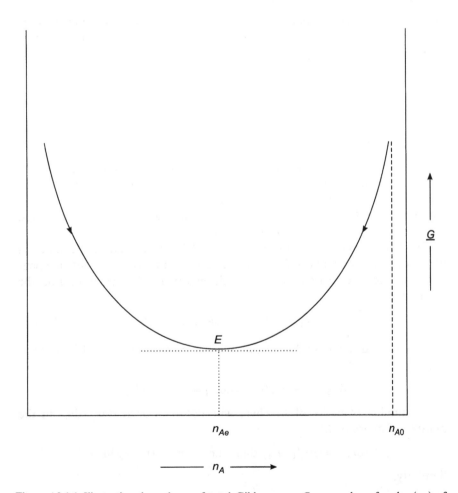

Figure A2.1.1 Illustrating dependence of total Gibbs energy $\underline{G}$ on number of moles (n_A) of limiting reactant as equilibrium is approached at constant temperature T and pressure P (see text) in a closed system. (n_{A0} is the number of moles of A initially charged to the system; n_{Ae} is the number of moles of A in the equilibrium mixture.)

Since temperature and pressure are constant, changes in the total Gibbs energy may be related as follows to changes in the number of moles of each component present:

$$dG = (\partial G/\partial n_A)_{T,P,n_B,n_C\ldots}dn_A + (\partial G/\partial n_B)_{T,P,n_A,n_C}dn_B + \ldots \quad (A2.1.1)$$

where $n_A, n_B, \ldots$ are the numbers of moles of components $A, B, \ldots$ present in the reactant/product mixture (it is assumed that the only work done on the system from outside arises from changes in the volume of the system). Quantities such as

$$(\partial G/\partial n_A)_{T,P,n_B,n_C} \quad \text{and} \quad (\partial G/\partial n_B)_{T,P,n_A,n_C}$$

are called the **chemical potentials** of components $A, B, \ldots$ (reference [7, p. 79]). In terms of chemical potentials the expression for dG can be written in the generalised form

$$\left(dG = \sum_i \mu_i dn_i \right)_{TP} \quad (A2.1.2)$$

where μ_i = chemical potential of component i
$$= (\partial G/\partial n_i)_{T,P,n_j}$$

(The subscript n_j following the differential coefficient indicates that the numbers of moles of all components except i remain constant during the differential operation.) The magnitude of the chemical potentials of the components in the mixture depend not only on the component considered but also on the composition of the mixture.

The changes $dn_A, dn_B, \ldots$ in equation (A2.1.1) which occur in the numbers of moles of components $A, B, \ldots$ as the reaction proceeds are not independent but are interrelated by the stoichiometry of the reaction. For the reaction

$$(-\nu_A)A + (-\nu_B)B \Leftrightarrow \nu_C C + \nu_D D \quad (A2.1.3)$$

for example, an increase dn_A in the number of moles of A will be accompanied by an increase

$$dn_B = [(-\nu_B)/(-\nu_A)]dn_A = (\nu_B/\nu_A)dn_A$$

in the number of moles of B. There will also be a decrease $(-dn_c)$ in the number of moles of C:

$$(-dn_c) = (\nu_C/(-\nu_A))dn_A \quad \text{i.e.} \quad dn_c = (\nu_C/\nu_A)dn_A$$

Similarly,

$$dn_D = (\nu_D/\nu_A)dn_A$$

In general, for a reaction written in the more general form of equation (2.6),

$$dn_i = (\nu_i/\nu_A)dn_A$$

where A is the limiting reactant.

Combining this result with (A2.1.1) or (A2.1.2)

$$d\underline{G} = (dn_A/\nu_A)(\mu_A\nu_A + \mu_B\nu_B + \dots)$$
$$= (dn_A/\nu_A)\sum_i(\mu_i\nu_i)$$

At the equilibrium point (in the absence of any discontinuity) $(d\underline{G}/dn_A)_{TP} = 0$ (see Figure A2.1.1). Hence,

$$\left[\sum_i \mu_i\nu_i = 0\right]_{\text{equilibrium}} \tag{A2.1.4}$$

(see also reference [7, p. 140]).

Equation (A2.1.4), which governs the relative magnitudes of the component chemical potentials at the equilibrium point, represents a very important first stage in the development of expressions for the thermodynamic equilibrium constant $\underline{K}$ and the form $(\underline{K}_p)$ to which this approximates under ideal gas conditions.

The second (and final) stage is to express these chemical potentials in terms of the component partial pressures (for a perfect gas phase) or component fugacities.

In a perfect gas phase, the components may be regarded as acting independently of one another and the chemical potential μ_i of a component i in the mixture with partial pressure p_i is equal to the chemical potential (and hence the molar Gibbs energy) of the pure gas i with **total pressure** p_i.

Accordingly, we can write

$$\mu_i = G_i^1 + RT \ln(p_i/p^\theta) \tag{A2.1.5}$$

where G_i^1 is the molar Gibbs energy of pure gas i at temperature T and 1 bar pressure and p^θ is 1 bar. $RT \ln(p_i/p^\theta)$ is the difference between the molar Gibbs energies of pure gas i at pressures p_i and p' respectively. Some authors define a perfect gas mixture as one in which the chemical potential of each of its components follows equation (A2.1.5). They then deduce the properties from this basis [7, p. 115].

Since the gas phase is perfect

$$G_i^1 = G_i^{\theta,\text{g}}$$

where $G_i^{\theta,\text{g}}$ is the molar Gibbs energy of pure component i normalised to an ideal gas state at 1 bar (since the gas phase is perfect no normalisation is necessary!).

Making this change in terminology and combining equations (A2.1.4) and (A2.1.5) gives, for a perfect gas phase,

$$\sum_i \nu_i G_i^{\theta,\mathrm{g}} + RT \sum_i \nu_i \ln(p_{ie}/p^\theta) = 0$$

that is,

$$-\sum_i \nu_i G_i^{\theta,\mathrm{g}} = RT \ln(\prod_i (p_{ie}/p^\theta)^{\nu_i}) \qquad (A2.1.6)$$

where p_{ie} is the partial pressure of component i in the equilibrium mixture. The term $\sum_i \nu_i G_i^{\theta,\mathrm{g}}$ in equation (A2.1.6) is equal to the Gibbs energy change for the reaction (equation (2.71)), i.e.

$$\sum_i \nu_i G_i^{\theta,\mathrm{g}} = \Delta_r G^{\theta,\mathrm{g}} \qquad (A2.1.7)$$

(This is best seen by taking a specific example such as a reaction of the form (A2.1.3). In this case

$$\sum_i \nu_i G_i^{\theta,\mathrm{g}} = (\nu_C G_C^{\theta,\mathrm{g}} + \nu_D G_D^{\theta,\mathrm{g}}) - (-\nu_A G_A^{\theta,\mathrm{g}} + \nu_B G_B^{\theta,\mathrm{g}})$$

$\qquad = \{\text{Gibbs energy of products in their standard states}\} -$
$\qquad \quad \{\text{Gibbs energy of reactants in their standard states}\}$
$\qquad = (\text{'increase in Gibbs energy which occurs when stoichiometric}$
$\qquad \quad \text{amounts of the products of the reaction are constituted from}$
$\qquad \quad \text{stoichiometric amounts of the reactants at specified tempera-}$
$\qquad \quad \text{ture, both products and reactants being in their standard}$
$\qquad \quad \text{states at temperature } T')$
$\qquad = \Delta_r G^{\theta,\mathrm{g}} \quad \text{(see equation (2.71))}$
Combining equation (A2.1.6) with (2.77),

$$(RT \ln \underline{K}_p = -\Delta_r G^{\theta,\mathrm{g}})_{\text{perfect gas}} \qquad (A2.1.8)$$

where

$$\underline{K}_p = \prod_i (p_i/p^\theta) \qquad (A2.1.8a)$$

as in equations (2.84) and (2.85) in the main text.

This result for perfect gas systems has been used extensively in the examples in Section 2.4 and the above notes are intended to provide a brief background to it. This background is not likely to be adequate in itself but does form a basis for further reading. Reference [7] should be helpful in this regard (it is still in print) but there are many other good treatments of this general area, including references [5], [6] and [8]. It is assumed that one such treatment is available to the reader.

The above treatment is, formally at least, readily extended to imperfect gas mixtures using the fugacity concept. The fugacity f_i of a component i in a mixture is defined [7] by the following relations:

$$\mu_i = G_i^{\theta,g} + RT \ln(f_i/f^\theta) \quad \text{and} \quad (f_i/p_i) \to 1 \quad \text{as } p \to 0 \qquad \text{(A2.1.9)}$$

where p is the total pressure and $f^\theta = 1$ bar. Under ideal gas conditions (A2.1.9) reduces to (A2.1.5). Based on the above definition and the properties of chemical potentials, properties of the pure component fugacities (such as the pressure dependence given by equation (2.104)) are readily obtained. This is true of component fugacities in mixtures also, though volumetric data for the mixtures are required to carry out the calculations. Combining equation (A2.1.4) with (A2.1.9) gives

$$\sum_i \nu_i G_i^{\theta,g} + RT \sum_i \nu_i \ln (f_{ie}/f^\theta) = 0$$

that is:

$$-\sum_i \nu_i G_i^{\theta,g} = RT \ln \left(\prod_i (f_{ie}/f^\theta)\right) \qquad \text{(A2.1.10)}$$

where f_{ie} is the fugacity of component i in the equilibrium mixture. As already noted (equation (A2.1.7)), $\sum \nu_i G_i^{\theta,g}$ is the free energy change for the reaction based on the hypothetical ideal gas standard state at 1 bar and is equal (equation (2.74a)) to $-RT \ln \underline{K}$, where $\underline{K}$ is the thermodynamic equilibrium constant. The latter depends only on temperature and composition; it is independent of pressure. Combining equation (A2.1.10) with (2.74a),

$$\underline{K} = \prod_i (f_{ie}/f^\theta)^{\nu_i} \qquad \text{(A2.1.11)}$$

where $f^\theta = p^\theta = 1$ bar.

Equation (A2.1.11) is more general in application than (A2.1.8a) since $\underline{K}$ is independent of pressure in all systems, ideal or non-ideal. Unfortunately, fugacities do not in general relate in a simple manner to system composition and this can lead to difficulties in applying the result under some conditions.

References

1. Angus, S., Armstrong, B. and de Reuck, K.M. (1976) *Carbon Dioxide, International Thermodynamic Tables of the Fluid State*, Vol. 3, Pergamon Press, Oxford.
2. Vargaftik, N.B., Vinogradov, Y.K. and Yargin, V.S. (1996) *Handbook of the Physical Properties of Liquids and Gases*, 3rd edition, Begell House, New York; Wallingford (UK).
3. Reid, R.C., Prausnitz, J.H. and Poling, B.E. (1987) *The Properties of Gases and Liquids*, 4th edition, McGraw-Hill, New York.
4. Smith, R. and Petela, E. (1992) *The Chemical Engineer*, 13 Feb., 24.
5. Bett, K.E., Rowlinson, J.S. and Saville, G. (1975) *Thermodynamics for Chemical Engineers*, Athlone Press, London.
6. Atkins, P.W. (1994) *Physical Chemistry*, 5th edition, Oxford University Press, Oxford.
7. Denbigh, K. (1981) *The Principles of Chemical Equilibrium*, 4th edition, Cambridge University Press, p. 65.

8. Smith, J.M. and Van Ness, H.E. (1987) *Introduction to Chemical Engineering Thermodynamics*, 4th edition, McGraw-Hill, New York, p. 118.
9. Gillespie, L.J. and Beattie, J.A. (1930) *Physical Reviews*, **36**, 1008.
10. Twigg, M.V. (ed.) (1996) *Catalyst Handbook*, 2nd edition, Manson Publishing. Co., London.
11. Freeman, R.D. (1982) *Bulletin of Chemical Thermodynamics*, **25**, 523.
12. Coulson, J.M., Richardson, J.F. and Sinnott, R.K. (1993) *Chemical Engineering, Vol. 6*, 2nd edition, Pergamon Press, Oxford, p. 857.
13. Newton, R.H. (1935) *Industrial and Engineering Chemistry*, **27**, 302.
14. Larson, A.T. and Dodge, R.L. (1924) *Journal of American Chemical Society*, **46**, 367.
15. Dodge, R.L. (1944) *Chemical Engineering Thermodynamics*, McGraw-Hill, New York.
16. Gillespie, L. J. and Beattie, J.A. (1930) *Physical Reviews*, **36**, 743, 1008.
17. Thomas, W.J. and Portalski, S. (1958) *Industrial and Engineering Chemistry*, **50**, 967.
18. Wade, R.B., Gengelback, R.B., Trumbley, J.L. and Houllbauer, W.C. (1981) in *Kirk Othmer Encyclopaedia of Chemical Technology*, Volume 15, 3rd edition, Wiley, New York, p. 398.
19. Chinchen, G.C., Denny, P.J., Jennings, J.R., Spencer, M.S. and Wangh. K.C. (1988) *Applied Catalysis*, **36**, 1.

3 Kinetics of homogeneous reactions and of reactions on solid catalyst surfaces

M.B. KING and J.M. WINTERBOTTOM

The first part of this chapter is concerned with homogeneous reactions, i.e. ones which involve only a single phase (normally either a gas phase or a liquid phase). In the second part of the chapter the situation is considered where the bulk reactant and product streams are homogeneous fluids but where the relevant reaction is predominately taking place on the surfaces of solid catalytic material with which the reactant/product mixture is contacted.

Catalytic reactors (particularly the fixed bed tubular reactor) are almost certainly the most important of the reactor classes, and the second part of this chapter leads directly forward to the sections on sizing and modelling fixed bed reactors in Chapter 4.

3.1 Reactions in homogeneous fluids

3.1.1 Some definitions

As in Chapter 2 we shall represent irreversible and reversible reactions by equations of the form

$$aA + bB + \ldots \rightarrow pP + qQ + \ldots$$
$$aA + bB + \ldots \rightleftharpoons pP + qQ + \ldots \tag{3.1}$$

respectively, where $A, B, \ldots$ are reactants and $P, Q, \ldots$ are products. The integers a, b, p, q, are the numbers of moles of components A, B, P, Q, respectively, entering into the reaction and are here called stoichiometric numbers.

As has been seen in Chapter 2, it is often convenient to write equation (3.1) in the form

$$0 = \nu_A A + \nu_B B + \ldots + \nu_Q Q + \ldots \tag{3.2}$$

ν_A, ν_B, ν_p and ν_Q are the **stoichiometric coefficients** of components A, B, P and Q. These are positive for the products and negative for the reactants, i.e.

$$\nu_A = -a, \quad \nu_B = -b, \quad \nu_P = p \quad \text{and} \quad \nu_Q = q$$

As in Chapter 2 we define the limiting reactant as being the one which would disappear first if the reaction proceeded sufficiently far.

The identity of the limiting reactant depends on the composition of the reactant mixture and the stoichiometry of the reaction. For example, if a moles of A react with b moles of B, and no other reactants are involved, A will be the limiting reactant if

$$\frac{\text{moles of } A \text{ fed}}{\text{moles of } B \text{ fed}} < \frac{a}{b}$$

It is usually convenient to choose the limiting reactant as the basis for calculations and we will adopt the convention of regarding component A as the limiting reactant. For example, the conversion X_A of this component given by

$$X_A = \frac{\text{moles of } A \text{ reacted}}{\text{moles of } A \text{ fed}} \qquad (3.3)$$

provides a convient measure of the degree to which the reaction as a whole has proceeded. Likewise, a convenient measure of the rate of the reaction is given by the rate at which limiting reactant A is being used up by the reaction per unit volume of reaction mixture.

The rate at which other components are being consumed or formed per unit volume may be calculated from the rate at which A is being consumed by considering the stoichiometry of the reaction.

We shall use the symbol r_i to designate the overall rate at which any component i is being formed in the reaction per unit volume of the homogenous phase and $\overrightarrow{r_i}$ to designate the rate of formation of i per unit volume by the forward reaction. ($\overrightarrow{r_i}$ is negative for reactants and positive for products.)

For an irreversible reaction, $r_i = \overrightarrow{r_i}$

For a reversible reaction $r_i = \overrightarrow{r_i} - (\overleftarrow{r_i})_i$

where $\overleftarrow{r_i}$ is the rate at which component i is being consumed in the back reaction.

In terms of the above terminology

$(-r_A) =$ net rate of consumption of limiting reactant A

in unit volume of reaction mixture $\qquad (3.4)$

The rates of formation and consumption of other components participating in the reaction are linked to the rate of consumption of the limiting reactant A by the ratio of the numbers of molecules of these components formed or destroyed to the number of molecules of A consumed. For the reaction shown in equation (3.1), for example,

$$r_p = (-r_A)(p/a) \qquad (3.5)$$

where r_p is net rate of formation of product p in the forward reaction in unit volume of reaction mixture.

In terms of the stoichiometric coefficients ν_p and ν_A in equation (3.2)

$$r_p = (\nu_p/\nu_A)r_A \qquad (3.6)$$

and, in general, for any component i

$$r_i = (\nu_i/\nu_A)r_A \qquad (3.7)$$

likewise

$$\overrightarrow{r_i} = (\nu_i/\nu_A)\overrightarrow{r_A} \qquad (3.8)$$

In an irreversible reaction there is no distinction between the forward and overall reaction rates and $-r_A = -\overrightarrow{r_A}$. The remainder of Section 3.1 is written with forward reactions primarily in mind. Reverse reactions follow similar mechanistic principles and the influence which these can have on the overall reaction rate is discussed in Section 3.2 where rate equations for reversible systems are briefly described.

3.1.2 Elementary, non-elementary and complex irreversible reactions

Single-line stoichiometric equations such as (3.1) show only the eventual outcome of the reaction and do not indicate whether the reaction takes place in a single stage or in a series of stages. Although some reactions do take place in a single stage, it is well known that a number of industrially important reactions proceed in many stages. *It is convenient to describe reactions which are believed to take place in one stage only (such as the reaction between nitric oxide and ozone in equation (3.9) below) as elementary reactions.* A reaction which takes place in a sequence of stages, the net result of which can be represented by a single stoichiometric equation of the form (3.1), is here described as a single but **non-elementary reaction**. A reaction of this type is that between nitrogen dioxide and fluorine (equations (3.10) and (3.11) below). These reactions are believed to proceed via the formation of highly reactive components (known as **active intermediates**). These have very short lifetimes and are not present in the final product. The stages by which a non-elementary reaction occurs are collectively known as the **mechanisim of that reaction**.

The following sections are concerned solely with **single reactions** both elementary and multi-stage. **Multiple reactions**, in which the reaction producing the desired product is accompanied by a variety of other usually unwanted reactions, will be considered briefly in later chapters. Multiple reactions cannot usefully be represented by a single stoichiometric equation. Together with non-elementary single reactions they are often referred to as 'complex'.

3.1.3   Single reactions and their mechanisms

An example of a reaction which is believed to be elementary is the gas phase reaction between nitric oxide and ozone.

$$NO + O_3 \rightarrow NO_2 + O_2 \tag{3.9}$$

For reasons which will be apparent later, elementary reactions rarely involve more than two molecules as reactants. (In other words, the **molecularity** of the reaction rarely exceeds 2.) An example of a single but non-elementary reaction is that between nitrogen dioxide and fluorine to form the reactive gas nitrogen fluoride. This has an overall molecularity of 3 and has the overall stoichiometry.

$$2NO_2 + F_2 \rightarrow 2NO_2F \tag{3.10}$$

The reaction is believed to take place in two stages via the formation of a short-lived intermediate (F). The proposed stages are

$$NO_2 + F_2 \rightarrow NO_2F + F$$
$$F + NO_2 \rightarrow NO_2F \tag{3.11}$$

The individual stages in non-elementary reactions such as the above are each taken to be elementary.

The first stage shown in equation (3.11) proceeds much more slowly than the second and determines the overall rate of the reaction. In situations such as the above, where the reaction proceeds in two or more stages, it is clear that the overall reaction cannot proceed faster than the slowest step. The slowest step in such a sequence is called **the rate-determining step**.

The hydrogenation of bromine, with the overall stoichiometric equation

$$H_2 + Br_2 \rightarrow 2HBr \tag{3.12}$$

is a further example of a single but non-elementary reaction. In this case the active intermediates (here H and Br free radicals) are continually regenerated within the reaction sequence, a situation which differs from that previously discussed. Reaction sequences of this type are called **chain reactions**. At about 300°C the following stages are believed to be involved:

Initiation stage $Br_2 \rightarrow 2Br\bullet$ (i)

Propagation $Br\bullet + H_2 \rightarrow HBr + H\bullet$ (ii)

Propagation $H\bullet + Br_2 \rightarrow HBr + Br\bullet$ (iii)

Termination $2Br\bullet \rightarrow Br_2$ (iv)

In the first stage (the **initiation** stage) molecular bromine Br_2 breaks up into two bromine free radicals ($Br\bullet$). (Free radicals are molecules or atoms having one or more unpaired electrons and are usually highly reactive. The unpaired electron is designated by a dot, as in Br•.) The second and third stages are **propagation** stages in which the final product of the reaction is produced and in which free radicals react with the formation of an equal number of other free radicals, thus propagating the reaction. During stage (ii) Br• reacts with molecular hydrogen to give HBr and a hydrogen free radical (H•). In stage (iii) the hydrogen free radical from stage (i) reacts with Br₂ to give HBr, regenerating at the same time the bromine free radical Br•. The net result of stages (ii) and (iii) is thus the formation of two molecules of HBr from one molecule of hydrogen and one of bromine without loss of the initiating bromine free radical. The final stage is the **termination stage** in which any unconsumed bromine free radicals recombine to give molecular bromine (a retardation stage may be present between (iii) and (iv) in which surplus H• reacts with HBr molecules to form Br• and molecular hydrogen). The above chain reaction is typical of many others propagated by free radicals [1].

(The comparatively complex reaction mechanism for the hygrogenation of bromine contrasts markedly with that for the hydrogenation of iodine:

$$H_2 + I_2 \rightarrow 2HI \tag{3.13}$$

This is believed to be an elementary reaction.)

The occurrence of chain reactions in polymer production is discussed in references [1,2]. Chain reactions are known (for example, that between O_2 and H_2) in which some of the stages (the so-called **chain-branching** stages) produce more free radicals than they consume. If conditions are such that the rate of formation of these excess free radicals exceeds their rate of recombination, the rate of reaction will increase as the reaction proceeds with explosive results[2].

3.1.4 *Reaction rates in homogeneous systems as functions of composition*

Good information about the relevant reaction rates and the manner in which these depend on composition and temperature is essential for reactor design. Elementary reactions will be considered first.

3.1.4.1 *Rate law for elementary reactions.* The rate law

$$(-r_A) = k\left[C_A^a C_B^b \ldots\right] \tag{3.14}$$

represents the reaction rates for **elementary reactions** (both in gases and in liquid solutions) quite well. C_A and C_B are the concentrations of reactants A and B and (in elementary reactions) a and b are the numbers of molecules of A and B which participate in each reaction event (see equation (3.1)). $(-r_A)$ is the rate at which reactant A is being consumed in the reaction per unit volume and k is the 'rate constant' which, although independent of concentration, is usually strongly temperature dependent.

(Some authors **define** an elementary reaction as one obeying equation (3.14). The present authors prefer the mechanistic definition given in Section 3.1.2).

Applied, for example, to the reaction (3.13) between hydrogen and iodine, equation (3.14) gives $(-r_{H_2}) = k[H_2][I_2]$ where $(-r_{H_2})$ is the rate at which hydrogen is being consumed in unit volume and $[H_2]$ and $[I_2]$ are the concentrations of hydrogen and iodine. (Here, as elsewhere throughout the book, the rate constant k is defined in terms of concentrations.)

3.1.4.2 *Physical background to the rate law for elementary reactions.*

Equation (3.14) has a long history going back to the formulation of the law of mass action in 1865, and even earlier than this. Indeed, it may be regarded as a modern statement of the law of mass action, though the early workers did not distinguish between elementary and non-elementary reactions.

Like some other important results in science it is based on a substantial body of experimental verification and on its physical reasonableness rather than on any single proof. It is physically reasonable for reactions involving two or more reactant molecules if one presumes that before the molecules can react they must first come together in the correct proportions. The probability of this occurring in a perfect gas mixture at given temperature will be proportional to the concentrations of the reactants raised to the powers of the appropriate stoichiometric numbers (see Section 3.1.6). As a good approximation the same would be expected to be true of reactions in solution also, though in this case it should be remembered that the nature of the solvent may profoundly influence the overall rate since the solvent molecules will impede the motions of the reacting species.

Consider the elementary reaction $A + B \rightarrow$ Products. We will presume that a reaction will occur only when molecules of types A and B collide due to thermal motions. Not every such collision will lead to a reaction but a proportion will do so. We will also presume that the proportion of collisions which lead to a reaction is constant for a given reaction in a given solvent (if one is present) at given temperature and pressure (the magnitude of the constant of proportionality is discussed in the next section). Then

$(-r_A)$ = rate of consumption of reactant A per unit volume

  = constant $\times$ (number of molecules of A per unit volume)

   $\times$ (probability that a given molecule of A will encounter a molecule of B in unit time)

The first term on the right-hand side of the above expression is directly proportional to the molar concentration of $A(C_A)$ while the last term will, if the encounters are random, be proportional to the number of molecules of B per unit volume and hence to the molar concentration (C_B) of B. Hence

$$(-r_A) = kC_A C_B \qquad (3.15)$$

where k is constant for a given reaction and solvent (if one is present) at given temperature and pressure.

If the elementary reaction were of the form

$$A + 2B \rightarrow \text{Products}$$

a necessary precondition for a reaction event to occur would be that a molecule of A should simultaneously encounter two molecules of B, a very unlikely event. The same arguments as those used above for bimolecular reactions would then lead to the result

$$(-r_A) = kC_A C_B^2$$

for the termolecular case. As pointed out in Section 3.1.3, truly termolecular elementary reactions are rare, if indeed they exist at all.

3.1.4.3 Reaction rates for single (non-elementary) reactions: reaction order. It is found experimentally that reaction rates for non-elementary single reactions often follow the general form of equation (3.14) but that the exponents a and b are not necessarily equal to the stoichiometric numbers for the overall reaction, nor are they always integers.

Thus typically, for a reaction between two components A and B

$$(-r_A) = kC_A^m C_B^n \qquad (3.16)$$

where C_A and C_B are the concentrations of A and B and $(-r_A)$ is the rate at which A is being consumed per unit volume.

A reaction obeying the above rate law is said to be of order m with respect to reactant A and of order n with respect to reactant B. The overall order is $(m + n)$. m and n are usually, though not invariably, integers or half integers. For example, the rate law for the bimolecular reaction between hydrogen and bromine (equation (3.12)) is, under suitable conditions,

$$(-r_{H_2}) = kC_{H_2} C_{Br_2}^{1/2}$$

The reaction is thus first order with respect to hydrogen and one half order with respect to bromine. The overall order is 3/2.

The gas phase oxidation of acetaldehyde (Example 3.1 below) is another example of a reaction with half integral kinetics.

In the special case of elementary reactions the overall order of the reaction is equal to the number of reactant molecules which participate in a reaction event. If, for example, the elementary reaction can be written as

$$A + B \rightarrow \text{Products}$$

the exponents m and n in equation (3.16) are both unity and the reaction is first order both with respect to A and with respect to B. The overall order is 2.

3.1.4.4 Pseudo-first-order reactions. A situation which sometimes arises is that of the **pseudo-first-order reaction**. This occurs when one reactant, reactant B say, in a bimolecular second-order reaction with a reactant A is present in considerable excess. The rate of consumption of component A is then given by

$$(-r_A) = kC_A C_B$$

where k is the second-order rate constant. However, since reactant B is present in large excess, its concentration will change very little during the entire reaction and the reaction rate will be approximately proportional to the concentration of component A. Under these conditions it is useful to define a **pseudo-first-order** rate constant k' such that

$$(-r_A) = k'C_A \tag{3.17}$$

where $k' = C_B k$. Hydrolysis reactions in water are frequently carried out under these conditions.

3.1.4.5 Determination of rate constants and reaction order from tests in a batch reactor. A few simple examples are given below which illustrate the principles involved. In these examples it is assumed that volume changes during the reaction are negligible. This is usually true as a good approximation for the liquid reaction systems typically used in batch reactors, but can break down badly in gas phase reactions in flow reactors if a change in the number of molecules occurs during the reaction. Corrections which should be applied to allow for volume changes when these occur are discussed in Section 3.2.2.

By definition, the reaction rate is given by

$$(-r_A) = -\frac{dn_A/dt}{V} \tag{3.18}$$

where $(-r_A)$ is the rate of consumption of the limiting reactant A per unit volume, V is the volume of reaction fluid considered and n_A is the number of moles of A present in it. The concentration C_A of A in the mixture is given by

$$C_A = n_A/V$$

and, accordingly,

$$\frac{dC_A}{dt} = \left(\frac{dn_A/dt}{V}\right) - \left(\frac{C_A}{V}\right)\frac{dV}{dt} \tag{3.19}$$

If volume change is negligible, then

$$(-r_A) = -\left(\frac{dC_A}{dt}\right) \tag{3.20}$$

Using equation (3.20) the order of a reaction can be inferred from observations of the way in which reactant concentrations vary with time.

For a first-order reaction, for example

$$(-r_A) = kC_A$$

Combining this result with (3.20) gives, for a constant volume system,

$$\frac{dC_A}{C_A} = -kdt$$

If the reaction is initiated at $t = 0$ when the concentration of A is C_{A0}, the above equation integrates to give

$$\ln\left(\frac{C_A}{C_{A0}}\right) = -kt$$

For a first-order reaction with negligible volume change in a batch reactor, therefore, a plot of $\ln C_A$ versus time should be linear.

As another example, the order of the reaction $A + B \rightarrow$ Products might be investigated. If this is an elementary reaction it should be second order, but this will not necessarily be the case if it is non-elementary. To investigate whether or not it is second order we might initiate the reaction with equal concentrations of A and B. If the reaction is second order then in the absence of volume changes,

$$r_A = \left(\frac{dC_A}{dt}\right) = \left(\frac{dC_B}{dt}\right) = -kC_A C_B = -kC_A^2$$

or

$$\left(\frac{dC_A}{C_A^2}\right) = -kdt$$

If the initial concentration of A is C_{A0}, the above expression integrates to give

$$\left(\frac{1}{C_A}\right) - \left(\frac{1}{C_{A0}}\right) = kt$$

in this case, therefore, a plot of reciprocal concentration should be linear in t.

If there is little prior knowledge about the order of the reaction it may be necessary to adopt a different approach. Suppose that, over at least part of the concentration range, the reaction rate can be written in the form

$$-\left(\frac{dC_A}{dt}\right) = (-r_A) = k^1 C_A^m \tag{3.21}$$

or

$$\ln\left(\frac{-dC_A}{dt}\right) = \ln k^1 + m \ln C_A \tag{3.22}$$

where C_A is the concentration of reactant A and m is the order of the reaction. The reaction may either be a true mth order reaction or a pseudo-mth-order reaction in which all reactant(s) apart from A are present in sufficient excess for fractional changes in their concentrations to be negligible as the reaction of A proceeds. (Pseudo-first-order reactions have been briefly discussed in Section 3.1.4.4.)

With the above proviso m and k' may be calculated from experimental measurements of C_A as a function of t. There are several techniques for doing this. Conceptually the most direct procedure at this stage is to evaluate dC_A/dt as a function of C directly from the experimental data (clearly closely spaced data points of good accuracy are required in this approach). m and $\ln k'$ may then be obtained graphically from a plot of $\ln(-dC_A/dt)$ versus $\ln C_A$ (the slope is m) or, alternatively, they may be obtained by fitting the data to equation (3.22) using a standard linear least-squares fitting procedure.

If equation (3.21) refers to a pseudo-mth-order reaction it will now be necessary to evaluate further parameters. If, for example, the full rate equation is

$$(-r_A) = kC_A^m C_B^n \tag{3.23}$$

the parameters k and n still remain to be determined. To obtain them, a further set of experimental data is required giving C_A and C_B as functions of time under conditions where C_B in equation (3.23) can no longer be regarded as constant. From equation (3.23)

$$\ln\left(\frac{(-r_A)}{C_A^m}\right) = \ln k + n \ln C_B \tag{3.24}$$

Since m is now known, $\ln((-r_A)/C_A^m)$ may be calculated from the new data set as a function of $\ln C_B$, enabling the parameters k and n to be determined from Equation (3.24).

Problem 3.1
Use the batch reactor data given below to show that the reaction is second order and determine the rate constant for the liquid phase reaction $2A \rightarrow$ Product.

Time (min)	0	2	4	6	8	15	30	50
Conc. (kmol m^{-3})	1.25	1.01	0.825	0.715	0.625	0.435	0.264	0.172

(*Answer*: 0.1 dm^{3}mol^{-1}min^{-1})

In the above examples it has been assumed that back reactions are negligible in extent.

If significant back reaction is present the tests should be carried out at low conversion of A. In extreme cases it is best to extrapolate the rate data back to zero conversion at a set of initial concentrations (C_{A0}) of the reactant A.

The initial reaction rates $(-dC_A/dt$ at $C_A = C_{A0})$ and initial concentrations can then be inserted in equation (3.21) and used to obtain values of k' and m for the forward reaction as previously described.

3.1.4.6 Specification of reaction rate constants: units. If the stoichiometric coefficients for the components in a reaction differ from one another, the magnitude of the rate constant will depend on the identity of the component from whose reaction rate k is calculated. In the present text the reaction rate and rate constant are taken to refer to the limiting reactant. However, **in cases of ambiguity this component should be specified.**

Consider for example the elementary reaction below:

$$A + 2B \xrightarrow{k} P \qquad (3.25)$$

for which $(-r_B) = 2(-r_A)$
k calculated from $(-r_A)$ is given by equation (3.14) to be

$$k(= k_A) = (-r_A)/C_A C_B^2$$

In terms of $(-r_B)$, however

$$k(= k_B) = (-r_B)/C_A C_B^2 = 2k_A$$

The units in which k is expressed depend on the order of the reaction and the units for time and concentration employed. Typically, k for a first-order reaction is given per second (s^{-1}) while for a second-order reaction the units are $dm^3 mol^{-1}s^{-1}$. However, many variants of the above are in use.

3.1.4.7 Rates of non-elementary reactions as determined by reaction mechanism. The use of equation (3.14) is not limited to single elementary reactions: when applied to each of the individual stages (when these are known) it can give an interpretation of observed reaction rates in non-elementary reactions also.

If, therefore, the mechanism for a reaction has been postulated but not confirmed, a useful check can be obtained by comparing the experimentally observed dependence of overall reaction rate on reactant concentrations with that obtained by applying equation (3.14) to each of the hypothesised individual stages. Example 3.1 is based on discussions in references [1] and [2].

Example 3.1
The vapour phase decomposition of acetaldehyde in the absence of air is found to be of the order 1.5.

The decomposition is quite a complex one. The dominant reaction is

$$CH_3CHO(g) \rightarrow CH_4(g) + CO(g)$$

Several by-products are formed together with trace amounts of ethane. The chain mechanism shown in equations (i) to (iv) below has been proposed to account for the main observed facts. It is oversimplified in that it does not account for the formation of by-products such as acetone or propanal.

$$\text{Initiation} \qquad CH_3CHO \xrightarrow{k_1} CH_3\bullet + \bullet CHO \qquad \text{(i)}$$

$$\text{Propagation(1)} \quad CH_3\bullet + CH_3CHO \xrightarrow{k_{P1}} CH_3\bullet CO + CH_4 \qquad \text{(ii)}$$

$$\text{Propagation(2)} \quad CH_3\bullet CO \xrightarrow{k_{P2}} CH_3\bullet + CO \qquad \text{(iii)}$$

$$\text{Termination} \qquad 2CH_3\bullet \xrightarrow{k_T} C_2H_6 \qquad \text{(iv)}$$

(The rate constants k_1 and k_{P1} are taken to be defined with respect to acetaldehyde and k_{P2} with respect to the $CH_3\bullet CO$ radical. k_T is defined with respect to the rate of ethane production.)

The initiation stage (see also Section 3.1.3) provides an initial source of methane radicals. This stage can be quite slow since these radicals are regenerated at the same rate as they are used up in the propagation stages. The initition stage is spontaneous and thermally initiated.

In propagation stage 1 a methane radical reacts with an acetaldehyde molecule to produce methane and a $CH_3\bullet CO$ radical.

In propagation stage 2, the $CH_3\bullet CO$ radical decomposes to form carbon monoxide, at the same time regenerating the methane radical.

In the termination stage surplus methane radicals recombine to form ethane.

In developing a rate equation consistent with the above hypothesised mechanism, *the net rate of production of intermediates (such as the free radicals $CH_3\bullet$ and $CH_3\bullet CO$) which do not appear in the final product is taken to be zero*, so that

$$d[CH_3\bullet CO]/dt = r_{CH_3\bullet CO}$$
$$= k_{P1}[CH_3CHO][CH_3\bullet] - k_{P2}[CH_3\bullet CO] \qquad (3.26)$$
$$= 0$$

and

$$d[CH_3\bullet]/dt = r_{CH_3\bullet}$$
$$= k_1[CH_3CHO] + k_{P2}[CH_3\bullet CO] - k_{P1}[CH_3CHO][CH_3\bullet] - 2k_T[CH_3\bullet]^2$$
$$= 0$$

$$(3.27)$$

Adding (3.26) to (3.27) and dividing by $2k_T$

$$[CH_3\bullet]^2 = (k_I/2k_T)[CH_3CHO] \qquad (3.28)$$

In the reaction scheme (i) to (iv), acetaldehyde is consumed at stages (i) and (ii).

Under conditions such that the initiation reaction (i) proceeds substantially more slowly than the propagation reactions, only the consumption in propagation step 1 need be considered and the rate of consumption of acetaldehyde is given by

$$(-r_{CH_3CHO}) = k_{P1}[CH_3\bullet][CH_3CHO] \tag{3.29}$$

Combining (3.28) and (3.29),

$$(-r_{CH_3CHO}) = \left(\frac{k_1 k_{P1}^2}{2k_T}\right)^{1/2}[CH_3CHO]^{1.5} \tag{3.30}$$

This result, together with the predicted presence of small amounts of ethane, agrees with the experimental findings summarised above.

3.1.5   Reaction rates as functions of temperature

The 'reaction rate constant' k in equation (3.14), although independent of the concentrations of the species involved in the reaction, will in general depend on temperature and, to a smaller degree, on pressure. The dependence on temperature can be particularly pronounced: usually (though not invariably) the rate increases strongly with this parameter. A doubling of reaction rate for a 10°C temperature rise from ambient is not uncommon and even larger increases occur. A few reactions are known [1] where the reaction rate is almost independent of temperature. These are exceptional, however. In the vast majority of cases reaction rates increase rapidly with increase in temperature.

In most cases the rate constants for elementary reactions can be correlated satisfactorily as functions of temperature using equations of the **Arrhenius type** according to which

$$k = Ae^{-E/RT} \tag{3.31}$$

where E is the molar activation energy and A is the so-called pre-exponential factor. Regarded as an empirical equation, equation (3.31), with constant values for A and E, can be very helpful for correlating and summarising reaction rate data.

Although there are a few exceptions [1] it does represent the temperature behaviour of most elementary reaction rate constants within experimental accuracy over a substantial temperature range and this is true also of some non-elementary reactions.

(It should be remembered, however, that the overall rate of a non-elementary reaction will not necessarily follow the Arrhenius equation even if each of the elementary steps by which it proceeds does.) Tables 3.1 and 3.2 show values of the Arrhenius constants A and E for some selected gas phase reactions and reactions in solution.

Table 3.1 Parameters in Arrhenius' equation (3.31) for some selected first-order reactions ($\bar{T}$ is mid-temperature of range over which A and E were fitted)

Reaction	Phase	$A \times 10^{-13}$ (s^{-1})	E $(kJ\,mol^{-1})$	$\bar{T}$ (K)	$(R\bar{T}e/Nh)^a$ $\times 10^{-13}(s^{-1})$
$Cl{\bullet}COO{\bullet}CCl_3 \rightarrow 2COCl_2$	g	1.4	61	285	1.6
$F_2O_2 \rightarrow F_2 + O_2$	g	0.6	72	231	1.3
$N_2O_5 \rightarrow N_2O_4 + (1/2)O_2$	g	4.6	103	306	1.7
$C_2H_5I \rightarrow C_2H_4 + HI$	g	2.5	209	–	–
$C_2H_5Br \rightarrow C_2H_4 + HBr$	g	9.1	218	666	3.8
$(CH_3)_2CO \rightarrow CH_4 + CH_2 = CO$	g	81	287	855	4.8
$N_2O_5 \rightarrow N_2O_4 + (1/2)O_2{}^b$	Bromine solvent	2.0	100		
$N_2O_5 \rightarrow N_2O_4 + (1/2)O_2{}^b$	Penta-chloro-ethane solvent	9.4	105		
$N_2O_5 \rightarrow N_2O_4 + (1/2)O_2{}^b$	HNO_3 solvent	65	118		
$(C_2H_5)_3SBr \rightarrow (C_2H_5)_2S + C_2H_5Br$	$C_2H_2Cl_4$ solvent	2600	127		
$(C_2H_5)_3SBr \rightarrow (C_2H_5)_2S + C_2H_5Br$	$C_6H_5NO_2$ solvent	0.006	120		

Data are from compilations in references [2], [4] and [5].
ᵃ $(R\bar{T}e/Nh)$ is the lead term in the transition state theory expression for A in first-order gas reactions. (see Section 3.1.7). $e = 2.718$.
ᵇ The parameters A and E for the decomposition of nitrogen pentoxide in each of the solvents nitromethane, carbon tetrachloride, ethylene chloride and chloroform fall between those for the solvents bromine and pentachloroethane shown above. Reaction rates at 20°C are accordingly very similar [5].

Table 3.2 Parameters in Arrhenius' equation (3.31) for some selected second-order reactions

Reaction	Phase	A $(dm^3mol^{-1}s^{-1})$	E $(kJ\,mol^{-1})$	$\bar{T}$ (K)	$(R\bar{T}e)^2/(Nhp^\theta)$ $(dm^3mol^{-1}s^{-1})$
$2NOBr \rightarrow 2NO + Br_2$	g	4.2×10^{10}	58	274	9.6×10^4
$2NOCl \rightarrow 2NO + Cl_2$	g	0.9×10^{10}	102	–	
$2NO_2 \rightarrow 2NO + O_2{}^a$	g	0.9×10^{10}	112	570	4.2×10^5
$H_2 + I_2 \rightarrow 2HI$	g	1.6×10^{11}	165	690	6.1×10^5
$2HI \rightarrow H_2 + I_2$	g	0.9×10^{11}	186	690	6.1×10^5
$H_2 + C_2H_4 \rightarrow C_2H_6$	g	1.24×10^6	180	–	–
$CH_3I + C_6H_5N(CH_3)_2 \rightarrow$ phenyltrimethyl ammonium iodide	$C_6H_5CH_2OH$ solvent	7.1×10^6	60.3		
$C_2H_5ONa + CH_3I \rightarrow$ $NaI + C_2H_5OCH_3$	C_2H_5OH solvent	2.4×10^{11}	81.6		
$C_2H_5Br + (C_2H_5)_2S \rightarrow$ $(C_2H_5)_3SBr$	$C_6H_5CH_2OH$ solvent	1.4×10^{11}	106.6		

Data are from compilations in references [2] and [4]. $p^\theta = 10^5\ N\ m^{-2}$.
ᵃ More recent values for A and E are:
$A = 0.2 \times 10^{10}$ dm³ mol⁻¹ s⁻¹ and $E = 111$ kJ mol⁻¹.

For some purposes it is convenient to rewrite equation (3.31) as

$$k = Ae^{-E'/k_BT} \tag{3.32}$$

where k_B is Boltzmann's constant, and E' is the molecular activation energy ($E' = E/N$) where N is Avogadro's constant.

A simple hypothesis leading to the form for equation (3.31) and equation (3.32) is that only collisions between molecules with an energy in excess of a certain minimum value (the molecular activation energy E' for the reaction) result in a reaction. The rate constant k for the reaction should thus be proportional to the fraction of the collisions for which this energy is exceeded, i.e. (from the Maxwell–Boltzmann distribution law) to $\exp(-E'/RT)$ [6].

In the case of gaseous reactions, these arguments have led to the **simple collision theory** for reaction rates.

3.1.6 Simple collision theory for gas reaction rates

According to this theory, as applied to the reaction $A + B \rightarrow$ Products, the reaction rate expressed as molecules of A reacting per unit volume per unit time is given by the product of the number of collisions between molecules of types A and B in unit volume in unit time (Z_{AB}) and the probability that a given collision will lead to a reaction event. This probability is in turn written as the product of the probability $(\exp(-E'/k_B T))$ that the colliding molecules will have the requisite energy and a steric factor p which is discussed further below.

Since $E'/k_B = E/R$, the reaction rate expressed as moles of A reacting per unit volume per unit time is given by

$$(-r_A) = (1/\mathbf{N}) \times (Z_{AB}) \times p \times \exp(-E/RT) \qquad (3.33)$$

where $\mathbf{N}$ is Avogadro's number
Z_{AB} is given by the simple kinetic theory of gases [2] as

$$Z_{AB} = \sigma_{AB}(8k_B T/\pi\mu)^{1/2}\mathbf{N}^2 C_A C_B \qquad (3.34)$$

where C_A and C_B are here the molar concentrations of A and B.
σ_{AB} is the collision cross-section

$$\sigma_{AB} = \pi[(d_A + d_B)/2]^2$$

where d_A and d_B are the diameters of molecules A and B, which are regarded as hard spheres.

μ is the so-called 'reduced mass' $= [(M_A M_B/(M_A + M_B)]$ where M_A and M_B are the masses of molecules A and B. Here p is the steric factor which, in terms of the theory, allows for the fact that molecules are not normally spheres and their orientation at collision must be correct if their kinetic energy is to be transformed into a form which can initiate the reaction.

Combining (3.15) with (3.33) and (3.34)

$$k = (pZ'/\mathbf{N})\exp(-E/RT) \qquad (3.35)$$

where

$$Z' = (Z_{AB}/C_A C_B) = \sigma_{AB}(8k_B T/\pi\mu)^{1/2}\mathbf{N}^2 \qquad (3.36)$$

In terms of simple collision theory, the pre-exponential factor A in equation (3.24) is thus given by pZ'/N and should, if p is taken to be independent of temperature, be proportional to $T^{1/2}$ (equations (3.35) and (3.36)). In practice it is very difficult to find experimentally whether or not this is the case since the exponential term is usually very much more temperature sensitive than this and the experimental data can be fitted equally well with and without the $T^{1/2}$ term.

A weakness in equation (3.35) is that the steric factor p is not usually predictable. This difficulty is overcome to some degree in the transition state theory discussed in the next section.

3.1.7 A note on the transition state theory of reaction rates

Transition state theory is more ambitious than the simple collision theory described above in that it aims to describe in more detail the changes in potential energy and Gibbs energy which take place during a collision event. The description given below is oversimplified but brings out the salient points. The theory starts from the premise that, when two reactant molecules come together as a result of thermal motions, an 'activated complex' may be formed. Whether or not this will occur will depend partly on their initial kinetic energies and partly on orientation and bond structure. The activated complex is taken to be in equilibrium with the reactants. It is unstable and the rate of formation of products is proportional to its concentration.

Figure 3.1 gives a simple representation of the potential energy of two reactant molecules as an exothermic reaction between them progresses. The progression of the reaction is indicated by the so-called 'reaction co-ordinate', which can be any property which increases progressively as a reaction event between the two molecules proceeds.

The pathway followed as the reaction develops will be such as to minimise the potential energy. The simple two-dimensional diagram in Figure 3.1 shows the potential energy as reaction proceeds along this pathway.

Initially, at the far left of the diagram, the reactant molecules, which are taken to have kinetic energy in excess of the activation energy, are widely separated and no reaction has yet occurred. As they come closer together there will be a slight fall in potential energy at first due to the intermolecular attractive forces, but when they get so close that their orbitals overlap, strong repulsion takes place and the potential energy rises sharply. According to the physical 'picture' underlying transition state theory, the initial kinetic energy is then progressively transformed into bond vibrational energy. As the peak of the potential energy curve is approached new bonds associated with the activated complex start to develop and these also vibrate.

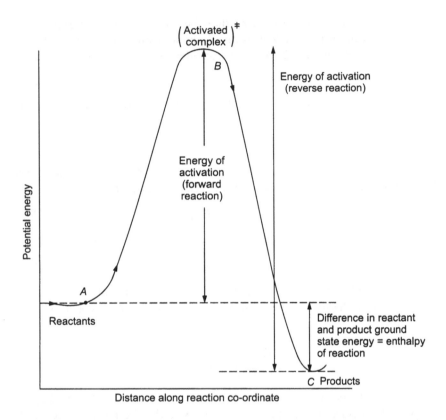

Figure 3.1 Potential energy changes accompanying a reaction event in a homogeneous phase.

As the transition state B is approached a complicated collective vibrational motion of all the atoms may be conceived [2], resulting in oscillation of the system along the reaction co-ordinate. According to the 'transition state' model, a complex is formed in the region close to B which is unstable to oscillations along the reaction co-ordinate. In this region most, if not all [2], of these oscillations will result in the system passing through the transition state B into the region BC, with consequent disintegration of the complex. If the oscillations are of frequency ν, the rate at which the complex disintegrates to form the products will be given by

$$(-r_{AB}^{\ddagger}) = \nu C_{AB}^{\ddagger} \tag{3.37}$$

where $C_{AB}^{\ddagger}$ is the concentration of the complex.

In the further development of the model it will be **assumed** that the complex (normally termed the 'activated complex') is formed in a rapid pre-equilibrium with the reactants. This is one of the fundamental postulates of the original model: it is supposed that the reaction does not appreciably disturb the equilibrium concentration of the activated complexes [5]. (For discussion, see Reference [2], p. 941.) Consider, for example, the reaction

$$A + B \rightarrow AB \tag{3.38}$$

According to the transition state theory, the reaction proceeds via the activated complex $(AB)^{\ddagger}$ as follows:-

$$A + B \Leftrightarrow (AB)^{\ddagger} \rightarrow AB \tag{3.39}$$

The rate of formation of the final product AB will be given by the rate of decomposition of the activated complex $(AB)^{\ddagger}$ so that (from (3.37))

$$(r_{AB}) = \nu C_{AB}^{\ddagger} \tag{3.40}$$

Likewise, the rate of consumption $(-r_A)$ of reactant A is given by

$$(-r_A) = \nu C_{AB}^{\ddagger} \tag{3.41}$$

If the activated complex is in equilibrium with the reactants A and B, as transition state theory requires, and **the reaction takes place under ideal gas conditions**,

$$\left(\frac{p_{AB}^{\ddagger} p^{\theta}}{p_A p_B} \right) = \underline{K} = \underline{K}_p \tag{3.42}$$

where $p_{AB}^{\ddagger}$ is the partial pressure of the activated complex and p_A and p_B are the partial pressures of reactants A and B. $\underline{K}$ is the thermodynamic equilibrium constant defined by (2.74a) and $\underline{K}_p$ is the value to which this reduces under ideal gas conditions (equation (2.77)). $p^{\theta} = 1$ bar $= 10^5$ Nm^{-2}. Since the gas is ideal, $C_{AB}^{\ddagger} = (p_{AB}^{\ddagger}/RT)$ with similar expressions for C_A and C_B, so, from (3.42)

$$C_{AB}^{\ddagger} = (C_A C_B)(\underline{K}RT/p^{\theta}) \tag{3.43}$$

Combining (3.41) with (3.43)

$$(-r_A) = \nu \underline{K} \left(\frac{RT}{p^{\theta}} \right) C_A C_B \tag{3.44}$$

where

$$\underline{K} = \exp(-\Delta G^{\ddagger,\,\theta,\mathrm{g}}/RT) \tag{3.45}$$

$\Delta G^{\ddagger,\,\theta,\,\mathrm{g}}$ is the Gibbs energy change for the transformation

$$A + B \rightarrow (AB)^{\ddagger} \tag{3.46}$$

when carried out under standard state conditions (in this case the ideal gas state at 1 bar). At this point it is well to remember that the activated complex is not an 'ordinary' compound (in statistical mechanical terms it lacks a vibrational mode). Bearing in mind this fact, a study of the partition functions contributing to $\underline{K}$ leads to the conclusions that this quantity should henceforth be regarded as a 'quasi-equilibrium constant' and that the decay frequency ν takes the value $(k_B T/h)$ [2,4] where k_B is Boltzmann's constant and h is Plank's constant.

Combining (3.44) with (3.45) and setting $\nu = (k_B T/h) = (RT/Nh)$

$$(-r_A) = C_A C_B \left(\frac{RT}{Nh}\right) \left(\frac{RT}{p^\theta}\right) \exp(-\Delta G^{\ddagger\theta,g}/RT) \qquad (3.47)$$

where N is Avogadro's number.

The rate constant k for the bimolecular reaction so far considered is obtained by comparing (3.47) with (3.15) and is given by

$$k = \left(\frac{RT}{Nh}\right) \left(\frac{RT}{p^\theta}\right) \exp(-\Delta G^{\ddagger\theta,g}/RT) \qquad (3.48)$$

The corresponding expression for a monomolecular gas phase reaction is [5]

$$k = \left(\frac{RT}{Nh}\right) \exp(-\Delta G^{\ddagger\theta,g}/RT) \qquad (3.49)$$

3.1.7.1 Comparison with the Arrhenius equation (3.31). The link between Equations (3.48) and (3.49) and the Arrhenius equation becomes clearer when the standard Gibbs energy change for the transformation shown in equation (3.46) is expressed in terms of the standard enthalpy and entropy changes as follows:

$$\Delta G^{\ddagger\theta,g} = \Delta H^{\ddagger\theta,g} - T\Delta S^{\ddagger\theta,g} \qquad (3.50)$$

(Equation (3.50) follows on applying the defining equation for the Gibbs energy $G = H - TS$ to the system before and after the transformation.

For a bimolecular reaction, for example, equation (3.48), with (3.50) gives

$$k = \left(\frac{RT}{Nh}\right)(RT/p^\theta)\left[\exp(\Delta S^{\ddagger\theta,g}/R)\right]\exp(-\Delta H^{\ddagger\theta,g}/RT) \qquad (3.51)$$

The standard enthalpy change $\Delta H^{\ddagger,\theta,g}$ is normally approximately equal [2] to the standard internal energy change and hence to the activation energy E. As an approximation, therefore, the final factor in equation (3.51) equates to the exponential term in the Arrhenius equation.

The exact relationship between $\Delta H^\ddagger$ and E depends on the molecularity of the gas reaction considered. It can be obtained by comparing the value of $(d \ln K/dT)$ given by the theory with the value given by the Arrhenius equation (3.31) according to which

$$\frac{\mathrm{d}\ln k}{\mathrm{d}T} = E/RT^2 \tag{3.52}$$

In the bimolecular case (equation (3.48)), for a gas phase reaction,

$$\frac{\mathrm{d}\ln k}{\mathrm{d}T} = \frac{2}{T} - \frac{1}{R}\frac{\mathrm{d}}{\mathrm{d}T}\left(\frac{\Delta G^{\ddagger}}{T}\right)$$

Applying the Gibbs–Helmholtz equation (2.79a),

$$\frac{\mathrm{d}\ln k}{\mathrm{d}T} = \frac{2}{T} + \frac{\Delta H^{\ddagger}}{RT^2} \tag{3.53}$$

Comparison of (3.52) with (3.53) then gives

$$E = \Delta H^{\ddagger} + 2RT \tag{3.54}$$

In the monomolecular case, similar differentiation of (3.49) and comparison with (3.52) gives

$$E = \Delta H^{\ddagger} + RT \tag{3.55a}$$

Inserting equation (3.54) into equation (3.51) gives, for the bimolecular gas phase reaction,

$$k = \frac{RTe^2}{Nh}\left(\frac{RT}{p^{\theta}}\right)\left[\exp\frac{(\Delta S^{\ddagger\theta,\mathrm{g}})}{R}\right]\left[\exp\left(\frac{-E}{RT}\right)\right] \tag{3.55b}$$

The corresponding equation for a monomolecular reaction (such as those listed in Table 3.1) is [5]

$$k = \frac{RTe}{Nh}\left[\exp\left(\frac{\Delta S^{\ddagger\theta,\mathrm{g}}}{R}\right)\right]\left[\exp\left(\frac{-E}{RT}\right)\right] \tag{3.56}$$

The term RTe/Nh in equation (3.56) is shown in Table 3.1 alongside the experimentally observed Arrhenius pre-exponential term A for some typical monomolecular reactions. In several cases agreement is very close, and in all cases except the last is within a factor of about 3, suggesting that the influence of the entropy term $\exp(\Delta S^{\ddagger}/R)$ is not large in these cases.

The term $(RTe)^2/Nhp^{\theta}$ in equation (3.55b) is shown alongside the experimentally observed value of A for some bimolecular reactions in Table 3.2. In every case shown, this term exceeds A by above four orders of magnitude. This behaviour is typical of bimolecular reactions. In terms of transition theory it implies that the term $\exp(\Delta S^{\ddagger}/R)$ is small ($\Delta S^{\ddagger}$ being a large negative number). This term can be regarded as allowing for the difficulty in transforming the kinetic energy of the reactant molecules into an appropriate form. (Entropy is a measure of disorder and a molecular orientation which is unlikely to occur corresponds to a very low entropy and a correspondingly negative value for the standard entropy change $\Delta S^{\ddagger}$) Broadly speaking, the term $\exp(\Delta S^\ddagger/R)$ plays a similar role to the factor p in collision theory. Unlike p it can, in principle at least, be calculated from the partition functions of the reacting molecules. In the few favourable instances where this calculation has been made, good agreement between observed and calculated pre-exponential factors has been reported [6].

Transition state theory predicts that the 'pre-exponential' term A should be temperature dependent. As in the case of collision theory, it is not easy to check this experimentally because of the normally strong temperature dependence of the term $\exp(-E/RT)$.

Although the equations given above were developed for reactions in an ideal gas phase, the concepts of transition state theory have been extended to reactions in solution [5] and in non-ideal gases [9].

3.1.8 Rates of reversible reactions

In previous sections detailed consideration has been given to the forward reaction only. Extension of the treatment to allow for the influence of appreciable backward reactions on the overall reaction rate $(-r_A)$ is straightforward provided kinetic data are available for the forward and reverse reactions.

The overall reaction rate of limiting reactant A is given by
$(-r_A) =$ (rate of consumption of A by the forward reaction)
$-$ (rate of production of A by the reverse reaction)

The rates of the forward and reverse reactions are equal at the equilibrium point and, in consequence, if these reactions are **elementary**, a simple form for the overall rate equation can be obtained which involves one rate constant only together with an equilibrium constant for the reaction. Consider, for example, the elementary reaction

$$A + B \underset{k_2}{\overset{k_1}{\rightleftharpoons}} C + D$$

Applying the rate law (3.14) to the forward and reverse reactions

$$(-r_A) = k_1 C_A C_B - k_2 C_C C_D$$

where C_A, C_B, C_C and C_D are the concentrations of reactants A and B and of products C and D respectively. At the equilibrium point $k_1 C_A C_B = k_2 C_C C_D$ and, as a consequence,

$$\frac{k_1}{k_2} = K_C = \frac{C_{Ce} C_{De}}{C_{Ae} C_{Be}}$$

where the subscript e is used to denote the equilibrium value (see Section 2.4.1.1)

Combining the above two equations,

$$(-r_A) = k_1 \left\{ C_A C_B - \left(\frac{C_C C_D}{K_C} \right) \right\} \tag{3.57}$$

Similarly if $A + B \rightleftharpoons 2C$

$$(-r_A) = k_1 \left(C_A C_B - \frac{C_C^2}{K_C} \right)$$

(where K_C is the equilibrium constant) and so on. Overall reaction rates for a number of reactions in solution and in the gas phase have been represented in this way. Example 4.19 in Section 4.8 is based on a 'pseudo-homogeneous' reaction of the form of equation (3.57).

3.2   Calculation of reactor volumes from reaction rate data: reaction rates as functions of conversion

3.2.1   Expressions giving the volumes of 'ideal' reactors which are required to meet specified values of throughput and conversion: significance of reaction rate $(-r_A)$ in these expressions

3.2.1.1   Batch reactors.   The overall throughput T_A of a liquid reactant A which can be achieved by processing a sequence of batches of material in a batch reactor is given by

$$T_A = \frac{n_{A0}}{t_r + t_0} = \frac{C_{A0} V}{t_r + t_0}$$

where n_{A0} = moles of A fed to reactor in each batch
   C_{A0} = initial concentration of A in reaction mixture
   V = volume of liquid fed to reactor. Assuming that no volume increase occurs during the reaction, V = design capacity of reactor.
   t_r = reaction time per batch required to achieve a desired conversion X_{AF} and
   t_0 = the time required to empty the reactor and to prepare it for the next batch.
   It will be shown in Section 4.1.1 (equation (4.4)) that

$$t_r = C_{A0} \int_0^{X_{AF}} \frac{dX_A}{(-r_A)}$$

where $(-r_A)$ is the rate of consumption of reactant A as a function of its conversion X_A. This equation presupposes that any volume changes in the reaction are negligibly small (it will be shown in chapter 4 that this assumption is a good one for reactions in batch reactors). Combining the above two equations

$$V(\text{batch}) = \frac{T_A}{C_{A0}} \left[C_{A0} \int_0^{X_{AF}} \frac{dX_A}{(-r_A)} + t_0 \right] \qquad (3.58)$$

X_{AF} is the specified conversion (see also Section 4.3.1)

3.2.1.2 Continuous flow reactors. It will also be shown in Chapter 4 that for ideal homogeneous tubular and continuous stirred tank reactors respectively,

$$V_r(\text{tubular}) = F_{A0} \int_0^{X_{AF}} \frac{dX_A}{(-r_A)} \tag{3.59}$$

(cf. equation (4.8)) and

$$V_r(\text{CSTR}) = \frac{F_{A0} X_{AF}}{(-r_A)} \tag{3.60}$$

(cf. equations (4.8) and (4.24) respectively).

In these equations F_{A0} is the molar flow rate of reactant A at the reactor inlet and X_{AF} is the required conversion at the reactor outlet. There is no restriction on volume changes during reaction in either case.

3.2.1.3 Calculation of reactor volumes from equation (3.58) to (3.60). The basic design equations (3.58) to (3.60) are derived and discussed with examples in Chapter 4.

For the present we simply note that before these equations can be evaluated expressions are required for the reaction rate $(-r_A)$ as a function of the conversion X_A of limiting reactant A within the reactor.

Reaction rate equations such as (3.14) or (3.23) give reaction rate as a function of the reactant concentrations, so at this stage relationships are required linking these concentrations to X_A.

If no volume change occurs during the reaction, the concentration C_A of limiting reactant A is directly linked to the conversion of A and the feed concentrations C_{A0} as follows:

$$C_A = n_A/V = \frac{n_{A0}(1 - X_A)}{V} = C_{A0}(1 - X_A) \tag{3.61}$$

n_{A0} is the number of moles of A present in the feed and n_A is the number of moles of A present in the volume V when the conversion of A is X_A (see equations (2.21) and (2.21a)). V is the volume of the feed which, since there is no volume change during the reaction, is the volume also of the product. (In a batch reactor the feed is the initial charge to the reactor; in a continuous flow reactor at steady state it is the input to the reactor over an interval specified in the basis for the calculation, as discussed in Section 2.2 and later sections.)

The corresponding equation for component i is

$$C_i = \left[n_{i0} - \left(\frac{\nu_i}{\nu_A}\right) n_{A0} X_A \right] / V = C_{i0} - \left(\frac{\nu_i}{\nu_A}\right) C_{A0} X_A \tag{3.62}$$

where ν_i and ν_A are the stoichiometric coefficients of components i and A (cf. equations (2.26) and (2.26a)). n_{i0} and n_i are, respectively, the numbers of moles of component i in the feed and in the resulting mixture in which the conversion of A is X_A.

Problem 3.2
Use the second-order rate constant calculated in Problem 3.1 (0.1 $dm^3 \, mol^{-1} \, min^{-1}$) to determine the liquid capacity required in a continuous stirred tank reactor for the conversion of 2000 kmol of A per 24 hours from an initial concentration of $1.25 \, kmol \, m^{-3}$ to $0.15 \, kmol \, m^{-3}$. Volume changes during the above reaction are negligible (Ans: $540 \, m^3$)

Equations (3.61) and (3.62) are usually true as a good approximation for liquid phase reactions and are applicable both to batch reactors and to flow reactors at steady state. They are also true for reactions in a perfect gas phase provided changes in temperature and pressure can be neglected and provided also that *the number of moles of products formed in the reaction is equal to the number of moles of reactants used up*, i.e. that $\sum_i \nu_i = 0$.

If these criteria are not obeyed, corrections to (3.61) and (3.62) are required, as described below.

3.2.2   Reactant concentrations as functions of conversion in ideal gas reactions in which appreciable volume changes occur

When volume changes occur during the reaction, the appropriate expressions for C_A and C_i become

$$C_A = \frac{n_A}{V} = \left(\frac{n_{A0}(1 - X_A)}{V_0}\right)\frac{V_0}{V} = C_{A0}(1 - X_A)\frac{V_0}{V} \qquad (3.63)$$

and

$$C_i = \frac{n_i}{V} = \left(\frac{n_{i0} - (\nu_i/\nu_A)n_{A0}X_A}{V_0}\right)\frac{V_0}{V} = \left(C_{i0} - \frac{\nu_i}{\nu_A}C_{A0}X_A\right)\frac{V_0}{V} \qquad (3.64)$$

where V_0 is the feed volume and V is the volume when the conversion is X_A. ν_i and ν_A are the stoichiometric coefficients of components i and A.

Because of the properties of the liquid state, the ratio (V_0/V) is frequently close to unity for liquid phase reactions.

In a gas phase reaction, however, V_0/V may depart significantly from unity, particularly if the number of moles of products produced differs from the number of moles of reactants consumed.

Suppose that, on the basis chosen, n_0 moles of all types are present in the feed and that n moles are present when the conversion is X_A. Suppose also that the feed temperature and pressure are T_0 and P_0 and that the corresponding values when the conversion is X_A are T and P respectively. Then, if the gas phase is ideal

$$\frac{V_0}{V} = \left(\frac{n_0}{n}\right)\left(\frac{T_0}{T}\right)\left(\frac{P}{P_0}\right) \tag{3.65}$$

Let Δn = increase in moles per mole of A which reacts. Then

$$n = n_0 + y_{A0}n_0 \, \Delta n X_A$$
$$= n_0(1 + y_{A0} \, \Delta n X_A)$$

where y_{A0} = mole fraction of A in feed. Let

$$\frac{n_0}{n} = \frac{1}{1 + \varepsilon_A X_A} \tag{3.66}$$

ε_A in equation (3.66) is given by

$$\varepsilon_A = y_{A0} \, \Delta n \tag{3.67}$$

Combining equations (3.63), (3.65) and (3.66) gives, for C_A,

$$C_A = \frac{C_{A0}(1 - X_A)}{1 + \varepsilon_A X_A}\left(\frac{T_0}{T}\right)\left(\frac{P}{P_0}\right) \tag{3.68}$$

Reactor pressure drops are normally small ($P/P_0 \approx 1$) and hence, for an **isothermal reactor**

$$C_A = \frac{C_{A0}(1 - X_A)}{1 + \varepsilon_A X_A} \tag{3.69}$$

with ε_A given by equation (3.67).

The corresponding equation for C_i is

$$C_i = \frac{C_{i0} - (\nu_i/\nu_A)C_{A0}X_A}{1 + \varepsilon_A X_A} \tag{3.70}$$

Under the above conditions, expressions similar to (3.69) and (3.70) hold also for the component mole fraction. For example, since $C_A = Py_A/RT$, where P and T are constants under the above conditions,

$$y_A = y_{A0}(1 - X_A) \tag{3.69a}$$

If temperature and pressure and the number of moles change during the reaction, equation (3.68) should be used for C_A. The corresponding equation for y_A is then

$$y_A = \frac{y_{A0}(1 - X_A)}{1 + \varepsilon_A X_A}$$

Δn in equation (3.67) is given by the stoichiometry of the reaction. For example, if $2A + B \rightarrow 2C$, then

$$\Delta n = [(2-3)/2] = -\frac{1}{2}$$

In general,

$$\Delta n = \left(\frac{\sum_i \nu_i}{-\nu_A}\right) \tag{3.71}$$

Examples of the use of the above equations are given in Chapter 4.

3.3   Reactions on catalyst surfaces

3.3.1   Definition of a catalyst

A catalyst is a substance which increases the rate at which a chemical reaction approaches equilibrium, without itself becoming permanently changed by the reaction or altering the equilibrium conditions.

There are some very important implications of the above definition. Firstly, since the catalyst does not alter the equilibrium conditions, the standard Gibbs energy change $\Delta_r G^\theta$ and the thermodynamic equilibrium constant K for the catalysed reaction must be the same as for the uncatalysed one. The uncatalysed reaction must therefore be thermodynamically feasible and proceed, even if only very slowly, in the absence of a catalyst. If substantial conversions are to be achieved, this implies that the standard Gibbs energy change for the reaction should be negative, i.e. that K should exceed unity (see equation (2.74)).

Since the equilibrium conditions are unchanged and since also the rates of the forward and reverse reactions are equal at equilibrium, it follows that the catalyst must increase equally the rates of the forward and reverse reactions.

Also, since

$$\Delta_r G^\theta = \Delta_r H^\theta - T\Delta_r S^\theta \tag{3.72}$$

and $\Delta_r G^\theta$ is the same for the catalysed and uncatalysed reactions, it follows that the standard enthalpy change $\Delta_r H^\theta$ and the standard entropy change $\Delta_r S^\theta$ must also be the same in the two cases.

3.3.2   Catalyst selectivity

A second property of a catalyst and one that is of supreme importance is its ability to selectively catalyse a reaction producing a desired product among a number of other reactions which may be thermodynamically feasible but which produce undesired products. A good example of this is provided by the following conversions of ethanol, all of which can occur at similar temperatures:

$$C_2H_5OH \xrightarrow[>300°C]{Al_2O_3} C_2H_4 + H_2O \qquad (3.73)$$

$$2C_2H_5OH \xrightarrow[>260°C]{Al_2O_3} C_2H_5OC_2H_5 + H_2O \qquad (3.74)$$

$$C_2H_5OH \xrightarrow[200-300°C]{Cu} CH_3 CHO + H_2 \qquad (3.75)$$

All the above reactions are thermodynamically feasible under similar conditions, yet copper catalyses the dehydrogenation reaction (3.75) while alumina catalyses the dehydration reactions (3.73) and (3.74). The homogeneous thermal decomposition of ethanol, predominantly to acetaldehyde and hydrogen, does not occur at a significant rate below 600°C.

3.3.3 Surface reaction rates and reaction rates per unit catalyst volume

Reaction rates on catalyst surfaces are directly expressed as rate of consumption of limiting reactant per unit surface area, which will be designated $(-r_{AS})$. The reaction rate per unit catalyst volume $(-r_{AV})$ is then given by

$$(-r_{AV}) = S_g \rho_p (-r_{AS}) \qquad (3.76)$$

where S_g is the surface area per unit mass of catalyst and ρ_p is the density of the catalyst material. In some cases (as in ammonia oxidation) the catalyst is present as a gauze, but in many other cases the catalyst will be present either in the form of porous particles or (more commonly) as a layer deposited within a porous carrier such as Al_2O_3 or SiO_2. The internal surface area presented by these porous particles (or 'pellets') can be large, several hundred square metres per gram of pellet being typical.

Spaced over this surface are sites onto which reactant molecules may be chemisorbed. It is believed that a necessary precursor for a catalysed reaction is that molecules of one or more of the reacting species should become adsorbed in this way.

3.3.3.1 Reduction in activation energy associated with catalysis.
As in the case of homogeneous reactions, surface-catalysed reactions are activated processes in that they require an activation energy E_S to proceed. Their rate constants k_σ are thus given by Arrhenius-type expressions of the form

$$k_\sigma = A_s \exp(-E_S/RT) \qquad (3.77a)$$

or

$$k_\nu = A_s S_g \rho_p \exp(-E_s/RT) \qquad (3.77b)$$

where k_σ is a surface rate constant and k_ν is the corresponding rate constant per unit volume of catalyst pellet. The term A_s is in some ways analogous to the pre-exponential term A in the Arrhenius equation (equation (3.31)) but, as seen below, it is more complex in form and can be composition dependent.

The temperature dependence of the rate constants is largely dictated by the exponential term in the above equations. A low value for E_s corresponds to a high value for this term and a correspondingly large contribution to the reaction rate.

As seen in the examples in Section 3.3.2, suitable catalysts have the ability to selectively enhance the rate of a desired reaction at given temperature, enabling this reaction to be carried out at a much lower temperature than would be the case in the absence of catalyst and also reducing the relative importance of any side reactions. In the light of equation (3.77) it may be deduced that in these cases the catalyst provides a lower energy pathway than the uncatalysed one and also that the catalysed reaction should have a lower activation energy than its homogeneous counterpart. This deduction is confirmed when experimental values for the activation energies of catalysed reactions are compared with those for uncatalysed ones as in Table 3.3.

The lower activation energies encountered in catalysed reactions are illustrated schematically in Figure 3.2 which shows the potential energy as a reaction event proceeds along the reaction co-ordinate in both the catalysed and uncatalysed cases (cf. Figure 3.1).

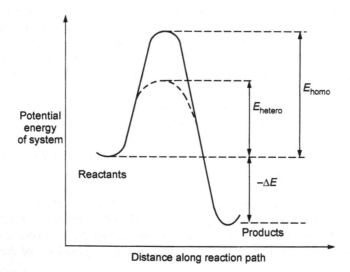

Figure 3.2 Comparison of potential energy changes associated with catalysed (—) and uncatalysed reaction events. $-\Delta E$ is closely related to the heat of reaction.

Table 3.3 Activation energies of uncatalysed and catalysed reactions

Reaction	E (kJ mol^{-1}) (uncatalysed)	E (kJ mol^{-1}) (catalysed)	Catalyst
$2HI \rightarrow H_2 + I_2$	175	105	Au
		59	Pt
$2N_2O \rightarrow 2N_2 + O_2$	236	122	Au
		137	Pt
2,2-Dimethylbutane pyrolysis	200	90	Nickel porphyrin

3.3.3.2 Rates of catalysed reactions compared with those of uncatalysed ones. For a catalysed reaction in a given volume to proceed at a faster rate than the uncatalysed one, the reaction rate per unit volume of catalyst pellet should be greater than the rate per unit volume of reaction fluid in the uncatalysed case, i.e.

$$k_v > k$$

It is evident from equation (3.77) that whether or not this is the case depends not only on the activation energy (which is normally lower for the catalysed case and hence tends to increase the catalysed rate) but also on the magnitude of the group $(A_s S_g \rho_p)$. Although exact comparison is difficult it is clear that this group is normally many orders of magnitude smaller than the Arrhenius parameter A for the uncatalysed reaction as given by equation (3.31). It has been calculated [5] that if a surface reaction and a homogeneous reaction had the same activation energies, a surface area of about 10^{14} m^3 per m^3 of catalyst pellet would be required in a typical case for the reaction rate per unit volume of pellet to equal the reaction rate per unit volume of homogeneous reaction mixture under the same conditions. For typical pellet material this would require a surface area of about 10^8 m^2 g^{-1}, which is substantially greater than the areas normally available. The implication is, that for a catalysed reaction to proceed as fast as its non-catalysed counterpart it should be necessary for the catalysed reaction to have a lower activation energy than the uncatalysed one. This is in accord with observed fact and as a 'rule of thumb' it is often taken that the lowering in the energy of activation should exceed about 70 kJ mol^{-1} if the catalyst is to produce any enhancement in rate in a typical case. The required reduction in activation energy in individual cases may deviate by as much as $\pm 40\%$ from the typical one and will be temperature dependent. Nevertheless the value of 70 kJ mol^{-1} is useful as an approximate guide when comparing activation energies in catalysed and uncatalysed reactions.

The low values for the pre-exponential term A_s in equation (3.77) can be explained qualitatively by extending the collision theory of chemical reactions (Section 3.1.6) to the catalysed situation. At this point it should be remembered that the catalyst surface is believed to contain, relative to the total number of atoms in its surface, only a small fraction of active or catalytic sites. In the homogeneous (uncatalysed) situation, the magnitude of the pre-exponential term A is dictated by the prerequisites that the reactant molecules should come together and that their orientations should be suitable. In the catalysed situation there is the additional prerequisite that at least one of the reactant molecules should collide with and then interact with an active site. The probability of this occurring is likely to be small, which probably accounts for the very small values for A_s in the catalysed case. In the iron-catalysed ammonia synthesis reaction, for example, it is found [7] that only about one N_2 molecule in 10^6 with sufficient energy to interact and dissociate on the iron surface actually does so. After dissociation of the reactants the reaction proceeds [6] according to the scheme

$$N_2 \rightarrow N \rightarrow NH \rightarrow NH_2 \rightarrow NH_3$$
$$H_2 \rightarrow H$$

In this case the weakness in the A_s term in equation (3.77) is compensated many times over by the reduction in activation energy, which is from 942 kJ mol^{-1} (uncatalysed) to about 13 kJ mol^{-1} (catalysed). The reaction rate per unit volume is enhanced by a factor calculated to be about 10^{48} [7].

3.3.4   Industrial importance of catalysis

A large number of well-established and very large-scale industrial processes involve catalysts. These include nitric acid production, with an annual output of 71 million tonnes world wide, ammonia synthesis (110 million tonnes), methanol synthesis, methanol oxidation and sulphur dioxide oxidation, to name but a few. Although consideration is given here only to reactions catalysed by solid surfaces (heterogeneous catalysis), there are also many important processes dependent on homogeneous catalysis or on enzyme catalysts. It is probably true to say that most new processes which will be discovered or developed in the future will involve not only catalysts but carefully designed catalysts capable of giving highly selective production of a given product. Such processes will be part of a 'clean technology' approach, reducing unwanted by-products, waste and pollutants.

3.3.5   Mechanisms involved in reactions catalysed by solid surfaces

The kinetics of reactions catalysed by solid surfaces are intrinsically more complex than those of their homogeneous counterparts. Such reactions normally depend on at least one reactant being adsorbed (usually chemisorbed) at an active site on the solid surface and then modified to a form in which it readily undergoes reaction. Migration from site to site may also occur.

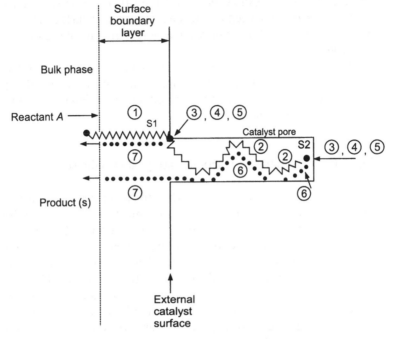

(1) Diffusional transport of reactant A from bulk phase to external catalyst surface

(2) Pore diffusion of reactant by molecular and Knudsen diffusion to internal surface site (S2)

(3) Adsorption of reactant at site S1 or S2

(4) Chemical reaction at sites S1 and S2

(5) Desorption of product(s) at sites S1 and S2

(6) Pore diffusion of product(s) to exterior surface

(7) Diffusional transport of products(s) to bulk phase across the surface boundary layer

Figure 3.3 Processes occurring when a species *A* reacts (a) at an outer surface site S1 and (b) at a site S2 which is deep in a catalyst pore.

In addition to the adsorption, reaction and desorption processes which will be taking place at the active sites themselves, a number of processes may be involved in the transport of the reactants to these sites and the removal of the products from them. As seen in Section 3.3.3, the catalyst is frequently present either in the form of porous particles or (more commonly) as a layer deposited within pellets of a porous carrier such as Al_2O_3 or SiO_2. Before the reactants can reach the active sites distributed throughout the pellet, they must diffuse first through a surface boundary layer surrounding the pellet and then through a network of pores within the pellet. These diffusional processes can be sufficiently slow to limit severely the rate at which reaction can proceed.

Figure 3.3 shows an idealised pore and the various processes undergone by a reacting species both when it reacts at a superficial surface site (S1) and when it reacts at a site (S2) deep down in a pore. Any one of the steps shown can be rate limiting.

In the immediately following sections (3.4 and 3.5) consideration will be given only to those steps involved with the surface reaction, i.e. adsorption, reaction and desorption. Reactions limited by diffusional resistance are considered later in Section 3.6.

3.4   The role of adsorption in reactions catalysed by solid surfaces

As noted in Section 3.3.5, it is evident that for a surface reaction to occur molecules of at least one of the reactants must first become adsorbed (usually chemisorbed) on 'active sites' on the surface. Several detailed mechanisms by which the reaction of interest then occurs have been proposed. In the Eley–Rideal mechanism a gas phase molecule then collides and reacts with the adsorbed molecule [2,3]. In the Langmuir–Hinshelwood model [2,3] the reaction takes place between molecules or fragments of molecules adsorbed on the surface. In each of these models adsorption plays a key role and the following sections are concerned with the interactions of gases with solid surfaces. These interactions must be understood before useful models for the rates of surface catalysed reactions can be developed and used for reactor design.

3.4.1   Types of adsorption onto solid surfaces: physical and chemical adsorption

There are broadly two types of adsorption on to surfaces. The first, defined as **physical adsorption**, arises because of **van der Waals'** interactions between the adsorbate molecules and atoms on the surface. The second, known as **chemical adsorption** or chemisorption, arises from chemical bonding between the adsorbate and atoms on the surface. In the process of heterogeneous catalysis both van der Waals' and chemical interactions are involved, although it is the latter which are the seat of catalytic activity and selectivity.

Physical adsorption results from comparatively weak (but long-range) interactions and is easily reversible. Chemisorption results from chemical bonding. Some criteria by which the two phenomena can be defined and distinguished are given in Table 3.4.

Physical adsorption involves no energy barrier in the forward direction, though an activation energy is required for desorption. Since only weak van der Waals' type forces are involved, the heat liberated during this form of adsorption $(-\Delta H_a)$ is comparatively low, being similar in

Table 3.4 Some criteria distinguishing physical from chemical adsorption

Parameter	Physical adsorption	Chemisorption
Heat of adsorption $(-\Delta H_a)$	Low $(< 20\,kJ/mol)$	High $(40 - 650\,kJ/mol)$
Energy of activation for adsorption	None	None in some cases but can be 40 kJ/mol or more in others
Number of layers formed	> 1	One only
Temperature at which effects are significant	Low $(<$ room temperature$)$	Sub-ambient to temperatures of 800 K and above in some cases

magnitude to the latent heat of condensation of the adsorbate. Typical values are around 20 kJ mol^{-1}. These small enthalpy changes are insufficient to lead to bond breaking and physically adsorbed molecules retain their identity [2]. Because of the similarity between the strengths of the adsorbate/surface and adsorbate/adsorbate interactions (evidenced by the similarity between the heat of adsorption and the heat of condensation) there is a tendency in physical adsorption for the formation of a succession of adsorbed layers above the adsorbent surface rather than the single layer which normally forms in chemisorption. A good qualitative representation of the partial pressure of the adsorbate in the gas space as a function of the mass of material adsorbed on the surface in the latter situation is given by the 'BET' equation described in the appendix to this chapter. Over a restricted range this equation is quantitatively accurate and, on the insertion of appropriate data, provides a measure of the number of molecules of a given adsorbate required to complete a single layer on a surface of interest. By using a standard gas (usually nitrogen) as adsorbate, for which the area excluded per molecule is known, the area of surface may be calculated. This technique is useful in determining the surface areas of porous materials, including catalyst pellets (Section 3.3.3).

Chemisorption involves the formation of a chemical bond and is usually an activated process. However, the activation energies for chemisorption onto a suitable catalytic surface are normally lower than those typically encountered in homogeneous reactions and, indeed, in some instances chemisorption can be a non-activated process (some gas adsorptions onto clean metal surfaces appear to be non-activated [2]). In most cases, however, chemisorption is an activated process with activation energies ranging up to about 40 kJ mol^{-1}. The potential energy changes occurring as a chemisorption event proceeds along the reaction co-ordinate with an appreciable activation energy E_a are shown schematically in Figure 3.4. The system in this figure consists of the adsorbate plus the surface; the initial state of the system is given by the minimum in the potential energy curve to the left of the figure, while the final state (after adsorption) is given by the minimum to the right of the figure.

There is a close, though not exact, correspondence between the heat of adsorption $(-\Delta H_a)$ and the difference $(-\Delta E)$ between the activation energies for the adsorption and desorption events.

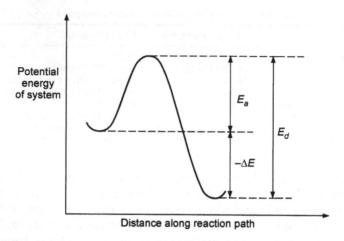

Figure 3.4 Energy profile for a chemisorption event in situation where there is an appreciable activation energy. E_a = activation energy for adsorption; E_d = activation energy for desorption.

A large heat evolution during adsorption is indicative of strong chemical bonding and is a measure of the minimum energy required to break the chemisorption bond.

Both physical adsorption and chemisorption are, except in very exceptional circumstances [2], exothermic, though the heats generated in the latter case are normally much larger than for physical adsorption.

Chemisorption is known to occur at very low temperatures though, in some cases, it can persist to very high temperatures. (Indirect evidence that chemisorption of the relevant species can persist to high temperatures is provided by the fact that many catalysed reactions occur at relatively high temperatures. Examples include catalytic cracking operations on zeolite (450°C) and the ammonia oxidation reaction catalysed by platinum rhodium gauzes at 850°C.)

During a chemisorption process the adsorbate molecules tend to dissociate prior to the formation of the chemisorbed species: indeed the presence of molecular fragments on the catalyst surface is probably an essential part of the catalysis process. An example is provided by the chemisorption of hydrogen onto nickel, the end result of which may be expressed as

$$H_2 + 2Ni = 2(Ni\text{—}H)$$

During the course of the formation of the chemisorbed species Ni—H, the molecular hydrogen is believed to dissociate, as illustrated in Figures 3.5 and 3.6.

Figure 3.5 shows the form of the potential energy changes which occur as a hydrogen molecule advances on the surface of the nickel and is eventually chemisorbed. Figure 3.6 shows diagrammatically stages which are believed

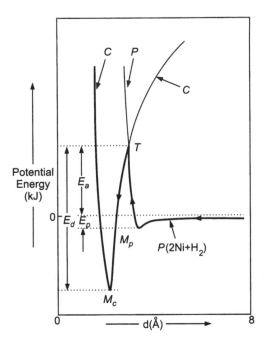

Figure 3.5 Potential energy changes as molecular hydrogen H_2 advances on surface of Ni atoms, decomposes at T and finally forms two chemisorbed species NiH (at M_c), ($d = 0$ at surface). Distance d from the surface is shown in angstrom units (Å).

to occur in the transition from molecular hydrogen to chemisorbed atomic hydrogen. In Figure 3.5, d is the distance of the hydrogen molecule or atom, as the case may be, from the 'surface', i.e. from a line drawn through the centres of the surface nickel atoms. P is the combined potential energy curve for the van der Waals' interactions between the hydrogen molecule and two adjacent nickel atoms on this surface. This is of the familiar 'Lennard–Jones' form with a minimum at M_p and arises from van der Waals' attractive and repulsive forces. The depth E_p of the potential energy well arising from these physical interactions is small, being of similar order of magnitude to the latent heat of hydrogen at its normal boiling point. The distance between the surface and the hydrogen molecule at which this minimum occurs is about 3 Å and is the sum of the van der Waals' radii of a hydrogen molecule and a nickel atom. C is the potential energy curve for the interaction between the two dissociated hydrogen atoms and the adjacent nickel atoms to which they are bonded, shown as a function of the distance between the centre of each nickel atom and the centre of its hydrogen atom.

At point T, curves C and P cross. To the right of this point the potential energy of molecular hydrogen approaching the surface is lower than that of the equivalent number of hydrogen ions; to the left of this point the reverse is true.

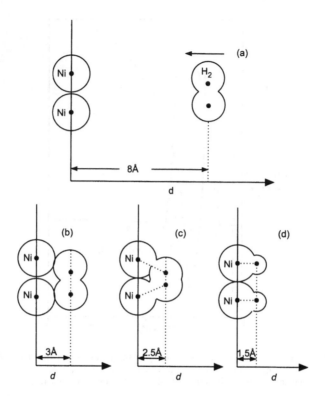

Figure 3.6 Idealised representation of stages in transition from molecular hydrogen to chemsorbed atomic hydrogen during chemisorption event (see Figure 3.5): (a) at large separation (e.g. at 8Å on Figure 3.5); (b) at point M_p; (c) at transition point T; (d) in chemisorbed state at point M_c.

Incoming hydrogen molecules/atoms will at all distances tend to follow a path of lowest potential energy. The appropriate sections of the curves P and C are shown as continuous thickened lines on Figure 3.5. At large distances from the surface the potential energy is close to zero and the path follows curve P down a track of gently decreasing potential energy until the shallow minimum at M_p is reached. Here, at sufficiently low temperature (not higher than about 100 K), hydrogen molecules would tend to remain in a state of physical adsorption. At the higher temperatures likely to be relevant in catalysis, however, molecules at the point M_p will rapidly acquire, by random thermal motions, sufficient energy either to leave the vicinity of the surface altogether or to come closer in to the nickel atoms, thus entering the repulsive part of the intermolecular force field with consequent transitory increase in the potential energy. Some will acquire sufficient energy to bring them to the point T (an activation energy E_a is required for this purpose). Stretching of the H—H bond in the hydrogen molecule now takes place accompanied by formation of Ni—H bonds. To the left of T the lowest potential energy is provided by curve C which represents the potential energy between the two by now dissociated H atoms and the nickel atoms. As the distance between the hydrogen atoms and the centres of the nickel atoms decreases further, so does the potential energy until finally the point M_c is reached. The magnitude of the activation energy E_a (about 30 kJ mol^{-1} in the present case) depends critically on the relative positions and shapes of the potential energy curves P and C.

Discussions of chemisorption in other systems, including hydrogen on carbon and oxygen on silver are given in reference [3].

3.4.2   The Langmuir isotherm and other adsorption isotherms

The Langmuir equation has played a very important role in the development of catalyst theory and is the basis of the Langmuir–Hinshelwood approach described in Section 3.5. The Langmuir isotherm gives the fraction θ_A of the area of surface which, at equilibrium, is covered by adsorbed species A as a function of the partial pressure p_A of the adsorbate.

There are two types of kinetic model which lead to an equation of the Langmuir form for the equilibrium adsorption isotherm. The first model is based essentially on physical adsorption but with the restriction that adsorption is limited to a single layer. In this model adsorption can take place at any point on the surface and the number of molecules which can be adsorbed is limited only by the area excluded by each molecule.

The second model is based on concepts which are closer to catalysis theory and is the one outlined below. This model is based on the following assumptions.

- Adsorption occurs at definite, localised sites on the surface
- Each site accommodates one adsorbed entity only
- All sites are energetically equivalent

(I) We first consider the adsorption of a single component (A) from the gas space on to the surface and we will suppose that no dissociation of this component occurs during the chemisorption process.

Since the formation of a chemical bond is involved it is convenient to represent the adsorption/desorption processes by an equation of the form:

$$A(\text{g}) + S \underset{k_d}{\overset{k_a}{\rightleftharpoons}} A - S \tag{3.78}$$

where $A(\text{g})$ = gas phase species
$\quad\quad S$ = surface site
$\quad A - S$ = chemisorbed species
$\quad k_a, k_d$ = rate constants for adsorption and desorption respectively

If n = number of sites (occupied + unoccupied) per unit area and s = number of unoccupied sites per unit area then, for a single adsorbate,

$$s = n(1 - \theta_A) \tag{3.79}$$

where θ_A = fraction of sites occupied by A.

At given temperature and a given gas (the 'adsorbate') and surface, the rate of adsorption per unit area (r_a) will be proportional to the following:

(1) The number of vacant sites per unit area (s).
(2) The active area δ_0^s around each vacant site into which an incoming adsorbate molecule from the gas space must land if adsorption is to occur. This area will be taken to be independent of θ_A and p_A and contributes to the constant of proportionality k_a in equation (3.80).
(3) r_a in the kinetic treatment is proportional also to the number of gas molecules impacting on unit area of surface per unit time. From simple gas kinetic theory this quantity will in turn be proportional to the partial pressure p_A of the adsorbate for a constant temperature.

Multiplying together factors (1) to (3) gives

$$\text{Rate of adsorption}: \quad r_a = k_a p_A s = k_a p_A n \theta_s \tag{3.80}$$

where θ_s is the fraction of sites unoccupied.
For a single adsorbate

$$r_a = k_a p_A n(1 - \theta_A) \tag{3.80a}$$

where k_a is the rate constant for adsorption at the given temperature.

Also, in the kinetic theory, the rate of desorption per unit area is proportional at a given temperature to the number of occupied sites, i.e.

$$\text{Rate of desorption}: \quad r_d = k_d n \theta_A \tag{3.81}$$

At equilibrium $r_a = r_d$ and hence

$$k_a p_A n(1 - \theta_A) = k_d n \theta_A \tag{3.82}$$

so that

$$\theta_A = \frac{k_a p_A}{k_d + k_a p_A} = \frac{b_A p_A}{1 + b_A p_A} \tag{3.83}$$

where

$$b_A = k_a/k_d. \tag{3.84}$$

b_A is the equilibrium constant for adsorption or 'adsorption coefficient' and is sometimes (as in Chapter 9) denoted K_A.

A large value of b_A indicates strong adsorption and conversely a low value indicates weak adsorption. The magnitude of this quantity is a function of the system type and the temperature.

(II) If a molecule **adsorbs dissociatively** by splitting into f fragments then

$$\theta = \frac{bp^{1/f}}{1 + bp^{1/f}} \qquad (3.85)$$

For example, in the case of the chemisorption of hydrogen onto nickel (Section 3.4.1)

$$H_2 + 2S = 2(H-S)$$

and

$$\theta_H = \frac{b_H p_H^{1/2}}{1 + b_H p_H^{1/2}} \qquad (3.86)$$

(In their analysis of the adsorption of hydrogen onto nickel and platinum, Halsey and Yeates [11] add an interaction term to (3.86).)

(III) If two species A and B are adsorbed competitively, non-dissociatively on the same type of site (only one molecule per site), then

$$\theta_A = \frac{b_A p_A}{1 + b_A p_A + b_B p_B} \qquad (3.87)$$

and

$$\theta_B = \frac{b_B p_B}{1 + b_A p_A + b_B p_B} \qquad (3.88)$$

In general, for i species,

$$\theta_A = \frac{b_A p_A}{1 + \sum b_i p_i} \qquad (3.89)$$

Similarly, for component B,

$$\theta_B = \frac{b_B p_B}{1 + \sum b_i p_i}$$

and so on.

Equation (3.89) is the **multi-component Langmuir equation**. The derivation of equations (3.87), (3.88) and (3.89) is similar to that of (3.83), except that the number of unoccupied sites per unit area is now

$$s = n(1 - \theta_A - \theta_B - \cdots)$$

so that the rate of adsorption of component $A(r_a(A))$ is now

$$r_a(A) = k_a(A)p_A n(1 - \sum \theta_i) = k_d(A)n\theta_A \qquad (3.90)$$

or $\theta_A = b_A p_A(1 - \sum \theta_i)$ where $k_a(A)$ and $k_d(A)$ are the rate constants for the adsorption and desorption of component A and

$$b_A = k_a(A)/k_d(A)$$

In general, for a component i

$$\theta_i = b_i p_i (1 - \sum \theta_i) = \frac{\theta_A b_i p_i}{b_A p_A}$$

So

$$\sum \theta_i = \left(\frac{\theta_A}{b_A p_A}\right) \sum (b_i p_i) \qquad (3.91)$$

Equation (3.89) follows directly on substituting (3.91) back into (3.90) and factoring out θ_A. For some purposes it is convenient to express the Langmuir equation in linear form. For example, equation (3.83) for the non-dissociative adsorption of component A can be rearranged to give

(i)
$$\frac{1}{b_A} + p_A = \frac{p_A}{\theta_A} \qquad (3.92)$$

or

(ii)
$$\frac{\theta_A}{p_A} + b_A \theta_A = b_A \qquad (3.93)$$

When evaluating the parameter b_A from an experiment in which the number of moles n_A of an adsorbate A adsorbed on a surface has been measured as a function of the partial pressure p_A of the adsorbate, the substitution

$$\theta_A = (n_A/n_s) \qquad (3.94)$$

is required where n_A is the number of moles of A adsorbed onto the given sample of solid in the experiment and n_s is the number of adsorption sites within the sample. Making this substitution in equation (3.92), for example, gives

$$\frac{p_A}{n_A} = \frac{p_A}{n_s} + \frac{1}{b_A n_s} \qquad (3.95)$$

A plot of (p_A/n_A) versus p_A, therefore, has a slope $(1/n_s)$ and an intercept $1/(b_A n_s)$ from which both b_A and n_s may be evaluated. Alternatively, numerical methods may be used. The area for adsorption within the sample is given by

$$\text{Area} = n_s \mathbf{N} \delta_0^s$$

where $\mathbf{N}$ is Avogadro's number and, in the above treatment, δ_0^s is the active area per site. In terms of the physical adsorption model (not given here) the same construction applies though δ_0^s is now the area excluded per molecule of adsorbate.

3.5 The kinetics of heterogeneously catalysed reactions

As seen in Section 3.1.4.3, the kinetics of **homogeneous** reactions, with the exception of chain reactions and free radical reactions, can be expressed in terms of simple rate expressions of the type

$$(-r_A) = k C_A^m C_B^n \tag{3.96}$$

where m and n are the respective orders of the reaction with respect to reactants A and B. Here C_A and C_B are the concentrations of these reactants. In elementary reactions, the exponents m and n are small integers such as 0, 1 or 2 though in free radical/chain reactions semi-integral values of reaction order such as 0.5 or 1.5 are frequently observed.

In the case of **heterogeneous reactions**, reaction rates can usually still be expressed in terms of the above power law type of expression, where C_A and C_B are now the concentrations of A and B in the fluid (usually pore fluid) adjacent to the catalytic surface. In this case, however, the exponents m and n can be markedly non-integral.

The reason for this is that the rate of reaction is dependent upon the concentrations of the surface (adsorbed) species. The latter must be related to the fluid or bulk phase concentrations via an appropriate adsorption isotherm. (Except in very unusual cases it is, of course, the bulk phase concentrations which are the only easily measurable parameters.)

The adsorption isotherm generally employed is the Langmuir isotherm since it provides a comparatively simple means of relating bulk phase concentrations to those on the surface.

In the model discussed below, the rates of the surface reactions are taken to be given by equations analogous to those for elementary reactions in a homogeneous phase but 'surface coverages' replace the volumetric concentrations employed for homogeneous reactions. The 'surface coverage' (sometimes called 'surface concentration') of an adsorbed component A is here identified with the fraction θ_A of the adsorption sites which are occupied by that component. For a surface reaction in which an adsorbed component A reacts to form products p, for example, the appropriate forward rate equation is

$$(-\vec{r}_{AS}) = k_{(1)} \theta_A \tag{3.97}$$

where $(-\vec{r}_{AS})$ is the rate at which the adsorbed component A is consumed by the forward reaction on unit area of surface and $k_{(1)}$ is the rate constant.

For a reaction between adsorbed components A and B the appropriate rate equation is

$$(-\vec{r}_{AS}) = k_{(2)} \theta_A \theta_B \tag{3.98}$$

and so on. The rate constants $k_{(1)}$ and $k_{(2)}$ have units mol s^{-1}m^{-2}.

The above expressions are clearly analogous to the first- and second-order reactions in a homogeneous phase. To avoid confusion, however, it is probably best to restrict the use of these terms to reactions which are first or second order with respect to the **fluid phase** concentrations or partial pressures. Using the Langmuir adsorption theory outlined previously, the surface coverages can be expressed in terms of fluid phase partial pressures, as described below.

The resultant rate equations are usually described as the Langmuir – Hinshelwood equations [12], though systematic application of this approach has also been made by Hougen and Watson [8, 13].

3.5.1   Langmuir–Hinshelwood–Hougen–Watson reaction models

3.5.1.1   Simple unimolecular reaction.   Consider the reversible reaction

$$A(g) \overset{\text{catalyst}}{\rightleftharpoons} R_{(g)}$$

$A_{(g)}$ and $R_{(g)}$ are gas phase species, though the reaction takes place on a solid surface.

A is first adsorbed onto this surface where it reacts to form R in adsorbed form. Finally R is desorbed and enters the gas space. According to this picture the reaction proceeds in the following sequence of three steps:

$$A(g) + S \underset{k_{-1}}{\overset{k_1}{\rightleftharpoons}} A - S \tag{3.99}$$

$$A - S \underset{k_{-2}}{\overset{k_1}{\rightleftharpoons}} R - S \tag{3.100}$$

$$R - S \underset{k_{-3}}{\overset{k_1}{\rightleftharpoons}} R_{(g)} + S \tag{3.101}$$

where S = surface site and A–S and R–S are adsorbed species.

If p_A, p_R are the gas phase partial pressures of A and R respectively, θ_A, θ_R are the surface 'concentrations' of A and R respectively and θ_s is the surface 'concentration' of vacant sites, the rate of each step can be written:

$$r_1 = k_1 p_A \theta_s - k_{-1} \theta_A \tag{3.102}$$

$$r_2 = k_2 \theta_A - k_{-2} \theta_R \tag{3.103}$$

$$r_3 = k_3 \theta_R - k_{-3} p_R \theta_s \tag{3.104}$$

Equations (3.102) and (3.104) are for the net rate of adsorption of A from the gas phase and the net rate of desorption of R into the gas phase respectively. The equations follow directly from the Langmuir kinetics outlined in Section 3.4.2. Equation (3.102), for example, is obtained by subtracting equation (3.81) for the rate of desorption of A from equation (3.80) for the rate of adsorption, and then writing

$$k_1 = nk_a$$

$$k_{-1} = nk_d$$

where n is the total number of sites per unit area.

θ_s in equations (3.102) and (3.104) is the fraction of adsorption sites which are vacant and is given by

$$\theta_s = \frac{S}{n}$$

where S is the number of vacant sites per unit area.

Equation (3.103) follows directly on applying equation (3.97) to the forward and reverse surface reactions respectively. Any one of the rates r_1, r_2 and r_3 in equations (3.102) to (3.104) can be sufficiently slow to control the overall rate of the reaction. For the present it will be assumed that the surface reaction (equation (3.100)) is much the slowest step and that its rate (given by equation (3.103)) controls the overall reaction rate. The adsorption and desorption steps (equations (3.99) and (3.101)) will therefore proceed sufficiently rapidly for the surface concentrations θ_A and θ_R of the adsorbed species A and R to achieve their equilibrium values corresponding to the partial pressure p_A and p_R of these species in the gas phase. Hence, from equation (3.89),

$$\theta_A = \frac{b_A p_A}{1 + \sum b_i p_i} = \frac{b_A p_A}{1 + b_A p_A + b_R p_R} \tag{3.105}$$

and

$$\theta_R = \frac{b_R p_R}{1 + \sum b_i p_i} = \frac{b_R p_R}{1 + b_A p_A + b_R p_R} \tag{3.106}$$

Combining these expressions for θ_A and θ_R with the rate equation (3.103) for the surface reaction gives

$$(-r_{AS}) = r_2 = \frac{k_2 b_A p_A}{1 + b_A p_A + b_R p_R} - \frac{k_{-2} b_R p_R}{1 + b_A p_A + b_R p_R} \tag{3.107}$$

where $(-r_{AS})$ is the net rate at which component A is being consumed in the reaction per unit catalyst area. For an **irreversible reaction**

$$(-r_{AS}) = r_2 = k_2 \theta_A = \frac{k_2 b_A p_A}{1 + b_A p_A + b_R p_R} \tag{3.108}$$

A number of limiting cases can be considered as follows:

(1) The product R is only weakly adsorbed or not adsorbed at all. b_R, which is a measure of the strength of adsorption of the product (equation (3.84)), will be virtually zero so that

$$(-r_{AS}) = r_2 = \frac{k_2 b_A p_A}{1 + b_A p_A} \qquad (3.109)$$

The order of the reactant overall reaction in this case depends on temperature and the partial pressure p_A of reactant A. The latter effect can be demonstrated as follows.

If p_A is very large,

$$1 \ll b_A p_A$$

Hence, from (3.109)

$$(-r_{AS}) = r_2 = k_2,$$

i.e. the reaction is zero order in p_A. Also, from (3.105), $\theta_A = 1.0$.

If p_A is very small

$$1 \gg b_A P_A$$

and

$$(-r_{AS}) = r_2 = k_2 b_A p_A = k_2' p_A$$

where $k_2' = k_2 b_A$. The reaction is then first order in p_A.

(2) The next situation to be considered is that when the product R is very strongly adsorbed. In this case

$$b_R p_R \gg 1 + b_A p_A$$

so that from equation (3.108),

$$r_2 = \frac{k_2 b_A p_A}{b_R p_R} = k_2'' \frac{p_A}{p_R}$$

where $k_2'' = k_2 b_A / b_R$. In this case the reaction is of order -1 in p_R and will be inhibited by high partial pressures of R.

If the product is not appreciably adsorbed, the overall reaction order is predicted to increase from zero to 1 as the partial pressure of A decreases, giving a progression of non-integral reaction orders.

This situation does not occur with homogeneous reactions.

It is a further consequence of the above analysis that, since the adsorption coefficients b_A and b_R are functions of temperature, the reaction order can also be temperature dependent.

Temperature dependence of the catalysed reaction rates in cases (1) and (2) above: activation energy of the surface reaction (equation (3.100))

The reaction rates in the model adopted above depend on the rate of the slowest step, which was taken to be that of the surface reaction (3.100). In the irreversible case the rate is accordingly given by (3.103) to be

$$(-r_{AS}) = r_2 = k_2\theta_A \qquad (3.108)$$

where $(-r_{AS})$ is the rate consumption of reactant A per unit catalyst area, k_2 is the rate constant for the surface reaction, θ_A is invariant at given temperature since it is the equilibrium coverage; however, it is a function of temperature since it depends on the Langmuir adsorption coefficients in the expressions for θ_A which are functions of temperature.

k_2 is also temperature dependent, being given by the Arrhenius expression

$$k_2 = A \exp\left(\frac{-E_S}{RT}\right)$$

where E_S is the activation energy for the surface reaction (which, as seen in Section 3.3.3.1, is normally lower than for the homogeneous case).

In the special case of the irreversible unimolecular reaction in which the product was only weakly adsorbed and p_A was very small (the first situation considered above), θ_A was unity and $r_2 = k_2$.

In this special case therefore the slope of a plot of the logarithim of the observed reaction rate against $(-1/T)$ will equal the activation energy for the surface reaction. In general, for unimolecular reactions the reaction rate depends both on k_2 and θ_A and the slope of the above plot gives

$$E_S - Q_A$$

where Q_A is the heat of adsorption of the reactant A.

3.5.1.2 Bimolecular surface catalysed reactions. If two or more reacting species are involved, the situation at the surface becomes much more complex. It may be, for example, that more than one type of surface site is involved [10]. In the present treatment it is assumed that only one type of site is involved in the reaction and that all the reacting species compete to a smaller or larger degree for that type of site.

If, in the reversible reaction

$$A + B \rightleftharpoons X + Y \qquad (3.110)$$

all reactants and products were adsorbed at finite rate and no one step in the adsorption/reaction sequence were rate controlling, quite a complex expression for the reaction rate would in general be obtained. This expression would incorporate finite rate terms for the adsorption/desorption of reactants A and B and for the desorption/adsorption of products X and Y in addition to a term or terms to take account of the surface reaction, at least five terms in all. The steps in the reaction sequence may be represented as:

$$A(g)+S \underset{k_{-1}}{\overset{k_1}{\rightleftharpoons}} A-S \tag{i}$$

$$B(g)+S \underset{k_{-2}}{\overset{k_2}{\rightleftharpoons}} B-S \tag{ii}$$

$$A-S+B-S \underset{k_{-3}}{\overset{k_3}{\rightleftharpoons}} X-S+Y-S \tag{iii}$$

$$X-S \underset{k_{-4}}{\overset{k_4}{\rightleftharpoons}} X+S \tag{iv}$$

$$Y-S \underset{k_{-5}}{\overset{k_5}{\rightleftharpoons}} Y+S \tag{v}$$

In the present treatment it is assumed that surface reaction (iii) is the rate-determining step and that the adsorption and desorption steps take place sufficiently rapidly for equilibrium to be achieved between the surface coverage of each species and the corresponding gas space partial pressure. In terms of gas space partial pressures, therefore, the 'surface coverages' of components A, B, X and Y are given in the present treatment by the Langmuir expression of the form of equation (3.89). The 'surface coverage' (or fractional occupancy of sites) for component A, for example, is given by

$$\theta_A = \frac{b_A p_A}{1 + b_A p_A + b_B p_B + b_X p_X + b_Y p_Y}$$

while for component B,

$$\theta_B = \frac{b_B p_B}{1 + b_A p_A + b_B p_B + b_X p_X + b_Y p_Y}$$

and so on.

The rate of the surface reaction in the forward direction is given by equation (3.98) to be

$$(-\vec{r}_{AS}) = k_3 \theta_A \theta_B$$

while the rate of the reverse reaction is $k_{-3}\theta_X\theta_Y$.

The net rate of the surface reaction is then

$$(-r_{AS}) = \frac{k_3 b_A b_B p_A p_B - k_{-3} b_X b_Y p_X p_Y}{(1 + b_A p_A + b_B p_B + b_X p_X + b_Y p_Y)^2} \tag{3.111}$$

where $(-r_{AS})$ is the net rate of consumption of reactant A per unit surface area, b_A, b_B, b_X and b_Y are the Langmuir adsorption coefficients for components A, B, X and Y respectively and p_A, p_B, p_X and p_Y are the partial pressures of the components in the gas space adjacent to the surface on which the reaction is taking place. Equation (3.111) is useful in its own right and in some instances is the best available form of the rate equation (see reference [14] for example). Also it is interesting to examine some limiting cases:

(1) If neither reactants nor products are strongly adsorbed, the denominator in equation (3.11) tends to unity so that

$$(-r_{AS}) = k_3 b_A b_B p_A p_B - k_{-3} b_X b_Y p_X p_Y = k_3' p_A p_B - k_{-3}' p_X p_Y \qquad (3.112)$$

This expression is the equivalent of that for a reversible second-order reaction in a homogeneous phase.

(2) If the reaction is irreversible so that $k_{-3} = 0$,

$$(-r_{AS}) = \frac{k_3 b_A b_B p_A p_B}{(1 + b_A p_A + b_B p_B + b_X p_X + b_Y p_Y)^2} \qquad (3.112a)$$

If, furthermore, reactant B is very much more strongly adsorbed than either reactant A or the products,

$$(-r_{AS}) = k_3 b_A b_B p_A p_B / (b_B p_B)^2 = k_3 (b_A/b_B)(p_A/p_B) = k_3''(p_A/p_B) \qquad (3.112b)$$

The reaction is thus of order -1 with respect to reactant B, so retardation by a reactant is predicted. There are in fact a number of examples of this type of behaviour [3] including the hydrogenation of pyridine on metal oxide catalysts which is retarded by high pyridine partial pressures.

The Langmuir–Hinshelwood model can quite readily be extended to include situations where reactants A and B are adsorbed onto different types of site or where the adsorption or desorption steps are rate controlling.

In conclusion, although the Langmuir–Hinshelwood model incorporates obvious oversimplifications and there are situations where equations based on other mechanisms, such as the Temkin equation ([15, 16, 24] and Chapter 7) or the Eley–Rideal mechanism [3] are preferable, it does provide a good overall interpretation of many general aspects of reaction rate behaviour on catalyst surfaces. Furthermore, the rate equations which it generates can, when the adjustable constants have been fitted, be very useful for fitting rate data [14].

Problem 3.3

The reaction $CO + Cl_2 \rightarrow COCl_2$ has been studied at 30°C and atmospheric pressure using an activated carbon catalyst, and the conclusion reached was that the overall rate was controlled by the surface reaction between adsorbed CO and adsorbed Cl_2. Parallel adsorption experiments showed that Cl_2 and $COCl_2$ were strongly adsorbed when the adsorption of CO was very weak, to the point that, although not zero, b_{CO} could be neglected relative to b_{Cl_2} and b_{COCl_2}.

The reaction is irreversible and the experiments were free from pore or film diffusion influence.

(a) Develop a rate equation on the basis of the above information.
(b) If, at 30°C, the initial rate for a 2:1 molar feed mixture of CO and Cl_2 was found to be 28% higher than that for a 1:1 molar feed mixture, calculate the approximate value of b_{Cl_2}.

Solutions

(a) $(-r_{CO}) = \dfrac{k_3 p_{CO} p_{Cl_2}}{\left(1 + b_{Cl_2} p_{Cl_2} + b_{COCl_2} p_{COCl_2}\right)^2}$ mol s^{-1}m^{-2}

(b) $b_{Cl_2} = 1.20$ atm^{-1}

3.6 Diffusional limitations on surface-catalysed reactions

The present section is principally concerned with situations where the catalyst material is present within a porous structure into which the reactant(s) must diffuse before reaction can occur. A common example of this arises when the catalytic material is dispersed within a porous support material such as alumina or silica, though there are other examples as described in Chapter 6. The porous particles within which the catalyst material is contained are here described as 'catalyst particles'. The best size for the catalyst particles used in 'fixed bed' reactors is dictated by a balance between pressure drop and mass transfer considerations. If the particles are too small the pressure drop through the bed becomes excessive and bed containment may be a problem. If, on the other hand, they are too large the catalyst material which is furthest from the outer surface of the particles will be largely ineffective since most of the reactant entering the particle will have reacted within the outer parts of the particle without reaching the deeper regions. The concentration of reactant in these regions will, accordingly, be lower than at the outer surface, leading to a reduction in reaction rate. A measure of this effect is provided by the **effectiveness factor** η.

3.6.1 Definition of effectiveness factor and intrinsic reaction rate

The effectiveness factor is defined as the ratio of the actual reaction rate per catalyst particle to the rate which would occur if the reactant concentrations throughout the particle were the same as at the surface. In consequence,

$$(-r_{AV}) = \eta(-r'_{AV}) \tag{3.113}$$

where $(-r_{AV})$ is the observed reaction rate per unit volume of catalyst particle. (This is an overall value being equal to the rate per particle divided by the volume of the particle.) $(-r'_{AV})$ is the 'intrinsic reaction rate', that is, the reaction rate per unit volume of particle which would occur if concentrations throughout the particle were the same as at the surface.

For design purposes values will be required both for η and for $(-r'_{AV})$ so that the actual achievable reaction rate can be calculated. Expressions for η for a simple form of surface reaction are developed below and, arising from these expressions, it is shown how effectiveness factors for particles of one size at given conditions may be estimated from data for other conditions and particle sizes.

Simple reaction form to be considered. To illustrate the principles upon which more elaborate treatments may be based, irreversible surface reactions in which the rate of consumption of reactant A follows the very simple first-order rate law

$$(-r_{AS}) = k_\sigma C_A \qquad (3.114)$$

will be considered. In equation (3.114), $(-r_{AS})$ is the rate at which reactant A is being consumed per unit area of catalytic surface, k_σ is the surface rate constant and C_A is the concentration of A in the pore fluid adjacent to the element of surface considered. If pore diffusion is slow, the steady-state value for C_A will vary with the depth of the pore beneath the outer surface of the particle (Figure 3.8). The case of ideal gas phase reactions will be considered so that $C_A = (p_A/RT)$ where p_A is the partial pressure of reactant in the gas space adjacent to the element of catalytic surface considered.

In terms of the Langmuir–Hinshelwood model, equation (3.114) would apply in the following situations:

(1) Unimolecular reactions, either with weak adsorption of both reactant and product(s) or weak adsorption of product and low concentration of A. Under these conditions $1 \gg b_A p_A$ and $k_\sigma = k_2 b_A RT$

(2) Irreversible bimolecular reactions in which the concentration (and hence partial pressure) of the second reactant B is substantially greater than that of A and components A and the product are only weakly adsorbed. Component B may either be weakly adsorbed (so that $b_A p_A + b_B p_B + b_x p_x + b_Y p_Y$ in equation (3.112a) $\ll 1$) or, alternatively, B may be very strongly adsorbed (so that (3.112b) applies). Because the partial pressure of B is substantially greater than that of A, changes in this during the reaction may be neglected (this is the analogy of the pseudo-first-order reactions discussed in Section 3.14.9).

In the first of the above cases k_σ in equation (3.114) would be given by

$$k_\sigma = k_3 b_A b_B p_B \, RT$$

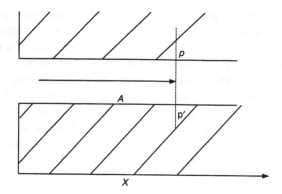

Figure 3.7 Schematic representation of pore (p–p' is stationary plane).

while, in the second,

$$k_\sigma = (k_3 b_A / b_B p_B) \, RT$$

3.6.1.1 Intrinsic reaction rate per unit volume of pellet and the intrinsic reaction rate constant.   In the absence of mass transfer limitations reaction rate per unit volume of pellet = intrinsic reaction rate $(-r'_{AV})$ where, for a reaction of the simple form of (3.114),

$$(-r'_{AV}) = \rho_p S_g k_\sigma C_{AS} \qquad (3.114a)$$

where k_a = first-order rate constant based on unit catalytic area.
   C_{AS} = concentration of A at surface (= that throughout pellet if no mass transfer limitations).
   ρ_p = pellet density (pellet mass/pellet volume).
   S_g = area of active surface per unit mass of pellet.

S_g is routinely evaluated experimentally, either using gas adsorption data or mercury porosimetry.

   It is convenient also to define an intrinsic first-order rate constant by the equation

$$(-r'_{AV}) = (-k'_V) C_{AS} \quad \text{where} \quad (-k'_V) = \rho_p S_g k_\sigma \qquad (3.114b)$$

In common with the intrinsic reaction rate, the intrinsic rate constant is based on unit volume of pellet material.

3.6.2   Diffusion into catalyst pores (see Figures 3.3 and 6.3)

The actual reaction rate per unit volume of catalyst pellet will normally be lower than the intrinsic value (η in equation (3.113) < 1) mainly due to slow diffusion through the pore system within the pellet.

The pores within the pellets will be of a range of sizes and the pore diffusional mechanism will range from 'normal' mutual or bulk diffusion in the large pores (i.e. pores with diameter d considerably greater than the mean free path λ of the reactant molecules) to Knudsen pore diffusion in the smallest pores (i.e. pores for which $d \ll \lambda$). In addition, there may be 'bulk flow' effects to consider. If the reaction results in an increase in the number of moles, the process of bulk diffusion into the large pores will take place against a counter current of outflowing gas, while if the reaction results in a decrease in the number of moles, it will take place in the same direction as bulk inflow.

3.6.2.1 Bulk diffusion. Rates of bulk diffusion are limited by the collision rate between the gas molecules and are given by Fick's law expressions such as equation (3.115). Values of binary diffusivities may be found in data tabulations or may be predicted from the more advanced forms of the kinetic theory of gases [20]. These are applicable provided the dimensions of the region in which the molecules move are all substantially in excess of the mean free path λ. If the gas phase can be regarded as near-ideal (Section 2.1.4.2), bulk diffusivities are inversely proportional to pressure and increase approximately as $T^{1.75}$.

Figure 3.7 is a schematic representation of a pore through which reactant A is diffusing.

$p - p'$ is a stationary plane within the pore, x is the distance up the pore measured from the mouth of the pore. It will be supposed that the pore is sufficiently wide for molecule to molecule collisions to be the principal barrier to molecular transport. For the moment it will be supposed also that the only component, apart from A, present in the pore is a component B so that D_{AB} is a binary diffusivity. The more general situation where more components are present is considered briefly at the end of the section. (There are several situations where pore fluid either is a binary mixture or may be regarded as such as a good approximation. For example, B may be a second reactant present in considerable excess or, as sometimes happens, an inert present in large amount. Also it may be the product of a unimolecular reaction.)

If there is no bulk flow, i.e. if there is no volume change resulting from the reaction and steady-state conditions prevail, the flux (j_A) of A across plane $p - p'$ in direction of increasing x is given by

$$j_A = -D_{AB} \frac{dC_A}{dx} \tag{3.115}$$

where C_A is the local concentration of A and the gradient (dC_A/dx) is taken at the plane $p - p'$. D_{AB} is the binary diffusivity of A into B.

If bulk flow is taking place,

$$j_A = \text{(flux due to diffusion)} + \text{(flux due to bulk flow)}$$

$$= -D_{AB} \left(\frac{dC_A}{dx} \right) + (j_A + j_B) y_A \tag{3.116}$$

where j_B is the flux of B across the plane $p - p'$ and y_A is the mole fraction of A. Equation (3.116) rearranges to give

$$j_A = \frac{-D_{AB}\left(\dfrac{dC_A}{dx}\right)}{1 - \left(1 + \dfrac{j_B}{j_A}\right)y_A} \tag{3.117}$$

The resistance to the transport of A up the pore arising from molecule/molecule collisions in the presence of bulk flow is given by

$$\Omega_A = \text{(driving force) divided by flux}$$
$$\text{flux} = -(dC_A/dx) \text{ divided by } j_A$$

$$j_A = (1/D_{AB})\left(1 - \left(1 + \frac{j_B}{j_A}\right)y_A\right) \tag{3.118}$$

if no volume change occurs during the reaction,

$$j_B = -j_A \quad \text{and} \quad j_A = -D_{AB}(dC_A/dx)$$

as in equation (3.115). This is also approximately true if y_A is very small.

3.6.2.2 Knudsen (or fine pore) diffusion.

In a fine pore, molecule/wall collisions may be more frequent than molecule/molecule ones and it is then molecule/wall collisions which provide the main resistance to diffusion. Diffusion under these conditions is termed **Knudsen diffusion**. According to kinetic theory

$$j_{KA} = \text{flux of } A \text{ in a Knudsen diffusion process} = -D_{KA}(dC_A/dx) \tag{3.119}$$

where

$$D_{KA} = \left(\frac{2\bar{r}}{3}\right)\left(\frac{8RT}{\pi M_A}\right)^{1/2} \tag{3.119a}$$

D_{KA} is the Knudsen diffusion coefficient for the transport of A into the pore, R is the molar gas constant, M_A is the molar mass of reactant A (in primary SI units of kg mol^{-1}) and $\bar{r}$ is the mean radius of the pores in m.

A measure of the resistance Ω_{KA} to the transport of A arising from molecule/wall collisions is given by

$$\Omega_{KA} = \frac{\text{driving force}}{\text{flux}} = \left(\frac{-dC_A/dx}{j_{KA}}\right) = \frac{1}{D_{KA}} \tag{3.120}$$

(An estimate for the mean pore radius $\bar{r}$ in equation (3.119a) may be obtained from experimentally determined values of the pellet porosity (ε_p), the pellet density (ρ_p) and the internal surface per unit mass of pellet (S_g) using the equation

$$\bar{r} = 2\varepsilon_p/(\rho_p S_g) \tag{3.121}$$

This result is obtained by regarding the pores as cylinders of equal radius).

Equations (3.117) and (3.119) respectively enable estimates to be obtained for diffusion rates in pores with diameters substantially larger or substantially smaller than the mean free path.

3.6.2.3 Transition from bulk diffusion to Knudsen diffusion. A combined equation for diffusion rates in both large and small pores, and also in pores of intermediate size, may be obtained by hypothesising that the overall resistance to flow, defined as (driving force)/(net flux of A) should be the sum of the resistances to flow due to molecule/molecule and molecule/wall collisions, i.e.

$$\frac{-dC_A/dx}{J_{A,\text{pore}}} = \frac{1}{D_{KA}} + \frac{1}{D_{AB}}\left[1 - \left(1 + \frac{j_B}{j_A}\right)y_A\right] \tag{3.122}$$

where $J_{A,\text{pore}}$ is the net flux of A into the pore.

Rearranging (3.122) gives

$$J_{A,\text{pore}} = \frac{-dC_A/dx}{(1/D_{KA}) + (1/D_{AB})(1 - (1 + (j_B/j_A)y_A)))} \tag{3.123}$$

where

$$J_{A,\text{pore}} = -D_{A,\text{pore}}\frac{dC_A}{dx} \tag{3.123a}$$

in which $D_{A,\text{pore}}$ is the net diffusivity of A into the pore. This quantity takes both bulk diffusion and Knudsen diffusion into account.

Equation (3.123) has been variously derived by several anthors [17–19].

3.6.2.4 Multi-component equations. Equation (3.123) is based on the presumption that only one component (component B) is present in the pore fluid in addition to reactant A so that the rate of bulk diffusion is given by equation (3.117). However, it has been generalised to the situation where any number of other components are present [19], the appropriate equations then being

$$J_{A,\text{pore}} = -D_A^M\left(\frac{dC_A}{dx}\right) \tag{3.124}$$

where D_A^M, the net diffusivity of A into the pore, is given by

$$\frac{1}{D_A^M} = \frac{1}{D_{KA}} + \sum_n\left[\frac{1}{D_{An}}\left(y_n - \left(\frac{j_n}{j_A}\right)y_A\right)\right] \tag{3.124a}$$

D_{KA} is given as before by equation (3.119a) and D_{An} is the binary bulk diffusivity in the system A/n; y_A and y_n are the mole fractions of A and n.

The above equations are based on the presumption that there is a continuous progression of pore sizes ranging from the smallest to the largest, i.e. that the system is monodisperse. It has been pointed out [17] that some pellet catalyst material can have a bidisperse pore structure with very small pores in the active particles and substantially larger pores in the support material. Equations for this situation have been developed [21].

3.6.3 *Effective diffusivity into catalyst pellets and effectiveness factor for spherical catalyst pellets*

The classical approach to the evaluation of the effectiveness factor for spherical pellets is to regard these as **homogeneous** spheres into and out of which the reactants and products uniformly diffuse. An effective diffusivity D_{eff} consistent with this model is attributed to each component.

This is based on the flux of each component A through unit area of pellet rather than through unit area of pore. It differs from the pore diffusivities partly because only part of any element of area within the pellet will be pore area and partly because the pores will not in general be uniform, nor will they be directed towards the centre of the pellet. For these reasons the effective diffusivity is usually expressed in terms of the net pore diffusivity by expressions of the form

$$D_{A,\text{eff}} = D_{A,\text{pore}}(\varepsilon_p/\tau) \tag{3.125}$$

where ε_p, the pellet porosity, is given as

$$\varepsilon_p = (\text{volume of pores})/(\text{volume of pellet}) \tag{3.126}$$

$D_{A,\text{eff}}$ and $D_{A,\text{pore}}$ in equation (3.125) are respectively the effective diffusivity of reactant A into the pellet and the net pore diffusivity of this component.

The factor ε_p allows for the fact that only a fraction ε_p of a given area within the pellet will be pore area. τ is the so-called 'tortuosity factor' and allows for the fact that pores are not normally directed towards the centre of the pellet but follow random and 'tortuous' paths. The effects of variations in pore area are often also lumped into this term [22] and this is done in the present treatment. In the present state of the art, τ cannot be evaluated from normally available data for catalyst pellets, and for this reason it is usually regarded as an adjustable constant. It should be remembered, however, that this quantity should be purely a function of the pore structure and should, for a given pellet material, be independent of parameters such as pressure (unless this is sufficiently high to deform the pellet), temperature and pellet radius.

τ (as defined in equation (3.125)) is typically about 4. However, values of 2 are not uncommon and values of 10 are not unknown. ε_p is typically about 0.4.

Problem 3.4

Estimate the Knudsen pore diffusion coefficient and also the effective diffusivity of thiophene at 660 K and a pressure of 30 bar in a catalyst pellet having an internal surface area of 180 $m^2 g^{-1}$, a pellet porosity of 0.4 and a pellet density of 1400 kg m^{-3}. The tortuosity factor τ is 2. The concentration of thiophene is very low and the pore fluid consists mostly of hydrogen. The mutual (or 'bulk') diffusivity of thiophene in hydrogen under the above conditions is 5.2×10^{-6} m^2 s^{-1}. (The RMM (or 'molecular weight') of thiophene is 84 and the molar gas constant R is 8.314 J mol^{-1} K^{-1})

Solution

$$D_{K,\text{Thiophene}} = 8.6 \times 10^{-7} m^2 s^{-1}; \quad D_{\text{eff}} = 1.5 \times 10^{-7} \ m^2 s^{-1}$$

3.6.4 Development of expression for effectiveness factor η for first-order reactions in isothermal spherical pellets (see equation (3.113))

The effectiveness factor is the ratio of the actual reaction rate per unit volume of pellet to the intrinsic reaction rate. For illustrative purposes we will first consider the simple form of surface reaction rate equation given by (3.114). The corresponding expression for the intrinsic reaction rate per unit pellet volume is then given by equation (3.114a). In the treatment below it is assumed that all parts of the pellet remain at the same temperature so that the surface reaction rate constant and the diffusivities do not vary within the pellet. To obtain the actual reaction rate per unit pellet volume by the route followed below we first obtain a differential equation relating reactant concentration to radial position in pellet (equation (3.130) below). To obtain this we take a steady-state mass balance on a spherical annulus of radius r and thickness δr, as shown in Figure 3.8.

Rate of consumption of A in annulus $= k_\sigma C_A x$ (surface area within annulus)

$$= k_\sigma C_A S_g \rho_p \text{ (volume of annulus)}$$
$$= k_\sigma C_A S_g \rho_p (4\pi r^2 \delta r) \qquad (3.127)$$

In steady state,
Rate of consumption of A in annulus $=$ rate at which A enters annulus by diffusion at $(r + \delta r)-$ rate at which A leaves annulus at r

$$= D_{A,\text{eff}}\left(4\pi(r + \delta r)^2\right)\left[\left(\frac{dC_A}{dr}\right)_{\text{at.} r} + \left(\frac{d^2 C_A}{dr^2}\right)\delta r\right] - D_{A,\text{eff}}(4\pi r^2)\left[\left(\frac{dC_A}{dr}\right)_{\text{at},r}\right]$$

$$= D_{A,\text{eff}}\delta r\left[4\pi r^2\left(\frac{d^2 C_A}{dr^2}\right) + 8\pi r\left(\frac{dC_A}{dr}\right)\right] + \text{(terms in}(\delta r)^2 \qquad (3.128)$$

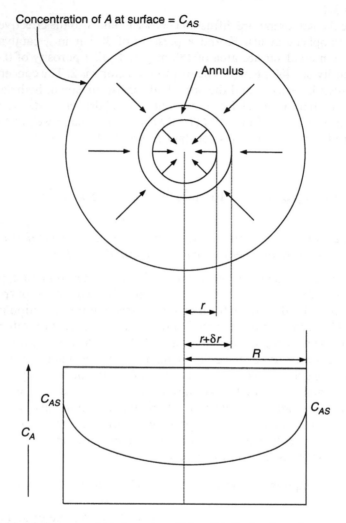

Figure 3.8 Mass balance on spherical annulus within pellet. Section through centre of pellet is shown. The concentration of A along this section is shown beneath.

We now equate (3.127) with (3.128), divide by $4\pi r^2 \delta r$ and let $\delta r \to 0$. This gives

$$k_\sigma \rho_p S_g (C_A) = D_{A,\text{eff}} \left[\frac{d^2 C_A}{dr^2} + \left(\frac{2}{r} \right) \frac{dC_A}{dr} \right]$$

that is,

$$\frac{d^2 C_A}{dr^2} + \left(\frac{2}{r} \right) \frac{dC_A}{dr} = \left(\frac{k_\sigma \rho_p S_g (C_A)}{D_{A,\text{eff}}} \right) \qquad (3.129)$$

It is convenient to express (3.129) in dimensionless form by writing

$$Z = (r/R)$$

$$C'_A = C_A/C_{AS}$$

then

$$Z\left(\frac{d^2 C'_A}{dZ^2}\right) + 2\left(\frac{dC'_A}{dZ}\right) = \phi^2_{S1} C'_A Z \qquad (3.130)$$

where

$$\phi_{S1} = R\left(\frac{k_\sigma \rho_p S_g}{D_{A,\text{eff}}}\right)^{1/2} \qquad (3.131)$$

ϕ is a dimensionless group, usually called Thiele's modulus. The subscripts S and 1 are used to show that, in this instance, the modulus is for spherical pellets (S) and a first-order surface rate equation. In subsequent calculations, it will usually be the reaction rate per unit volume of catalyst pellet (rather than the rate per unit catalyst surface) which is of primary interest and equation (3.131) will be expressed in the form

$$\phi_{S1} = R\left(\frac{k^I_\nu}{D_{A,\text{eff}}}\right)^{1/2} \qquad (3.132)$$

where k^I_ν is the intrinsic first-order rate constant for unit pellet volume defined as in equation (3.114b). Under given conditions of temperature and pressure, and for a given reaction, and pellet composition and structure, k^I_ν and $D_{A,\text{eff}}$ will be constant so that

$$\phi_{S1} \propto R \qquad (3.133)$$

(As will be seen from equation (3.142), the effectiveness factor can be calculated directly from ϕ_{S1} so equation (3.133) can often be useful for estimating variations of effectiveness factor with pellet size.)

Before equation (3.133) can be solved, two boundary conditions are required. The statement of these will depend on the form of the differential equation to be solved. Two conditions which the final solution must satisfy are given by equations (3.134) and (3.135) below. The first of these (which is a consequence of the symmetric nature of the pellet) states (see Figure 3.8) that

$$\text{at } r = 0, \quad dC_A/dr = 0 \qquad (3.134)$$

The second boundary condition is that,

$$\text{at } r = R, \quad C_A = C_{AS} \qquad (3.135)$$

C_{AS} is the concentration at the outer surface of the pellet. (If gas film resistance around the pellet is negligible, C_{AS} will be equal to the concentration of A in the bulk gas space.) On inserting appropriate boundary conditions, the solution to (3.130) is (as shown in Appendix 3.1) found to be

$$C_A^r = \frac{\sinh(\phi_{S1} Z)}{Z \sinh(\phi_{S1})} \tag{3.136}$$

Differentiating with respect to Z and then setting $Z = 1$,

$$\left(\frac{dC_A^r}{dZ}\right)_{Z=1} = \phi_{S1}\left(\frac{1}{\tanh \phi_{S1}}\right) - 1 \tag{3.137}$$

that is,

$$\left(\frac{dC_A}{dr}\right)_{r=R} = \left(\frac{\phi_{S1} C_{AS}}{R}\right)\left[\left(\frac{1}{\tanh \phi_{S1}}\right) - \left(\frac{1}{\phi_{S1}}\right)\right] \tag{3.138}$$

Actual reaction rate per pellet = rate at which reactant can diffuse into interior of pellet across outer boundary where $r = R$

$$= 4\pi R^2 D_{A,\text{eff}}\left(\frac{dC_A}{dr}\right)_{r=R} \tag{3.139}$$

with dC_A/dr given by (3.138)

Combining (3.139) with (3.138) and dividing by the pellet volume,

Reaction rate per unit volume of pellet $= (-r_{AV}) = \dfrac{\phi_{S1} D_{A,\text{eff}}}{R^2}$

$$(3C_{AS})\left(\frac{1}{\tanh \phi_{S1} - \frac{1}{\phi_{S1}}}\right) \tag{3.140}$$

The intrinsic reaction rate per unit volume of pellet is given by equation (3.114b) to be

$$(-r_{AV}^I) = k_V^I C_{AS}$$

Hence, **the effectiveness factor**, η, is given by

$$\eta_{S1} = \frac{(-r_{AV})}{(-r_{AV}^I)} = \frac{3\phi_{S1} D_{A,\text{eff}}}{R^2 k^I}\left(\frac{1}{\tanh \phi_{S1}} - \frac{1}{\phi_{S1}}\right) \tag{3.141}$$

But from equation (3.132), $(R^2 k_\nu^I / D_{A,\text{eff}}) = \phi_{S1}^2$, so

$$\eta_{S1} = \frac{3}{\phi_{S1}}\left(\frac{1}{\tanh \phi_{S1}} - \frac{1}{\phi_{S1}}\right) \tag{3.142}$$

The way in which effectiveness factor η_{S1} varies with Thiele's modulus ϕ_{S1} for an irreversible first-order reaction in a spherical pellet is shown in Figure 3.9 and Table 3.5.

N.B. The expression for Thiele's modulus differs in the two cases shown being given in general by

$$\phi_{Sn} = R\left(\frac{k_{\nu}^I (C_{AS})^{n-1}}{D_{A,\mathrm{eff}}}\right)^{\frac{1}{2}}$$

where n is the order of the reaction (see Section 3.6.4.5).

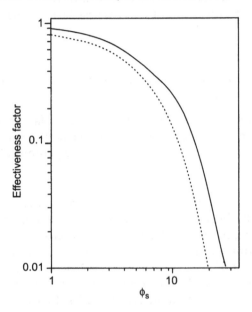

Figure 3.9 Effectiveness factor as function of Thiele's modulus: (a) for first-order irreversible reactions in spherical catalyst pellets (continuous curve); (b) for second-order reaction (dotted curve).

Table 3.5 Effectiveness factor η_{S1} as function of Thiele's modulus ϕ_{S1} for first-order irreversible reaction in spherical catalyst pellet
Below the range covered in the table, equation (3.142) should be used.
Above the range shown the following approximations may be made:

$$40 \geqslant \phi_{S1} \geqslant 4: \quad \eta_{S1} = \left(\frac{3}{\phi_{S1}}\right)\left(1 - \left(\frac{1}{\phi_{S1}}\right)\right) \quad \text{or} \quad \phi_{S1} = \frac{1}{0.5 - (0.25 - (\eta_{S1}/3))^{\frac{1}{2}}}$$

$$\phi_{S1} > 40: \quad \eta_{S1} = \frac{3}{\phi_{S1}} \quad \text{or} \quad \phi_{S1} = 3\eta_{S1}$$

ϕ_{S1}	η_{S1}	ϕ_{S1}	η_{S1}
1.20	0.916	2.80	0.696
1.40	0.890	3.00	0.672
1.60	0.862	3.20	0.647
1.80	0.834	3.40	0.624
2.00	0.806	3.60	0.603
2.20	0.777	3.80	0.583
2.40	0.750	4.00	0.563
2.60	0.722		

Small values of ϕ (which would be obtained for example with small pellets or high pore diffusivities) correspond to effectiveness factors close to unity since, under these conditions, the effects of poor pore diffusion are negligible. High values of ϕ correspond to low effectiveness factors and are obtained with large pellets and/or low pore diffusivities.

A very thorough discussion of effectiveness factor behaviour is given in reference [22].

3.6.4.1 Use of effectiveness factor and intrinsic reaction rate in predicting catalytic reaction rates

When the intrinsic reaction rate and effectiveness factor are known at a given temperature and system composition, the actual reaction rate can at once be calculated at the same conditions from equations of the form

$$(-r_{AV}) = \eta(-r_{AV}^I) \qquad (3.143)$$

where $(-r_{AV})$ is the actual reaction rate per unit volume of catalyst pellet and $(-r_{AV}^I)$ is the intrinsic rate. The intrinsic rate may be deduced as a function of temperature from laboratory reaction rate tests on small-diameter particles (for which pore diffusional limitations are small) or from tests on two or more beds in which the particle diameter is substantially different (see Problem 3.7 below). Observed reaction rates for systems with first order intrinsic kinetics can be represented as follows.

$$(-r_{AV}) = k_V^{\text{apparent}} C_{AS} \qquad (3.144)$$

also

$$(-r_{AV}^I) = k_V^I C_{AS} \qquad (3.145)$$

where k_V^I is the intrinsic first-order rate constant and k_V^{apparent} is the apparent first-order rate constant for the pore diffusion limited reaction, as given by equation (3.144). Both k_V^I and k_V^{apparent} refer to the unit volume of pellet.

Combining (3.143) with (3.144) and (3.145),

$$k_V^{\text{apparent}} = \eta_{S1} k_V^I \qquad (3.146)$$

3.6.4.2 Dependence of η_{S1} on pellet radius.
One important use of equation (3.132) with (3.142) is in predicting the way in which effectiveness factor varies with pellet diameter and system conditions for an irreversible first-order reaction in spherical pellets. (The pore structure of the pellets is taken to be independent of radius.) Reactions of other orders are considered briefly at the end of the section.

If, for example, Thiele's modulus ϕ_{S1} is known at given temperature for one pellet diameter, it can, using equation (3.132), be calculated for other pellet diameters at the same temperature. The effectiveness factor can then be calculated from ϕ_{S1} for each diameter using equation (3.142). Assuming that the intrinsic rate is known, the actual reaction rate may then be obtained from (3.143). Likewise, the apparent first-order rate constant may be calculated from the intrinsic rate constant using equation (3.146). There is a range of possible ways in which ϕ_{S1} might be obtained for pellets of one diameter. If, for example, the intrinsic first-order rate constant k_V^I is known (possibly from tests with fine particles) and also k_V^{apparent} has been measured for pellets of one diameter, η for that diameter can be found from (3.146) as

$$\eta_{S1} = \frac{k_V^{\text{apparent}}}{k_V^I} \tag{3.147}$$

Knowing η_{S1}, ϕ_{S1} for that diameter can be back-calculated using equation (3.142). ϕ_{S1} at any other pellet diameter of interest can then be obtained from (3.132) and used to obtain effectiveness factors using (3.142).

Alternatively, if isothermal reaction rate data (and hence values of k_V^{apparent}) have been obtained for pellets of two different sizes, ϕ_{S1} can be evaluated for each of these sizes, as in Problem 3.7. Furthermore, a value for the intrinsic rate constant can also be evaluated from the data, enabling effectiveness factors and reaction rate data to be predicted for other pellet sizes. There are many other possibilities, some of which are illustrated in the examples and problems below.

Example 3.2
The ratio of the intrinsic first-order constant (k_V^I) to the observed (or 'apparent') value (k_V^{apparent}) for an irreversible first-order reaction in spherical catalytic pellets is known to be 1.38 at temperature t and ambient pressure. The pellet diameter is 2×10^{-3} m.

(a) What is the effectiveness factor and Thiele's modulus for the 2 mm pellets?
(b) Estimate the effectiveness factor for pellets with the same pore structure but a diameter of 3×10^{-3} m.

Solution
Effectiveness factor for the 2 mm pellet $= 0.725$
Thiele's modulus (from Table 3.5) $= 2.58$
Effectiveness factor for the 3 mm pellet corresponds to $\phi = 3.87$ and is (from Table 3.5) 0.576.

Example 3.3
(a) Evaluate effectiveness factor (η) for a first-order irreversible reaction at three values of Thiele's modulus (ϕ) in the range $\phi = 0.8$ to $\phi = 1.2$ and

plot η versus ϕ in this range. Assume spherical catalyst pellets. (This graph will be useful in part (b) of this question.)

(b) Laboratory tests involving a first-order catalysed reaction have shown that Thiele's modulus for near-spherical pellets with an equivalent diameter of 3.0 mm is 10.0. How fine would you have to grind the catalyst to obtain a reaction rate which was 95% of the intrinsic reaction rate at the same reaction conditions? Why might pellets of this size be unsuitable for use in a full-scale packed bed reactor?

Solution to part (b)
From information provided, required effectiveness factor = 0.95
From the graph drawn in part (a), required value of Thiele's modulus = 0.89. Since ϕ is proportional to diameter for given pellet structure, required diameter = $3 \times 0.89/10 = 0.27$ mm.

Problem 3.5
The possibility is being investigated of changing the size of the near-spherical catalyst pellets used in a fixed bed tubular reactor in which an irreversible first-order reaction is taking place. The pellets currently in use are 4×10^{-3} m in diameter and Thiele's modulus for these is known to be 3.0. Temperature gradients within them are negligible. It is proposed to change from the 4 mm pellets to pellets which are 1 mm in diameter but whose properties are otherwise identical.

Calculate the ratio of the apparent first-order rate constant with the modified pellets to that with the original pellets.

Solution
1. Effectiveness factor for original pellets corresponds to Thiele's modulus of 3.0. From Table 3.5 this is 0.672.
2. From equation (3.133), Thiele's modulus for the 1 mm pellets will be one-quarter of this, i.e. 0.75.
3. From equation (3.142), η for these pellets = $(3/0.75)\,[1.574 - 1.333] = 0.963$.

Hence: Required ratio of apparent first-order rate constant with the modified pellets to that with original pellets = $(0.963k_v^I)/(0.672k_v^I) = 1.433$.

Problem 3.6
Tests have shown that the effectiveness factor for 8 mm diameter catalyst pellets catalysing an irreversible first-order gas reaction at 1 bar pressure and maintained at 400°C is 0.248, the corresponding value of Thiele's modulus being 11.0.

Estimate the effectiveness factor for these pellets if the reaction was carried out at 2 bar pressure and 450°C. The intrinsic first-order rate constant at 400 and 450°C are 3.0 and 4.5 s^{-1} respectively. The pore diffusivity is virtually equal to the bulk diffusivity and **is proportional to $T^{1.8}$ and P^{-1}.**

Solution
From equation (3.132), new value for Thiele's modulus is:

$$(\text{old value}) \times (4.5/3.0)^{1/2} \times (2)^{1/2} \times [(400 + 273)/$$
$$(450 + 273)]^{0.9} = 11.0 \times 1.225 \times 1.41 \times 0.937 = 17.80$$

New effectiveness factor $= (3/17.80)[1 - 0.056] = 0.159$

3.6.4.3 Approximate forms of equation (3.142) valid for high values of Thiele's modulus. Although the exact equation (3.142) is required to calculate η at low values of Thiele's modulus (up to about 4), approximations can be made at higher values than this (see Table 3.5).

(i) If Thiele's modulus is greater than about 4, $1/\tanh \phi$ in equation (3.142) may be set equal to unity, giving

$$\eta_{S1} = (3/\phi_{S1})[1 - (1/\phi_{S1})] \tag{3.148}$$

(at $\phi_{S1} = 4$, $(1/\tanh \phi)$ is 1.0007 and η is 0.563).
(ii) If ϕ_{S1} is greater than about 40 it becomes reasonable to neglect $(1/\phi_{S1})$ with respect to unity, in which case (3.142) reduces to

$$\eta_{S1} = (3/\phi_{S1}) \tag{3.149}$$

This result is valid to within 3% at $\phi_{S1} = 40$ and becomes progressively more accurate as ϕ_{S1} rises above this value. We shall describe systems in which ϕ_{S1} exceeds 40 and hence which follow equation (3.149) as being **strongly pore diffusion limited**. The pore limitation may arise because the pellet is large, so that the diffusion path to the centre is long, and/or because the effective diffusivity is low.

Problem 3.7
Experimental reaction rates measured for an irreversible first-order isothermal reaction using spherical catalyst pellets were as follows:

• With 8 mm diameter pellets, the reaction rate was 120 mol per kg of catalyst per hour.
• With 4 mm diameter pellets, and similar conditions of temperature, pressure and composition the reaction rate was 216 mol per kg of catalyst per hour.

Appreciable diffusional limitation was observed in both cases. Determine the intrinsic reaction rate, i.e. the rate at which the reaction would go in the absence of diffusional limitation, and predict what the observed reaction rate would be using 1 mm diameter spherical catalyst pellets.

Solution
First we note that

$$\frac{\eta_8}{\eta_4} = \left(\frac{\text{observed reaction rate for 8 mm pellets}}{\text{observed reaction rate for 4 mm pellets}}\right) = \frac{120}{216} = 0.5556$$

where η_8 and η_4 are the effectiveness factors for the 8 mm and 4 mm pellets.

Since appreciable diffusional limitation has been observed it seems reasonable to express η_{S1} in terms of ϕ_{S1} for each pellet size using the approximation given by equation (3.148) and check the validity of this at the end by evaluating tanh ϕ_{S1} for the values of ϕ_{S1} obtained. Let ϕ_8 = Thiele's modulus for the 8 mm pellets, then Thiele's modulus for 4 mm pellets = $(\phi_8/2)$ and, from equation (3.148),

$$0.5556 = 0.5\left[1 - \left(\frac{1}{\phi_8}\right)\right]\left[1 - \left(\frac{2}{\phi_8}\right)\right]^{-1}$$

from which $\phi_8 = 11.0$, $\phi_4 = 5.5$. (As check, tanh (11) = 1.000 = tanh (5.5), so use of equation (3.148) was correct at this stage.)

The effectiveness factor for the 8 mm pellets is given by (3.148) to be (3/11)(0.909) = 0.248. The intrinsic reaction rate = (observed rate)/η_{S1} = 120/0.248 = 484 mol per kg catalyst per hour. For the 1 mm pellets $\phi = 11/8 = 1.375$. This is outside the range of equation (3.148) and the full equation (3.142) must be used (or Table 3.5) to calculate the corresponding effectiveness factor.

From equation (3.142), the effectiveness factor for 1 mm pellets = 0.893. The observed reaction rate with these pellets is predicted to be

$$484 \times 0.893 = 432 \text{ mol per kg catalyst per hour.}$$

3.6.4.4   Reaction rates of systems with strong pore diffusional limitation (i.e. which follow equation (3.149)). Combining equation (3.149) with (3.143) and (3.145) gives, for a system in which pore diffusion is strong,

$$(-r_{AV} = \frac{3}{\phi_{S1}}(k_v^1 C_{AS})$$

$$= \frac{3}{R}\left(\frac{D_{A,\text{eff}}}{k_v^I}\right)^{\frac{1}{2}}(k_v^1 C_{AS})$$

$$= \frac{3}{R}(D_{A,\text{eff}}k_v^1)^{\frac{1}{2}}C_{AS} \tag{3.150}$$

where $(-r_{AV})$ is the rate of consumption of reactant A (i.e. the rate of reaction) per unit pellet volume, C_{AS} is the concentration of A at the outer surface of the pellet, R is the radius of the pellet and $D_{A,\text{eff}}$ is the effective diffusivity of A in the pellet; k_v^I is the intrinsic first-order rate constant which will normally increase sharply with increase in temperature. $D_{A,\text{eff}}$ will increase a little with increase in temperature, but not to the same extent as k_ν^I The reaction rate in the strongly diffusion limited system thus increases with temperature approximately as $(k_\nu^I)^{1/2}$. In a system with negligible pore resistance, however, $(-r_{AV}) = k_\nu^I C_{AS}$ and reaction rate increases proportionally to k_ν^I.

In consequence, reaction rates for systems with strong pore diffusional limitation are less temperature dependent than those for systems with negligible pore limitation.

3.6.4.5 Other orders of reaction.
Up to this point first-order surface reactions for which

$$(-r_{AS}) = k_\sigma C_A \qquad (3.114)$$

have been considered. The mathematics for other reaction orders parallels that for the first-order case. If, for example, the nth-order expression

$$(-r_{AS}) = k_{\sigma,n}(C_A)^n \qquad (3.151)$$

is used in place of equation (3.114) in the derivation in Section 3.6.4, the rate of consumption of A in the annulus (equation (3.127)) becomes

$$k_{\sigma,n} C_A^n S_g \rho_p 4\pi r^2 \, dr$$

Consequently $(C_A)^n$ replaces C_A on the right-hand side of the differential equation (3.129). Equations (3.130) and (3.131) then become

$$Z\left(\frac{d^2 C_A^r}{dZ^2}\right) + 2\left(\frac{dC_A^r}{dZ}\right) = \phi_{S_n}^2 (C_A^r)^n Z \qquad (3.130a)$$

where

$$\phi_{S_n} = R\left(\frac{k_{\sigma,n}\rho_p S_g C_{AS}^{n-1}}{D_{A,\text{eff}}}\right)^{\frac{1}{2}} = R\left\{\frac{k_\nu^I(C_{AS})^{n-1}}{D_{A,\text{eff}}}\right\}^{1/2} \qquad (3.131a)$$

where ϕ_{S_n} is Thiele's modulus for the nth-order reaction. C_{AS} is the concentration of reactant A in the fluid phase at the outer surface of the pellet.

The solution to equation (3.130a) for concentration as a function of radial position in the pellet clearly differs from the solution to (3.130) and this leads in the further development of the derivation to a different form of variation of effectiveness factor with Thiele's modulus, as shown in Figure 3.9 for a second-order reaction. Effectiveness factor falls more rapidly with Thiele's modulus for higher order reactions than for low. Furthermore, as seen from equation (3.131a), Thiele's modulus in the general n^{th}-order case contains the term C_{AS}^{n-1}. Consequently, for reaction orders greater than 1, effectiveness factor decreases with increasing reactant concentration at the outer surface of the pellet.

For very large values of Thiele's modulus (strongly pore diffusion-limited system), the limiting result

$$\eta_{s_n} = \left(\frac{2}{n+1}\right)^{\frac{1}{2}}\left(\frac{3}{\phi_{s_n}}\right) \tag{3.152}$$

is obeyed, where η_{s_n} is the effectiveness factor and ϕ_{s_n} is Thiele's modulus for the nth-order reactions.

In the first-order case, the apparent reaction rate equation had the same order as the intrinsic rate equation (both were unity). In general, however, this is not true and the apparent reaction order n^{apparent} is related to the intrinsic order n by

$$n^{\text{apparent}} = \frac{n+1}{2}$$

3.6.4.6 Non-isothermal pellets. Up to this point it has been assumed that the interior of the pellets is at a uniform temperature. However, if the thermal conductivity of the pellets is low and/or substantial heat generation takes place during reaction, the temperature at the centre of each pellet may differ substantially from that at the surface. For exothermic reactions this may lead to effectiveness factors greater than 1.

3.6.5 Surface film resistance to reaction rate (interphase resistance)

Before it can enter a catalyst pellet, a reactant A must diffuse across a surface film surrounding the pellet (Figure 3.10). The resistance encountered at this stage can be significant, and even control the reaction rate if the gas flow rate is low (leading to low film mass transfer coefficients) and the temperature high.

Let C_{AB} = concentration of A in bulk gas space
$\quad\quad C_{AS}$ = concentration of A in fluid phase at surface of pellet.
In steady state, the rate at which reactant A is being used up in the reaction within the pellet is equal to the rate at which it diffuses through the surface film, that is:

$$[(4/3)\pi R^3](-r_{AV}) = 4\pi R^2 k_{CA}(C_{AB} - C_{AS}) \tag{3.153}$$

where $\quad R$ = pellet radius
$\quad (-r_{AV})$ = rate of consumption of reactant A per unit pellet volume
$\quad\quad\quad = k^{\text{apparent}} C_{AS}$
$\quad\quad k_{CA}$ = average mass transfer coefficient of A in film around pellet
$(C_{AB} - C_{AS})$ = concentration driving force across the surface film

From equation (3.153)

$$(-r_{AV}) = (3/R)k_{CA}(C_{AB} - C_{AS}) = k_v^{\text{apparent}} C_{AS} \tag{3.154}$$

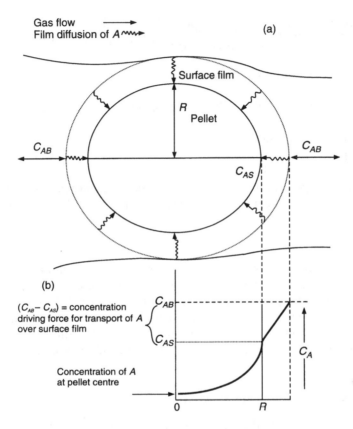

Figure 3.10 Illustrating interphase transport of a reactant A: (a) spherical pellet with surface film (idealised); (b) concentration profile in situation where film resistance is appreciable though not controlling.

From (3.154)

$$\frac{C_{AB} - C_{AS}}{C_{AS}} = \left(\frac{k_v^{\text{apparent}}}{k_{CA}}\right)\frac{R}{3} \qquad (3.155)$$

If k_{CA} is large, $C_{AS} \cong C_{AB}$. If k_{CA} is small, $C_{AB} - C_{AS} \cong C_{AB}$ (3.156)

In the latter case the reaction rate is **controlled** by the rate at which reactant A can diffuse across the film. We then have **interphase transport control**. Combining equation (3.156) with (3.154) gives the reaction rate as

$$(-r_{AV})_{\text{Film control}} = (3/R)k_{CA}C_{AB} \qquad (3.157)$$

k_{CA} in equation (3.157) is conveniently given by a correlation due to Thoenes and Kramers [23].

$$\frac{k_{CA}d_p}{D_{AB}}\left(\frac{\varepsilon_b}{1-\varepsilon_b}\right)\left(\frac{1}{\gamma}\right) = \left(\frac{ud_p\rho}{\mu(1-\varepsilon_b)\gamma}\right)^{\frac{1}{2}}\left(\frac{\mu}{\rho D_{AB}}\right)^{\frac{1}{3}} = (Re')^{\frac{1}{2}}(Sc)^{\frac{1}{3}} \qquad (3.158)$$

The above correlation is valid in the range

$$0.25 < \varepsilon_b < 0.5, \quad 40 < Re' < 4000, \quad 1 < Sc < 4000$$

d_p = pellet diameter (m)
γ = shape factor = 1 for sphere
u = superficial gas velocity (m s^{-1})
ε_b = bed voidage = (volume of space between pellets)/(bed volume)
ρ_p and μ = density (kg m^{-3}) and viscosity (kg m^{-1}s^{-1}) of the reactor fluid, respectively.

Problem 3.8
The apparent first-order rate constant per unit pellet volume for the dehydrogenation of ethylbenzene to styrene in a reactor containing 4 mm (effective) diameter porous pellets is found to be 2.71 s^{-1}. Using a correlation, it is found that the mass transfer coefficient for ethylbenzene across the outer film is 0.15 m s^{-1}. What is the ratio of the concentration of ethylbenzene at the periphery of each pellet to that in the bulk gas space?

Solution
From equation (3.155) and data given:

$$C_{AS}/C_{AB} = [(2 \times 10^{-3}/3)(2.71/0.15) + 1]^{-1} = 0.988$$

3.7   Reaction rate regimes

(1) **Surface reaction control** occurs when k_σ is very small. The reaction rate at the catalytic surface is then very low and hence rate controlling. Because of the very low transport rate required to feed the reaction, neither pore diffusion nor surface film diffusion provide appreciable limitation ($\eta \cong 1.00$) and ($C_{AB} - C_{AS}) \cong 0$). For a first-order reaction, the rate is thus given by

$$(-r_{AV}) = k_v^I C_{AS}$$

with similar expressions for other orders. This situation tends to occur at very low temperatures. The activation energy is typical of that for a catalysed reaction.

(2) **Strong pore diffusion limitation**. The effectiveness factor is then very low and given by

$$\eta_{s1} = 3/\phi_{s1}$$

for 1st-order reaction.

Because of the very low rate of transport across the surface film,

$$C_{AB} - C_{AS} \approx 0$$

The reaction rate achieved when the intrinsic reaction is first order is given by equation (3.150) to be

$$(-r_{AV}) = (3/R)(D_{A,\text{eff}}k_v^I)^{1/2}C_{AS}$$

Reaction rates are not as temperature sensitive as those for surface reaction-controlled systems, the activation energy being 50% that for the intrinsic reaction (see Section 3.6.4.4). This situation tends to occur at temperatures between those at which the surface reaction controls the reaction rate and those at which interphase resistance controls the rate.

(3) **Interphase resistance control.** The reaction rate is given by equation (3.157) to be

$$(-r_{AV}) = (3/R)k_{CA}C_{AB}$$

for all orders of reaction. The rate is determined and controlled by the film mass transfer coefficient. This situation occurs at high temperatures (the reaction rate inside the pellet is then very high) and low Reynolds numbers (leading to low film mass transfer coefficients).

Appendix 3.1

Solution of equation

$$Z\left(\frac{d^2 C_A^r}{dZ^2}\right) + 2\left(\frac{dC_A^r}{dZ}\right) = \phi_{S1}^2 C_A^r Z \qquad (3.130)$$

subject to boundary conditions appropriate to the centre of the pellet and its outer surface. (In equation (3.130), $Z = r/R$), where $r =$ radial position in pellet and $R =$ pellet radius. Also $C_A^r = (C_A/C_{AS})$, where $C_A =$ concentration of A in pore fluid within the pellet and $C_{AS} =$ concentration of A in pore fluid at outer surface of pellet.)

Solution

Set $(C_A^r Z) = u$, then

$$\frac{du}{dZ} = C_A^r + Z\left(\frac{dC_A^r}{dZ}\right)$$

$$\frac{d^2u}{dZ^2} = 2\left(\frac{dC_A^r}{dZ}\right) + Z\left(\frac{d^2C_A^r}{dZ^2}\right)$$

Substituting in (3.130),

$$\frac{d^2u}{dZ^2} - \phi_{S1}^2 u = 0 \qquad\qquad (A3.1.1)$$

Assume a solution of the form $u = A_m e^{mZ}$ (see reference [24, p.46]). The auxiliary equation is $(m^2 - \phi_{S1}^2) = 0$, that is $m = \pm\phi_{S1}$, so

$$u = A_1 \exp(\phi_{S1}Z) + A_2 \exp(-\phi_{S1}Z) \qquad\qquad (A3.1.2)$$

We now evaluate A_1 and A_2 from the boundary conditions

(1) **At centre of pellet** $Z = 0$, so $(ZC_A^r) = u = 0$.
Substituting in (A3.1.2), $A_1 = A_2$
Also, from (A3.1.2),

$$\begin{aligned}(C_A^r) &= \frac{1}{2}A_1[\exp(\phi_{S1}Z) - \exp(-\phi_{S1}Z)] \\ &= \left(\frac{2A_1}{Z}\right)\sinh(\phi_{S1}Z)\end{aligned} \qquad\qquad (A3.1.3)$$

(2) **At outer surface of pellet,** $(C_A/C_{AS}) = 1$ and $r = R$, that is:

$$C_A^r = 1 \quad \text{when} \quad Z = 1$$

Inserting this result in (A3.1.3) gives

$$A_1 = \frac{1}{Z\sinh(\phi_{S1})} \qquad\qquad (A3.1.4)$$

so

$$C_A^r = \frac{\sinh(\phi_{S1}Z)}{Z\sinh(\phi_{S1})} \qquad\qquad (3.136)$$

This is the required result, giving C_A^r as a function of radial position in the pellet.

Appendix 3.2   The BET equation

This equation was developed by Brunauer, Emmett and Teller in 1938 (*J. Amer. Chem. Soc.* **60**, 309) to represent data in systems where multi-layer physical adsorption takes place. The equation gives the equilibrium pressure p of an adsorbate vapour over a solid surface at given temperature as a function of the number of moles of vapour (n_a) adsorbed on the surface (i.e. it gives the 'adsorption isotherm' for the system). According to this equation:

$$\frac{x}{n_a(1-x)} = \frac{1}{cn_m} + \frac{(c-1)x}{cn_m} \qquad \text{(A3.2.1)}$$

where x is the ratio of the adsorbate pressure in the vapour space to the saturation vapour pressure

$$x = (p/p^0)$$

If only one component (the adsorbate) is present in the vapour space, p is the system pressure. If an inert component is present, the adsorption of which is negligible under the conditions studied, p is the **partial pressure** of the adsorbate in the vapour space.

n = number of moles of adsorbate on the surface

n_m = number of moles of adsorbate which (in the BET model) are required to form a complete monolayer over the surface

c = ratio of the equilibrium constant for the adsorption of vapour molecules onto the base surface to the equilibrium constant for the adsorption of vapour molecules onto an existing adsorbed layer of these molecules.

If data are available for the number of moles of adsorbate n_a adsorbed on a surface as a function of p at given temperature, the parameters n_m and c (if required) can be evaluated.

Although the BET model is qualitatively successful in predicting the general forms of adsorption isotherm which can be obtained it is quantitatively only reliable in the rather restricted range

$$0.05 < (p/p^0) < 0.3$$

Within this range, however, it provides a useful and widely used method for determining surface areas of porous samples including **catalysts**. For this purpose a standard adsorbate (usually nitrogen) is contacted with the sample of interest at a temperature close to the normal boiling point of the adsorbate (77 K for N_2) for a range of adsorbate partial pressures which fall within the range of validity of the BET equation. By fitting the data to equation (A3.2.1), n_m for the sample is deduced. The area exposed by the sample to the adsorbate vapour can then be estimated from the equation

$$\text{area} = n_m \sigma^0 N$$

where n_m = moles of adsorbate required to form a complete monolayer over the surface

σ^0 = area of surface excluded by each adsorbate molecule (16.2×10^{-20} m^2 for N_2)

N = Avogadro's number

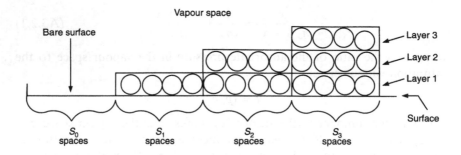

Figure A3.2.1 Idealised diagram of layer formation envisaged in BET model.

The area σ^o excluded by each molecule of adsorbate on the surface is usually deduced from liquid phase density data for the adsorbate. The value obtained is checked by carrying out adsorption tests on non-porous material of known area. The procedures have been discussed in detail by S.J. Gregg and K.S.-W. Sing in *Adsorption, Surface Area and Porosity*, 2nd edition, Academic Press (1982).

Automated equipment is available for carrying out area determinations in the above manner. In this equipment, N_2 is the adsorbate normally employed and the determinations are normally carried out in the presence of helium as inert gas.

Derivation of the BET equation

The derivation is an extension of that used to arrive at the Langmuir equation in Section 3.4.2. In obtaining the Langmuir equation (equation (3.83)) the rate of desorption from the surface, k_d (number of occupied sites) was balanced against the rate of adsorption onto the surface, $k_a p$ (number of unoccupied sites) to give the equilibrium relationship

$$\theta = \frac{n_a}{n} = \frac{bp}{1 + bp} \tag{A3.2.2}$$

in which n_a is the number of moles adsorbed and n the total number of adsorption sites on the surface considered. The so-called equilibrium constant for adsorption b is the ratio of the adsorption rate constant k_a to the desorption rate constant k_d

$$b = k_a/k_d$$

(Equation (A3.2.2) differs slightly from (3.83) in that subscripts A, which referred to a specific component A, have been omitted.)

Vapour space

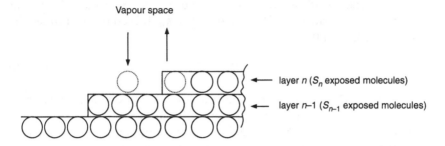

layer n (S_n exposed molecules)

layer n–1 (S_{n-1} exposed molecules)

Figure A3.2.2 The equilibrium situation. To maintain the numbers of molecules in each layer constant, as many molecules must evaporate from layer n as condense on layer $(n-1)$.

In the BET model allowance is made for the formation of layer upon layer of adsorbed molecules rather than just a single layer. The condition for equilibrium is then that (rate of evaporation from layer n) = (rate of condensation on layer $(n-1)$ (cf. Figure A3.2.2).

It is presumed that evaporation and condensation take place only from and onto parts of these layers which are directly exposed to the vapour space, the appropriate rates being given by the Langmuir-type arguments used in Section 3.4.2 so that, for $n > 1$,

$$S_n k_d = S_{n-1} k_a p \qquad (A3.2.3)$$

where p = equilibrium pressure of adsorbate vapour
S_n = number of exposed molecules in layer n
S_{n-1} = number of exposed molecules in layer $(n-1)$
k_a and k_d = rate constants for adsorption and desorption respectively.

From (A3.2.3),

$$S_n / S_{n-1} = bp \qquad (A3.2.4)$$

where $b = k_a/k_d$ = equilibrium constant for adsorption of an adsorbate molecule onto an existing layer of adsorbed molecules.

It is a postulate of the theory that b is independent of n provided $n > 1$.

The rate of evaporation from the first adsorbed layer $(n = 1)$ must, for equilibrium, be balanced against the rate of condensation on the bare surface $(n = 0)$. Again Langmuir-type arguments are used to evaluate these rates and, in direct analogy to (A3.2.3), we obtain

$$S_1 k_d^* = S_0 k_a^* p \qquad (A3.2.5)$$

where S_1 = number of molecules in layer 1 which are in contact with the vapour space
S_0 = number of exposed spaces on the surface.

Since adsorbate/surface interactions would not be expected to be similar to adsorbate/adsorbate ones, $k_d^* \neq k_d$ and $k_a^* \neq k_a$. It is convenient to rewrite equation (A3.2.5) in the form

$$S_1/S_0 = b^*p \qquad (A3.2.6)$$

where $b^* = k_a^*/k_d^* =$ equilibrium constant for adsorption of an adsorbate molecule directly onto the surface.

Combining equations (A3.2.4) and (A3.2.6) we obtain the following general expression for the number of molecules in layer n which are exposed to the vapour space:

$$\begin{aligned} S_n &= (bp)S_{n-1} \\ &= (bp)^{(n-1)}S_1 \\ &= (b^*p)(bp)^{(n-1)}S_0 \\ &= c(bp)^n S_0 \end{aligned} \qquad (A3.2.7)$$

where $c = b^*/b =$ ratio of equilibrium constant for the adsorption of adsorbate molecules onto the base surface to the equilibrium constant for absorption of adsorbate molecules onto an existing adsorbed layer.

In the theory, c is a constant for a given system at a given temperature. It is convenient to denote the product (bp) in equation (A3.2.7) by the symbol x, giving

$$S_n = cx^n S_o \qquad (A3.2.8)$$

To develop the model further we need to reflect that the number of exposed molecules in layer n will equal the number of surface spaces which are covered by n molecules (see Figure A3.2.1). The total number of adsorbed molecules (N_a) is thus given by

$$N_a = \sum_{n=1}^{\infty}(nS_n) \qquad (A3.2.9)$$

Also the number of molecules plus the number of bare spaces exposed to the vapour remains constant throughout the adsorption process, being equal to $\sum_{n=0}^{\infty} S_n$. This number is also equal to the number of spaces initially available for adsorption onto the surface and to the number of molecules N_m required to form a monolayer over the surface, i.e.

$$N_m = \sum_{n=0}^{\infty} S_n \qquad (A3.2.10)$$

From equations (A3.2.8), (A3.2.9) and (A3.2.10) therefore,

$$\frac{N_a}{N_m} = \frac{\sum_{n=1}^{\infty}(nS_n)}{\sum_{n=0}^{\infty}(S_n)} = \frac{cS_0\sum_{n=1}^{\infty}(nx^n)}{S_0 + cS_0\sum_{n=1}^{\infty}(x^n)} = \frac{c\sum_{n=1}^{\infty}(nx^n)}{1 + c\sum_{n=1}^{\infty}(x^n)} \qquad \text{(A3.2.11)}$$

However, from the binomial theorem,

$$\sum_{n=1}^{\infty}(nx^n) = x + 2x^2 + \cdots = \frac{x}{(1-x)^2}$$

$$\sum_{n=1}^{\infty}(x^n) = x + x^2 + \cdots = \frac{x}{1-x}$$

Inserting these results in equation (A3.2.11),

$$\frac{N_a}{N_m} = \frac{cx/[(1-x)^2]}{1 + [cx/(1-x)]} \qquad \text{(A3.2.12)}$$

where $x = bp$

If $x = 1, b = 1/p$.

If we allow x in equation (A3.2.12) to tend to unity

$$(N_m/N_a) \to \frac{1}{1-x} \to \infty \qquad \text{(A3.2.13)}$$

i.e. the number of molecules in the adsorbed layer is becoming very large and we have for all intents and purposes a liquid. Physically p would then be expected to approximate to the liquid vapour pressure p^0.

According to (A3.2.13), therefore,

$$b = 1/p^0 \quad \text{and} \quad x = p/p^0$$

The ratio of the number of moles adsorbed (n_a) to the moles required to form a complete monolayer (n_m) is of course identical to the ratio of the molecules adsorbed to the molecules in a complete monolayer. In conclusion, according to the BET model

$$\frac{n_a}{n_m} = \frac{cx/[(1-x)^2]}{1 + [cx/(1-x)]} \qquad \text{(A3.2.14)}$$

where $x = (p/p^0)$

Evaluation of parameters in BET equation

If evaluating the parameters n_m and c in the BET equation graphically, it is convenient to express the equation in the form shown in equation (A3.2.1) and to plot $[x/n_a(1-x)]$ against x.

The intercept (i) is given by

$$i = \frac{1}{cn_m} \qquad \text{(A3.2.15)}$$

The slope (S) is given by

$$S = \frac{c-1}{cn_m} \qquad \text{(A3.2.16)}$$

Solving these two equations simultaneously gives

$$n_m = \frac{1}{S+i} \qquad \text{(A3.2.17)}$$

and

$$c = \left(\frac{S}{i}\right) + 1 \qquad \text{(A3.2.18)}$$

If i is very small (as in the following problem)

$$n_m = 1/S \qquad \text{(A3.2.19)}$$

c (if required) can be evaluated by substituting n_m obtained as above, together with an observed value of n_a at given x into the BET equation.

Problem

The following values have been obtained for the number of moles (n_a) of nitrogen adsorbed on a sample of rutile at 75 K as a function of the nitrogen pressure P.

P (torr)	1.17	45.8	128	164
n_a	2.68×10^{-2}	3.67×10^{-2}	4.67×10^{-2}	5.12×10^{-2}

Estimate the surface area of the sample by a graphical method. (The vapour pressure of N_2 at 75 K is 573 torr; σ^0 for $N_2 = 16.2(\text{Å})^2$; $N = 6.02 \times 10^{23}$ mol^{-1}.)
Solution: 3500m^2

References

1. Nauman, E.B. (1987) *Chemical Reactor Design*, Wiley, New York.
2. Atkins, P.W. (1994) *Physical Chemistry*, 5th edition, Oxford University Press, Oxford.
3. Adamson, A.W. (1990) *Physical Chemistry of Surfaces*, 5th edition, Wiley, New York.
4. Moelwyn-Hughes, E.A. (1961) *Physical Chemistry*, 2nd edition, Pergamon, Oxford.
5. Glasstone, S., Laidler, K.J. and Eyring, H. (1941) *The Theory of Rate Processes*, McGraw-Hill, New York.
6. Mahan, B.M. and Myers, R.J. (1987) *University Chemistry*, 4th edition, Benjamin/Cummings Publishing Co., Menlo Park, California.

7. Spencer, M.S. (1996) Chapter 1 in *Catalyst Handbook* (M. V. Twigg, ed.), Manson Publishing Co., London.
8. Hougen, O. A. and Watson, K.M. (1943) *Chemical Process Principles, Part 3, Kinetics and Catalysis*, Wiley, New York
9. Hamann, S.D. (1963) *High Pressure Physics and Chemistry*, Vol. 2 (ed. R. S. Bradley), Academic Press, New York.
10. Weale, K.E. (1967) *Chemical Reactions at High Pressure*, E & FN Spon, London.
11. Halsey, G.D. and Yeates, A.T. (1979) *J. Phys. Chem.* **83**, 3236.
12. Hinshelwood, C.N. (1941) *Kinetics of Chemical Change*, Oxford University Press, Oxford.
13. Hougen, O.A. and Watson, K.M. (1943), *Ind. Eng. Chem.* **35**, 529.
14. Lloyd, L., Ridler, D.E. and Twigg, M.V. (1996), Chapter 6 in *Catalyst Handbook* (ed. M.V. Twigg), Manson Publishing Co., London.
15. Temkin, M.I. and Pyzhev, V. (1940), *Acta Physiochim. (USSR)* **12**, 327.
16. Temkin, M.I., Moroov, N.M. and Shapatina, E.N. (1963) *Kinet. Cata (USSR)* **4**, 565.
17. Wakao, N. and Kaguei, S. (1982) *Heat and Mass Transfer in Packed Beds*, Gordon & Breach, New York
18. Evans, R.B., Watson, G.M. and Masson, E.A. (1961) *J. Chem. Phys.* **35**, 2076
19. Rothfield, L.B. (1963), *Amer. Inst. Chem. Eng. J.* **9**, 19.
20. Reid, R.C., Prausnitz, J.H. and Poling, B.E., (1987) *The Properties of Gas and Liquids*, 4th edition, McGraw-Hill, New York.
21. Wakao, N. and Smith, J.M. (1962) *Chem. Eng. Sci.* **17**, 825.
22. Satterfield, C.N. (1981) *Mass Transfer in Heterogeneous Catalysis*, Robert E Krieger Publishing Co.
23. Thoenes, D. and Kramers, H. (1958) *Chem. Eng. Sci.* **8**, 271.
24. Jenson, V.G. and Jeffreys, G.B. (1977) *Mathematical Methods in Chemical Engineering*, 2nd edition, Academic Press, New York.

4 Simple reactor sizing calculations

J.M. WINTERBOTTOM and M.B. KING

The sizing of a chemical reactor starts with a consideration of the material balance equation for the reactor. This applies equally to batch and continuous flow systems. In the first part of this chapter, material balance equations will be developed for the principal classes of reactor on the assumption that flow conditions in the reactor are 'ideal'. (The 'ideal' situation is 'plug flow' for tubular reactors or 'complete mixing' for CSTR or batch reactors.) Isothermal operation of the reactors will then be considered, followed by non-isothermal operation. The effects of non-ideal flow behaviour are considered in Chapter 5.

4.1 The material balance equations for batch and continuous reactors

A molar balance applied to a reactant species A and taken around a volume element within a reactor or reactor system can be written as follows:

Molal flow of A into volume element in unit time		Molal flow A out of volume element in unit time		Rate of removal of A by reaction in volume element in unit time		Rate of accumulation of A in volume element in unit time	
	=		+		+		
Input	=	Output	+	Consumption	+	Accumulation	
(1)		(2)		(3)		(4)	(4.1)

(cf. equation (2.15) and Figure 2.4).

If A is generated by the reaction, the consumption term is negative.

There is an analogous heat balance equation but this can be ignored for isothermal operation. For reasons explained earlier, reactant A is most conveniently taken to be the limiting reactant, i.e. the reactant which is used up first as the reaction proceeds.

4.1.1 The batch reactor

The ideal batch reactor is assumed to be perfectly mixed so that there are no concentration gradients within it. Furthermore, because no material enters or leaves the reactor during operation, the material balance for any component may be applied to the whole reactor. Taking a molar balance on component A, as in equation (4.1),

Input rate (1)　　　　　= output rate (2) = 0
Rate of consumption (3) = −rate of accumulation (4)

where

$$\text{rate of consumption} = (-r_A)V_r \text{ kmol s}^{-1}$$

in which $(-r_A)$ = rate of reaction of A (kmol m^{-3} s^{-1})
　　　　V_r = reaction volume (m^3), which in this case is the volume of the contents of the reactor.

In the case of a gas phase batch reactor, the actual reactor volume and the volume of the reactants will be identical, but for a liquid phase reaction there will be a head-space above the liquid surface so that $V_r < V_t$, where V_t is the total volume of the reactor shell.

The accumulation term is given by the rate of change of moles A per unit time, i.e. rate of accumulation = dn_A/dt(kmol s^{-1})
Therefore

$$dn_A/dt = -(-r_A)V_r$$

or

$$-d(C_A.V_r)/dt = (-r_A).V_r$$

where $-r_A$ is the rate of consumption of A per unit volume of reaction mixture and C_A is the concentration of A (kmol m^{-3}), $(C_A = n_A/V_r)$.

For reasons outlined in Section 3.2.1, volume changes accompanying batch reactions are normally very small and the treatment below is based on a constant volume system.

For a constant volume system

$$-dC_A/dt = (-r_A) \tag{4.2}$$

For many purposes it is convenient to write equation (4.2) in terms of the fractional conversion X_A of component A, defined as in equation (3.47). For a constant volume system,

$$C_A = C_{A0}(1 - X_A) \quad \text{and} \quad dC_A = -C_{A0}dX_A$$

Combining with equation (4.2),

$$C_{A0}dX_A/dt = (-r_A)$$

and

$$dt = C_{A0}dX_A/(-r_A) \tag{4.3}$$

Integrating (4.3) from zero time, at which the reactor was charged, to time t gives

$$t = C_{A0} \int_0^{X_{AF}} (\mathrm{d}X_A/(-r_A)) \qquad (4.4)$$

where $(-r_A) = kf(C)$ in which k is the reaction rate constant and $f(C)$ is a function of the concentrations of the reacting and (for reversible systems) the back-reacting components (see Chapter 3).

Provided the form of the rate equation giving $(-r_A)$ is known, equation (4.4) allows the reaction time required to obtain a given conversion to be calculated.

4.1.2 The continuous reactor

It will be assumed that the continuous reactor considered is operating under steady-state conditions (this is not the case at start-up or, of course, for semi-batch operation). In this case, there is no accumulation term in equation (4.1) and

$$\text{input} = \text{output} + \text{consumption}$$

4.1.2.1 The plug flow reactor. Reactors in which the flow conditions approximate to plug flow are normally tubular and may or may not contain a bed of catalyst. The equations for the two situations are very similar. The case where there is no catalyst bed (i.e. the homogeneous plug flow reactor) will be considered first.

If the reactor is operating under ideal plug flow conditions, then

(a) there is no radial variation in composition, temperature or flow rate
(b) all fluid elements spend an equal time within the reactor, which implies that
(c) there is no axial mixing between each fluid element.

In stating the above, it must be remembered that for a **gas phase reaction**, where a change occurs in the total number of moles as the reaction proceeds, the volume of the reactant product mixture will differ from that of the reactants from which it was constituted and the design of the reactor should allow for this (see Chapter 3, Section 3.2.2.1).

4.1.2.2 Homogeneous plug flow reactors. In order to develop the material balance equation for the homogeneous plug flow reactor we consider a differential volume element $\mathrm{d}V_r$ (Figure 4.1).

In the absence of a catalyst bed the volume of reaction mixture contained in this element will also be $\mathrm{d}V_r$. In this case the moles of reactant A used up in element $\mathrm{d}V_r$ in unit time will be $\mathrm{d}V_r(-r_A)$.

Here, as elsewhere throughout this book,

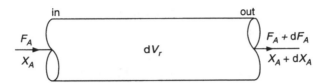

Figure 4.1 Flow of reactant A through volume element dV_r of plug flow reactor. (F_A = molar flow rate of reactant A; X_A = fractional conversion of A.)

$(-r_A)$ = rate of consumption of reactant A per unit volume of reaction mixture. (4.5)

Carrying out the molar balance on reactant A over the element dV_r gives

$$F_A \quad = \quad F_A + dF_A \quad + \quad (-r_A)dV_r \qquad (4.6)$$

| Moles of A entering element per unit time | moles of A leaving element per unit time | moles of A used up in reaction per unit time |

that is,

$$-dF_A = (-r_A)dV_r$$

but

$$F_A = F_{A0}(1 - X_A)$$

where F_{A0} is the molar flow rate of A at the reactor inlet where $X_A = 0$. In terms of X_A

$$dF_A = -F_{A0}dX_A$$

So

$$F_{A0}dX_A = (-r_A)dV_r$$

or

$$dV_r/F_{A0} = dX_A/(-r_A) \qquad (4.7)$$

Integrating over the length of the reactor and inserting the boundary condition $X_A = 0$ when $V_r = 0$,

$$V_r/F_{A0} = \int_0^{X_{AF}} dX_A/(-r_A) \qquad (4.8)$$

where V_r is the reactor volume required to achieve conversion X_{AF} at the reactor outlet.

For some purposes it is convenient to write equation (4.8) in terms of the volumetric flow rate of the feed (v_0) and the concentration of reactant A in the feed (C_{A0}) using the identity

$$F_{A0} = v_0 C_{A0} \qquad (4.9)$$

Combining (4.8) and (4.9),

$$V_r/v_0 = \tau = C_{A0} \int_0^{X_{AF}} \mathrm{d}X_A/(-r_A) \qquad (4.10)$$

The parameter τ has dimensions of time and is called the **space time**. It is a measure of the reaction rate and can be a useful parameter for correlating reactor characteristics. It can be determined experimentally from tests in which reactor volume and inlet conditions are known and outlet conversion can be measured. If density changes accompanying the reaction are negligible, τ is equal to the residence time of a given fluid element within the plug flow reactor.

Residence times in 'ideal' reactors are discussed in Section 4.2.

4.1.2.3 Heterogeneous plug flow reactors: mass balance equations and alternative expressions for reaction rate. When considering heterogeneous reactors (i.e. reactors containing a particulate bed of catalyst material) equation (4.8) and other equations developed for homogeneous systems cannot be used directly with the terms defined previously. However, equations of identical form can be obtained for the heterogeneous case by replacing the reaction rate per unit volume of reaction mixture $(-r_A)$ in equations (4.7) to (4.10) by the reaction rate per unit volume of catalyst bed $(-\underline{r}'_A)$. (Reactor volume V_r is also replaced by bed volume V_b, though apart from end effects these two volumes will be equal anyway.)

If, for example, a mass balance is taken on volume element $\mathrm{d}V_b$ of catalyst bed, the equation

$$\mathrm{d}V_b/F_{A0} = \mathrm{d}X_A/(-\underline{r}'_A) \qquad (4.11)$$

is obtained where

$(-\underline{r}'_A)$ = rate of consumption of reactant A per unit volume of catalyst bed

(4.12)

Equation (4.11) is directly analogous to equation (4.7) for a homogeneous system. Integration of equation (4.11) with boundary conditions $X_A = 0$ when $V_b = 0$ gives

$$V_b/F_{A0} = \int_0^{X_A} \mathrm{d}X_A/(-\underline{r}'_A) \qquad (4.13)$$

where V_b is the catalyst bed volume required to give a conversion X_A of component A.

The reaction rate per unit bed volume $(-\underline{r}'_A)$ in equation (4.13) may be calculated from the rate per unit volume of pellet (expressions for which were obtained in Chapter 3) using the bed voidage ε_b. This is defined as

$$\begin{aligned}\varepsilon_b &= \text{bed voidage} \\ &= \text{(volume of space between the catalyst pellets)/(volume of bed)}\end{aligned} \qquad (4.14)$$

From the above definition (since volume of bed = (volume of pellets) + (volume of space between pellets)

$$(1 - \varepsilon_b) = \text{(volume of pellets)/(volume of bed)} \qquad (4.15)$$

By definition,

$$(-\underline{r}_{AV}) = \text{rate of consumption of reactant } A \text{ per unit volume of catalyst pellet} \qquad (4.16)$$

So, from (4.12) and (4.15),

$$(-\underline{r}'_A) = (1 - \varepsilon_b)(-r_{AV}) \qquad (4.17)$$

Another measure of reaction rate in catalytic systems which is often employed is the reaction rate per unit catalyst mass

$$(-\underline{r}_{AW}) = \text{rate of consumption of reactant } A \text{ per unit mass of catalyst pellets} \qquad (4.18)$$

In terms of this quantity, $(-\underline{r}'_A)$ is given by

$$(-\underline{r}'_A) = \rho_p(1 - \varepsilon_b)(-\underline{r}_{AW}) \qquad (4.19)$$

where ρ_p is the pellet density.

An equation equivalent to (4.13), but expressed in terms of mass of catalyst rather than in terms of the catalyst bed, may be obtained by taking a molar balance on reactant A across an element of catalyst mass dW. This gives

$$-dF_A = (-\underline{r}_{AW})dW$$

or

$$F_{A0}dX_A = (-\underline{r}_{AW})dW \qquad (4.20)$$

Equation (4.20) may be rearranged and integrated to give

$$W/F_{A0} = \int_0^{X_A} (dX_A/(-\underline{r}_{AW}))$$

(4.21)

where W is the catalyst mass required to give conversion X_A of reactant A and F_{A0} is the molar flow rate of A at reactor inlet where $X_A = 0$.

4.1.2.4   The well-mixed or continuous stirred tank reactor.   In this case the 'ideal' situation is that of perfect mixing. The effluent from the reactor tank then has the same composition as that of the bulk fluid at all points in the tank. The material balance can be taken over the whole reactor giving (Figure 4.2)

$$F_{A0} = F_A + (-r_A)V_r$$

(4.22)

where F_{A0} = molar flow rate of reactant A entering reactor
     F_A = molar flow rate of A leaving reactor
     V_r = volume of reactant/product mixture contained in the reactor.

$(-r_A)$ is the rate at which reactant A is used up per unit volume of reactant/product mixture. Equation (4.22) may be expressed in terms of the conversion X_A of component A by writing

$$F_A = F_{A0}(1 - X_A)$$

(4.23)

Making this substitution and rearranging,

$$V_r/F_{A0} = X_A/(-r_A)$$

(4.24)

where X_A is the conversion achieved in the reactor. **If no volume changes** accompany the reaction,

$$C_A = C_{A0}(1 - X_A)$$

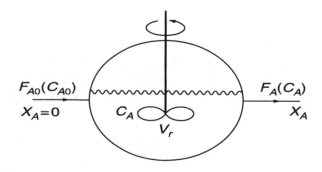

Figure 4.2 Flow of reactant A through continuous stirred tank reactor of volume V_r. (F_{A0} and C_{A0} are the molar flow rate and concentration of A in the feed. The outlet molar flow rate and concentration of A are F_A and C_A.)

and
$$X_A = (C_{A0} - C_{A1})/C_{A0}$$

where C_{A0} = concentration of reactor A entering reactor
         C_{A1} = concentration of A leaving the reactor.
Since also
$$F_{A0} = \nu_0 C_{A0}$$

it follows from (4.24) that

$$
\begin{aligned}
C_{A0} &= C_{A1} + (-r_A)(V_r/\nu_0) \\
&= C_{A1} + (-r_A)\tau
\end{aligned}
\tag{4.24a}
$$

where ν_0 = volumetric flow rate of fluid entering reactor (which in this case is also the volumetric flow rate of stream leaving the reactor)
         τ = 'space time' for the reactor

$$(\tau = V_r/\nu_0) \tag{4.24b}$$

If the CSTR is not in fact perfectly mixed, the 'ideal' equations (4.22) and (4.24) should be used with care. Some effects of imperfect mixing are discussed in Chapter 5.

The same general comments apply to the CSTR as to the batch reactor discussed in Section 4.1.1. Continuous stirred tank reactors in the form considered here are normally used industrially for liquid phase reactions only. Their use with gases is largely restricted to kinetic and fundamental studies carried out in the laboratory. Having said this, however, it should be remembered that a perfectly good, though rather specialised, example of a well-mixed continuous reactor employed industrially for gas phase reactions is the fluidised bed reactor, the design of which is discussed in Chapter 8.

Examples based on the design equations for batch and continuous reactors developed above will be given in Section 4.3.

4.2   Residence times in 'ideal' continuous flow reactors

The residence time (t) of any element of fluid in a reactor is the time spent by that element in the reactor. In an ideal plug flow reactor the residence times of all elements of fluid for given flow rate are the same, i.e. there is a single residence time. **If no volume changes occur during the reaction**, this residence time is given by

$$t = (V_r/\nu_0)_{\text{constant density}} \tag{4.25}$$

where V_r is the reactor volume and ν_0 is the volumetric flow rate at the inlet (in this situation ν_0 is also the flow rate at all points in the reactor tube).

In an 'ideal' continuous stirred tank reactor (CSTR), the fluid elements leaving the reactor will have spent a range of times within the reactor. These residence times range theoretically from zero (the most probable time) to infinity. The mean residence time in the constant density situation is given by

$$\bar{t} = V_r/\nu_0 \tag{4.26}$$

However, because of the highly 'skewed' nature of the distribution of these times (Figure 5.3a) 63.2% of the fluid spends less than this time in the tank.

The residence time distributions for a number of situations of ideal and non-ideal flow in reactors are analysed in Chapter 5, where the importance of the information that can be derived from measurements of the residence time distributions (RTDs) is emphasised. In the examples considered in Chapter 5, volume changes occurring during the reaction are negligible.

For reasons already stated, substantial volume changes are unlikely to occur in the liquid reactions typically carried out in continuous stirred tank reactors. However, when plug flow reactors (PFRs) are employed with gas phase reactions in which the number of moles changes during the reaction, volume and density changes occurring in a given element of fluid as it passes through the reactor can be quite substantial. Even under these conditions, a single residence time should result if the conditions for plug flow can be maintained, but its value will not be given by equation (4.25).

The magnitude of the residence time t can be investigated as follows: Consider a stationary section of a PFR of length dl, distance l from the feed point, as shown in Figure 4.3.

Also shown in Figure 4.3 as a dotted line is a filament of fluid within the main stream which enters the reactor section at time t and leaves at $t + dt$.

Time (dt) taken for the filament to cross length $dl = dl/u$

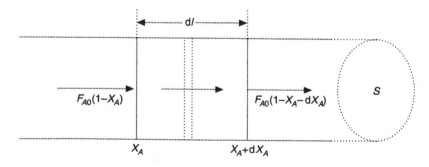

Figure 4.3 Illustrating passage of filament of fluid across section of plug flow reactor of length dl. (The filament is shown by broken lines.)

where u = average fluid velocity within the section dl of the reactor.

Volume of reactor section considered $(dV_r) = Sdl$

where S is cross-section area of reactor.
Hence,

$$dt = dl/u = Sdl/v = dV_r/v \qquad (4.26a)$$

where

v = volumetric flow rate of fluid passing through the reactor section
$= F_t \times v'$

$$(4.26b)$$

F_t and v' are average values of the total molar flow rate and molar volume within the section of reactor considered

Taking a material balance across the reactor section,

$$F_{A0}dX_A = (-r_A)dV_r$$

where F_{A0} is the molar flow rate of A at reactor inlet. Combining this result with (4.26a) and (4.26b),

$$dt = F_{A0}dX_A/F_t v'(-r_A) \qquad (4.26c)$$

Integrating (4.26c) with boundary conditions $t = 0$ when $X_A = 0$

$$t = F_{A0} \int_0^{X_A} \frac{dX_A}{F_t v'(-r_A)} \qquad (4.27)$$

where t is the time required for filament of fluid to traverse the reactor, i.e. 'residence time'. If no density changes accompany the reaction, the product $(F_t v')$ (i.e. the volumetric flow rate) will be the same at all points in the reactor, being equal to the volumetric flow rate v_0 at the inlet. Combining (4.26b) and (4.27) and setting $v = v_0$ then gives, for constant fluid density,

$$t = \frac{F_{A0}}{v_0} \int_0^{X_A} \frac{dX_A}{(-r_A)} = C_{A0} \int_0^{X_A} \frac{dX_A}{(-r_A)}$$

where C_{A0} = concentration of A (mol m^{-3}) at the reactor inlet.

This is identical with the expression for space time (τ) given by equation (4.10), so

$$(t = \tau)_{\text{constant } V}$$

This result only applies if the density (and hence the volumetric flow rate) of the reactor fluid remains constant as the reaction proceeds.

For reasons outlined above, this is usually a valid approximation for liquid reactions and it will be true for isothermal gaseous reactions in which the number of moles does not change in the reaction.

If the number of moles decreases during a gaseous reaction, $t > \tau$. If the number of moles increases, $t < \tau$.

To illustrate this we shall consider the first-order gaseous reaction

$$A \rightarrow 2B$$

which is carried out in an isothermal plug flow reactor, the conversion of B being zero at the inlet. We first evaluate the residence time t using equation (4.27) which we will rewrite in terms of the concentration of A using the equivalence

$$C_A = (F_A/F_t)(1/v')$$

where v' is the molar volume of a mixture within the reactor where the concentration of A is C_A and the molar flow rate of A is F_A. Making this substitution and writing

$$F_A = F_{A0}(1 - X_A)$$

gives

$$t = \int_0^{X_A} \frac{C_A dX_A}{(1 - X_A)(-r_A)}$$

Since the reaction is first order, $(-r_A) = kC_A$ and

$$t = \frac{1}{k}\int_0^{X_A} \frac{dX_A}{1 - X_A} = -\frac{1}{k}\ln(1 - X_A) \qquad (4.28)$$

For comparison, we now evaluate the space time from equation (4.10) and for this purpose we require the expression for the concentration of A, (C_A) as a function of conversion given by equation (3.69). (We assume ideal gaseous conditions so that the volume occupied by a given element is proportional to the number of moles contained in it at a given temperature and pressure.)

From equation (3.69)

$$C_{A0} = C_A \left(\frac{1 + \varepsilon_A X_A}{1 - X_A}\right)$$

where

$$\varepsilon_A = y_{A0}\Delta n$$

Δn = increase in moles per mole of A reacting = 1 in present instance
y_{A0} = mole fraction of A at inlet which is also unity

Hence

$$\varepsilon_A = 1 \quad \text{and} \quad C_{A0} = C_A\left(\frac{1 + X_A}{1 - X_A}\right)$$

Substituting in equation (4.10) and setting $(-r_A) = kC_A$ gives

$$\tau = \frac{1}{k}\int_0^{X_A} \frac{1 + X_A}{1 - X_A}\,dX_A$$

that is,

$$\tau = 1/k[-2\ln(1 - X_A) - X_A]$$
$$= \text{space time}$$
$$= V_r/v_0$$

Comparing this result with equation (4.28) we obtain, for a conversion of 0.9,

$$\frac{t}{\tau} = -1/k\ \ln(1 - 0.9)/\{1/k[-2\ln(1 - 0.9) - 0.9]\}$$
$$= 2.3/3.7$$
$$\text{i.e. } \tau = 1.61(t)$$

When non-ideal flow conditions prevail, the residence time does not take a single value for a tubular reactor at given conditions since different elements of fluid will remain in the reactor for different time periods. In this case a distribution of residence times is obtained, as described in Chapter 5.

4.3 Determination of reactor volumes required for given production rate and throughput specifications: homogeneous reactions under isothermal conditions

4.3.1 The batch reactor

Production rates that can be achieved with batch reactors depend on a number of factors in addition to the required reaction time given by equation (4.4), which will henceforth be termed t_r. These other factors are operational. A reactor must be charged, heated, cooled, emptied and cleaned each time a batch is processed. The total production time for a given batch is thus given by

$$t_T = t_r + t_0 \tag{4.29}$$

where t_r = reaction time
t_0 = operational time

Example 4.1

A first-order isomerisation reaction $A \rightarrow B$ is irreversible and is carried out in the liquid phase in a batch reactor. Calculate the reactor volume and also the number of batches required for an annual production of 1×10^4 kg of B. (The plant is operational for 300 days (7200 h) in the year.)

Data

Reaction temperature $= 150°C$
k (first-order rate constant) $= Ae^{-E/RT}$
  $= 2.6 \times 10^{14} e^{-125,000/RT}$ h^{-1}
X_{AF} $= 0.9$
t_0 (min) $= 10$ (filling) $+ 16$ (heating) $+ 14$
      (emptying) $+ 30$ (cleaning)
  $= 70$ min
Density $= 900$ kg m^{-3}
(The densities of the feed and product are similar.)

Solution

From equation (4.4),

$$t_r = C_{A0} \times \int_0^{X_{AF}} dX_A/(-r_A)$$

where C_{A0} is the initial concentration of reactant A.

Reaction rate per unit volume of reactor mixture

$$(-r_A) = kC_A = kC_{A0}(1 - X_A)$$

$$t_r = C_{A0} \times \int_0^{X_{AF}} dX_A/kC_{A0}(1 - X_A)$$

$$= -1/k \ \ln(1 - X_{AF})$$

$$k = 2.6 \times 10^{14} \exp[-125,000/(8.314 \times 423)]$$

$$= 9.4 \times 10^{-2} h^{-1}$$

For $X_A = 0.9$,

$$t_r = 24.5 \, h$$

From equation (4.29),
  Total time per batch $= 24.5 + 70/60$
      $= 24.5 + 1.17$
      $= 25.7 \, h$
  Total number of batches which can be processed annually
      $= 7200/25.7 = 280$
  Mass of A which must be processed per batch
      $= 1 \times 10^4/(280 \times 0.9) = 39.7 \, kg$

Volume of mixture contained in reactor
$$= 39.7/900 = 0.044\,m^3$$
$$= 44\,dm^3$$
In practice, the actual reactor shell volume would probably be about 20% larger, giving

$$\text{Total reactor volume} = 53\,dm^3$$

4.3.2 The plug flow reactor

The relevant equations have been given in Sections 4.1 and 4.2 and are best illustrated by the following examples.

Example 4.2
The chlorination of dichlorotetramethylbenzene in acetic acid occurs at 30°C according to the equation

$$C_6(CH_3)_4Cl_2 + Cl_2 = HCl + C_6(CH_3)_3CH_2ClCl_2$$

The following data were obtained for the above reaction in a well-stirred batch reactor.

Time, t (s)	0	48	85	135	171	223	257
Fractional conversion	0	0.21	0.32	0.44	0.52	0.60	0.64

If the initial concentrations of dichlorotetramethylbenzene (B) and chlorine (A) were respectively $C_{B0} = 34.7\ mol\ m^{-3}$ and $C_{A0} = 19.2\ mol\ m^{-3}$.

(a) use the batch reactor data to show that the reaction is second order
(b) calculate the plug flow reactor volume required to give 90% chlorine conversion for a volumetric flow rate of $1.5 \times 10^{-4}\ m^3\ s^{-1}$ using the same initial concentrations as above.

Solution
(a) If the reaction is second order, then

$$(-r_A) = k_2 C_A C_B \qquad (A = Cl_2, B = \text{organic reactant})$$

Since $C_{A0} \neq C_{B0}$,

$$(-r_A) = k_2 C_{A0}^2 (1 - X_A)(M - X_A), \quad \text{where } M = C_{B0}/C_{A0}$$

or

$$C_{A0}(dX_A/dt) = k_2 C_{A0}^2 (1 - X_A)(M - X_A)$$

Rearrangement and integration gives

$$\ln M - X_A/M(1 - X_A) = (M - 1)C_{A0}k_2 t$$

Hence if the reaction is second order, a plot of $\ln(M - X_A)/M(1 - X_A)$ vs t should be linear with a slope $(M - 1)C_{A0}k_2$. In fact the plot is linear, hence the reaction is second order (show this) and $k_2 = 1.47 \times 10^{-4}$ $m^3\,mol^{-1}\,s^{-1}$

(b) The required volume for the plug flow reactor may now be obtained using equation (4.8)

$$V_r/F_{A0} = \int_0^{X_A} dX_A/(-r_A) = \int_0^{X_A} dX_A/k_2 C_{A0}^2 (1 - X_A)(M - X_A)$$

Carrying out the integration,

$$
\begin{aligned}
V_r &= \left[\left(\frac{F_{A0}}{k_2 C_{A0}^2}\right)\left(\frac{1}{M - 1}\right)\right] \ln\left[\frac{M - X_A}{M(1 - X_A)}\right] \\
&= \left[\frac{V_0}{k_2 C_{A0}(M - 1)}\right] \ln\left[\frac{M - X_A}{M(1 - X_A)}\right] \\
&= \left[\frac{1.5 \times 10^{-4}}{(1.48 \times 10^{-4}) \times (19.2 \times 0.81)}\right] \times \ln\left[\frac{1.81 - 0.9}{1.81(1 - 0.9)}\right] \\
&= 0.105\,m^3
\end{aligned}
$$

Example 4.3
Acetaldehyde decomposes homogeneously according to the equation

$$CH_3CHO \rightarrow CH_4 + CO$$

It is required to decompose $0.01\,kg\,s^{-1}$ of acetaldehyde at $520°C$ and at 1 atm in a PFR. The reaction is irreversible, second order and the rate constant k_2 is $0.43\,m^3\,kmol^{-1}\,s^{-1}$.

Calculate (a) the reactor volume for 90% conversion, (b) the respective space time for 90% conversion (τ) and (c) the residence time (t) for 90% conversion. (Gas constant $R = 0.082\,m^3\,atm\,kmol^{-1}K^{-1}$ and relative molecular mass (or molecular weight) of acetaldehyde = 44.)

Solution
(a) **Reactor volume**
According to equation (4.8)

$$\frac{V_r}{F_{A0}} = \int_0^{X_A} \frac{dX_A}{(-r_A)} \tag{i}$$

where $(-r_A)$ is the rate of consumption of reactant A per unit volume of reactant/product mixture and F_{A0} is the molar flow rate of reactant A at the inlet to the reactor. V_r is the desired reactor volume.

The reaction is second order in A, so

$$\frac{V_r}{F_{A0}} = \int_0^{X_A} \frac{\mathrm{d}X_A}{k_2 C_A^2} \tag{ii}$$

where C_A is the concentration of A.

To proceed further we need an expression for C_A as a function of the conversion X_A of this component. Proceeding as in the illustration at the end of Section 4.2.1, and applying equation (3.69) with $\Delta n = 1$ and $y_{A0} = 1$, we find

$$C_A = C_{A0}(1 - X_A)(1 + X_A)^{-1} \tag{iii}$$

At the conditions stated, ideal gas relationships will be closely followed so that $C_{A0} = (P/RT)\,\mathrm{kmol\,m^{-3}}$ where $P = 1$ atm. Combining this result with (iii) and (ii) gives

$$\frac{V_r}{F_{A0}} = \left(\frac{RT}{P}\right)^2 \frac{1}{k_2} \int_0^{X_A} \left(\frac{1 + X_A}{1 - X_A}\right)^2 \mathrm{d}X_A$$

$$= \left(\frac{RT}{P}\right)^2 \frac{1}{k_2} \left(\frac{4}{1 - X_A} - 4 + 4\ln(1 - X_A) + X_A\right)$$

Also, since $F_{A0} = (0.01/44)\,\mathrm{kmol\,s^{-1}}$,

$$V_r = (0.01/44)(0.082 \times 793)^2(1/0.43) \times 27.7 = 61.9\,\mathrm{m^3}$$

(b) The space time (τ)
By definition,

$$\text{Space time} = \text{reactor volume} / v_0$$

where v_0 = volumetric flow rate at entry to reactor.
In this case

$$v_0 = (0.01/44) \times (0.082 \times 793/1)$$
$$= 0.0148\,\mathrm{m^3\,s^{-1}}$$

So
$$\tau = 61.9/0.0148$$
$$= 4.18 \times 10^3\,\mathrm{s}$$

(c) The residence time (t)
Proceed as in Example 4.2 but modify the treatment to allow for the reaction being second order rather than first order. From equation (4.28),

$$t = \int_0^{X_A} (C_A/((1 - X_A)(-r_A)))\mathrm{d}X_A \tag{iv}$$

In the present case

$$(-r_A) = k_2 C_A^2$$

where $C_A = C_{A0}(1 - X_A)(1 + X_A)^{-1}$ (equation (iii)) and $C_{A0} = RT/P$. Making these substitutions in (iv)

$$
\begin{aligned}
t &= RT/k_2 P \int_0^{X_A} [(1 + X_A)/(1 - X_A)^2] dX_A \\
&= RT/k_2 P[(2/(1 - X_A)) - X_A - \ln 1/(1 - X_A)] \\
&= 0.082 \times 793/0.43[16.8] \\
&= 2540 \, s
\end{aligned}
$$

So the residence time (t) is about 60% of the space time (τ) in this case.

4.3.3   The continuous stirred tank reactor

The design equation (4.24) for the CSTR is very simple in form and a single example will suffice.

Example 4.4
As in Example 4.2, we consider the chlorination of dichlorotetramethyl-benzene (DTMB). In Example 4.2 we calculated the volume of a PFR required to achieve 90% conversion of DTMB at 30°C for a specified feed. In the present example we calculate the volume hold-up of the reactor contents (V_r) in a CSTR required to effect 90% conversion of DTMB using the same conditions and feed as those used in Example 4.2.

According to equation (4.24),

$$V_r/F_{A0} = X_A/(-r_A)$$

where F_{A0} is the molar flow rate of component A entering the reactor, X_A is the outlet conversion of reactant A and $(-r_A)$ is the rate at which A is being consumed for unit volume of reaction mixture. So,

$$V_r/F_{A0} = V_r/v_0 C_{A0} = X_A/k_2 C_{A0}^2(1 - X_A)(M - X_A)$$

and

$$V_r = v_0 X_A (k_2 C_{A0}(1 - X_A)(M - X_A))^{-1}$$

with symbols as in Example 4.2.

Inserting numerical values,

$$V_r = (1.5 \times 10^{-4} \times 0.9)/(1.48 \times 10^{-4} \times 19.2(1 - 0.9)(1.81 - 0.9)) = 0.52 \, m^3$$

Note that the volume of a plug flow reactor for the same duty is only 0.105 m³ (Example 4.2).

($M = C_{B0}/C_{A0}$, v_0 is the volumetric flow rate of the feed and C_{A0} and C_{B0} are the concentrations of A and B in the feed. k_2 is the second-order rate constant.)

4.4 Performance of series and parallel combinations of homogeneous plug flow reactors under isothermal conditions

4.4.1 Plug flow reactors in series

Consider n plug flow reactors in series (Figure 4.4). For the ith reactor (volume V_{ri}), application of equation (4.8) gives

$$\frac{V_{ri}}{F_{A0}} = \int_{X_{A(i-1)}}^{X_{Ai}} \left(\frac{1}{-r_A}\right) dX_A$$

where X_{Ai} is the conversion of A in the stream leaving reactor i.

The combined volume of all the reactors in the group V_r (total) is given by

$$V_r(\text{total}) = \sum^n V_{rn} = F_{A0}\left\{ \int_0^{X_{A1}} \frac{dX_A}{-r_A} + \ldots + \int_{X_{A(i-1)}}^{X_{Ai}} \frac{dX_A}{-r_A} + \ldots \right.$$

$$\left. + \int_{X_{A(n-1)}}^{X_{An}} \frac{dX_A}{-r_A} \right\} = F_{A0} \int_0^{X_{An}} \frac{dX_A}{-r_A}$$

So n PFRs in series give the same performance as a single PFR of the same total volume.

4.4.2 Plug flow reactors in parallel (Figure 4.5)

If $V_{r1} = V_{r2} = \ldots = V_r$ and the flow is split equally between each reactor then the conversion for each reactor will be the same and will, of course, be equal to the conversion achieved by the entire reactor assembly. In this case, as in the previous one, the performance of the combination of reactors is the same as that of one reactor of the same total volume.

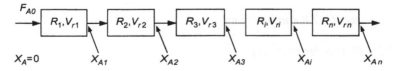

Figure 4.4 Illustrating performance of n homogeneous plug flow reactors ($R_1, R_2, \ldots,$ $R_i, \ldots, R_n$) in series. (V_{ri} and X_{Ai} are the volume and outlet conversion for reactor i.)

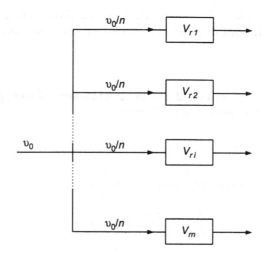

Figure 4.5 Illustrating performance of group of n identical plug flow reactors assembled in parallel. ν_0 is volumetric flow rate of feed to group manifold.

If the reactor tubes are of different size, then it follows that if the flow is split in proportion to the volume of the tubes, then again the conversions will be identical for each tube.

4.5 Performance of series and parallel combinations of continuous stirred tank reactors under isothermal conditions

4.5.1 Parallel CSTR combinations

From equation (4.24), the conversion X_A of reactant A obtained from a single CSTR is given by $X_A = (-r_A)V_r/F_{A0}$.

Since $(-r_A)$ is constant for given X_A, it follows from this equation that:

(a) if flow is split equally between two reactors each with liquid hold-up V_r, then the conversion will be identical to that of a single tank of volume $= 2V_r$;

(b) if a second tank of volume V_r is placed in parallel to another identical tank, the throughput is doubled for the same conversion.

4.5.2 Series CSTR combinations

4.5.2.1 First-order reactions

(a) *Tanks of equal volume* $(= V_r)$. The volumetric flow rate of reactant/product mixture (ν_0) is taken to be approximately the same for each reactor

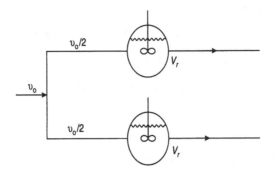

Figure 4.6 Illustrating performance of two CSTRs of equal volume (V_r) mounted in parallel (see Figure 4.5).

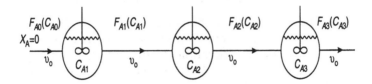

Figure 4.7 Illustrating performance of series of CSTRs of equal volume (V_r) mounted in series. F_{A0} and F_{Ai} are molar flow rates of reactant A at entry and leaving reactor i respectively. C_{Ai} is concentration of A in reactor i.

in the chain. (This is a reasonable approximation for most liquid phase reactions.)

The molar flow rates (F_A) and concentrations (C_A) of reactant A are as shown in Figure 4.7.

The material balance for tank (1) is given by equation (4.24a) to be

$$C_{A0} = C_{A1} + (-r_A)V_r/v_0 = C_A + (-r_A)\tau$$

where $\tau = V_r/v_0 =$ 'holding time' or 'space-time' for each reactor.

For first-order reaction

$$(-r_A) = k_1 C_{A1} \tag{4.30}$$

Therefore

$$C_{A0} = C_{A1} + k_1 C_{A1}\tau$$

Analogous equations can be written for tanks (2) and (3):

$$C_{A1} = C_{A2} + k_1 C_{A2}\tau$$

$$C_{A2} = C_{A3} + k_1 C_{A3}\tau$$

Rearranging each equation

$$C_{A0}/C_{A1} = 1 + k_1\tau; \quad C_{A1}/C_{A2} = 1 + k_1\tau; \quad C_{A2}/C_{A3} = 1 + k_1\tau$$

Multiplying the three results together gives

$$C_{A0}/C_{A3} = (1 + k_1\tau)^3 \qquad (4.31)$$

For n tanks

$$C_{A0}/C_{An} = (1 + k_1\tau)^n$$

The total holding time $\tau_t = n\tau$. Therefore,

$$C_{A0}/C_{An} = (1 + k_1\tau_t/n)^n \qquad (4.32)$$

Equation (4.32) can be expanded as a binomial series and as $n \to \infty$

$$C_{A0}/C_{An} \to \exp(k_1\tau_t) \quad \text{or} \quad \tau_t \to 1/k_1 \ln(C_{A0}/C_{An})$$

This result may be compared with the expression $\tau = 1/k_1 \ln(C_{A0}/C_{An})$ for a PFR in which a first-order reaction with no volume change is taking place. (This expression follows from equations (4.10) and (4.30) on applying the constant density result $dX_A = -(dC_A/C_{A0})$, where C_{A0} is the concentration of A in the fluid entering the reactor.) Hence an infinite series of CSTRs would be required to simulate a single plug flow reactor of the same total volume giving the same conversion.

Some extensions of the above results are as follows:

(b) *Different sized tanks.* Suppose tanks 1, 2 and 3 in the sketch have volumes V_{r1}, V_{r2} and V_{r3} and that (as previously) the volumetric flow rate v_0 through each tank is the same. The 'holding times' (or 'space times') in the tanks will be written

$$\tau_1, \tau_2, \tau_3$$

Taking mass balances as before:
for tank 1,

$$C_{A0} = C_{A1} + kC_{A1}\tau_1; \text{ therefore } C_{A0}/C_{A1} = 1 + k\tau_1$$

for tank 2,

$$C_{A1} = C_{A2} + kC_{A2}\tau_2; \text{ therefore } C_{A1}/C_{A2} = 1 + k\tau_2$$

for tank 3,

$$C_{A2} = C_{A3} + kC_{A3}\tau_3; \text{ therefore } C_{A2}/C_{A3} = 1 + k\tau_3$$

Hence (cf. equation (4.31))

$$C_{A0}/C_{A3} = (1 + k\tau_1)(1 + k\tau_2)(1 + k\tau_3) \qquad (4.33)$$

(c) *Equal sized tanks, different temperatures, same volumetric flow rate* (ν_0) *through each.* If the tank temperatures differ, the first-order rate constants will not be the same so, if $k_{(1)}$, $k_{(2)}$ and $k_{(3)}$ are the rate constants in tanks 1, 2 and 3, $k_{(1)} \neq k_{(2)} \neq k_{(3)}$

Proceeding as above, we then obtain

$$C_{A0}/C_{A1} = (1 + k_{(1)}\tau)(1 + k_{(2)}\tau)(1 + k_{(3)}\tau) \qquad (4.34)$$

where τ is the space time for each tank ($\tau = V_r/\nu_0$)

4.5.2.2 Second-order reactions in series arrangements of CSTR tanks

(a) *Analytical solution: two tanks only with equal volumes and space times.* Analytical solutions are only possible in very simple situations. In this illustration we suppose that the reaction rate can be expressed as the square of the concentration of reactant A only:
Consider a rate equation

$$(-r_A) = k_2 C_A^2 \qquad (4.34a)$$

For tank 1, $C_{A0} = C_{A1} + (-r_A)\tau$ (cf. equation (4.24a)), i.e.

$$C_{A0} = C_{A1} + k_2 C_{A1}^2 \tau \qquad (4.34b)$$

where C_{A0} and C_{A1} are the concentrations of reactant A in the streams entering and leaving tank 1.

For tank 2,

$$C_{A1} = C_{A2} + k_2 C_{A2}^2 \tau \qquad (4.34c)$$

In this very simple situation, the following algebraic solution applies

$$C_{A1} = (-1 \pm \sqrt{1 + 4k_2\tau C_{A0}})/2k_2\tau$$

and

$$C_{A2} = (-1 \pm \sqrt{1 + 4k_2\tau(-1 \pm \sqrt{1 + 4k_2\tau C_{A0}})/2k_2\tau})/2k_2\tau$$

If the number of tanks exceeds two and/or more general forms for the rate equation apply, analytical solutions become increasingly cumbersome or impossible and graphical solutions are necessary.

(b) *Graphical solution.* Consider the general form of the material balance equation around a single continuous stirred tank reactor, assuming no volume change on reaction. Then (see Section 4.1.2.2, equation (4.24a)),

$$C_{A0} = C_{A1} + (-r_A)\tau$$
$$= C_{A1} + (-r_A)V_r/v_0$$

Rearranging

$$C_{A0} - C_{A1} = (-r_A) \cdot V_r/v_0$$

or

$$(-r_A)/(C_{A0} - C_A) = v_0/V_r = 1/\tau \tag{4.35}$$

where

$$(-r_A) = kf(C_A) \tag{4.36}$$

If the kinetic equation is known, then $(-r_A)$ can be calculated and plotted as a function of C_A as in Figure 4.8.

On Figure 4.8, the ratio $(-r_A)/(C_{A0} - C_{A1})$ in equation (4.35) is represented by the ratio of the lengths BC to AC.

We thus can write

$$(1/\tau) = BC/AC = v_0/V_r = -(\text{slope of line } AB) \tag{4.37}$$

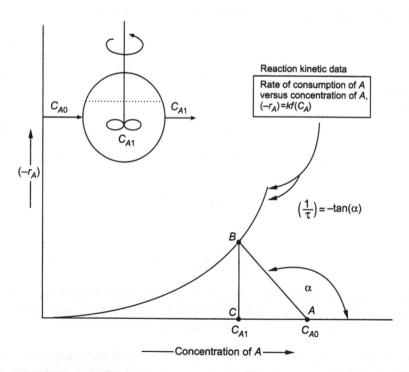

Figure 4.8 Construction based on reaction kinetic data giving space time (V_r/v_0) for CSTR if inlet and outlet reactant concentrations are known, or vice versa.

To illustrate the use of this construction we now consider various possibilities.

(c) *n Tanks each with the same feed rate and of the same size.* A construction similar to that in Figure 4.8 can be drawn for each of the reactors in the series leading to a set of lines A_1B_1, A_2B_2, A_3B_3, etc. equivalent to the line AB in Figure 4.8. Since v_0 and V_r are the same for each tank, $1/\tau$ will be the same for each and so, according to the construction in Figure 4.8, will the slopes of the lines A_1B_1, A_2B_2, etc. A set of parallel construction lines will thus be obtained as in Figure 4.9.

(d) *n Tanks of different sizes.* In this case the slopes of the lines (A_1B_1), (A_2B_2), (A_3B_3), etc., in Figure 4.9 will be different.

(e) *Tanks at different temperatures.* A different rate curve will be needed for each tank but a solution for a series of tanks can still be obtained.

Example 4.5
Acetic acid hydrolysis is carried out in a series of three equal-sized tanks. The reaction is first order and the rate constant $k_1 = 0.158\ \mathrm{min}^{-1}$. What is the fractional conversion X_A if the volume of each tank is $1.8\ \mathrm{dm}^3$ and the volumetric flow rate is $0.6\ \mathrm{m}^3\ \mathrm{min}^{-1}$?

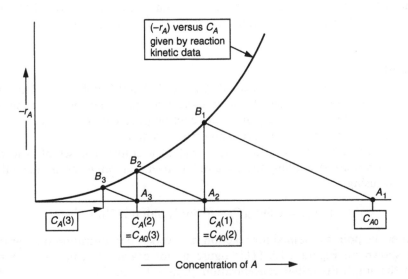

Figure 4.9 Construction showing outlet concentrations $C_A(1)$, $C_A(2)$ and $C_A(3)$ from a series of CSTR tanks of same size (see also Figure 4.8).

Solution

From equation (4.31), the ratio of the concentration C_{A0} of reactant A in the feed in the first tank to the concentration C_{A3} of A in the product from the third tank is given by

$$C_{A0}/C_{A3} = (1 + k_1\tau)^3 = (1 + 0.158 \times 1.8/0.6)^3$$

where k_1 is the first-order rate constant and τ is the space time. ($\tau = V_r/v$, where V_r is the volume hold-up and v the volumetric flow rate.) Therefore

$$C_{A0}/C_{A0}(1 - X_A) = (1.474)^3 = 3.2$$

where X_A is the overall conversion achieved by the three tanks in series. Therefore

$$1 - X_A = 1/3.2 = 0.31 \text{ and Fractional conversion} = 0.69$$

Example 4.6

Ethyl acetate (ethyl ethanoate) was hydrolysed by an aqueous solution of caustic soda according to

$$C_2H_5COOCH_3 + NaOH \rightarrow C_2H_5OH + CH_3COONa$$

The reaction was carried out in a series of three equal-sized tanks (CSTRs) at 25°C. The initial concentrations (C_{A0}) of both reactants were 1×10^{-2} mol dm^{-3} and a total conversion of 60% was required. Calculate the volumetric flow rate required if the volume of each tank was 130 dm^3. The second-order specific rate constant k_2 is 6.86 dm^3 mol^{-1} min^{-1} at 25°C.

Solution

The initial concentrations are equal $(C_{A0} = C_{B0})$ and 1 mole of ethyl acetate is consumed for each mole of caustic soda used up, so the concentrations of the two reactants remain equal as the reaction proceeds $(C_A = C_B)$.

A graphical solution and trial and error is required.

First calculate the reaction rate $(-r_A)$ at a convenient set of reactant concentrations straddling the required conversion range as in the table below, using

$$(-r_A) = 6.86 C_A C_B = 6.86 C_A^2 \text{ mol min}^{-1} dm^{-3}$$

Enter the points obtained for reaction rate versus concentration on a graph as shown on Figure 4.10. (This figure is not drawn to scale.) Also enter on the graph the initial concentration $(C_{A0} = 1 \times 10^{-2}$ mol dm$^{-3})$ and the concentration C_{AF} when 60% conversion has been achieved $(C_{AF} = (1.0 \times 10^{-2}) \times (1 - 0.6) = 0.4 \times 10^{-2}$ mol dm^{-3}.

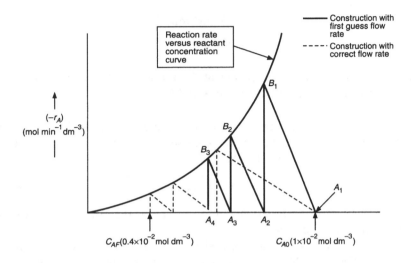

Figure 4.10 Trial and error procedure for determining flow rate required to give desired conversion in series of three CSTRs (see also Figure 4.8).

$C_A(\times 10^2)(\text{mol dm}^{-3})$	1.0	0.8	0.6	0.4	0.2
$(-r_A) = k_2 C_A^2 (\times 10^4)(\text{mol min}^{-1}\text{dm}^{-3})$	6.86	4.39	2.47	1.10	0.27

Trial and error is now required.

Guess a value $v_0(1)$ for the unknown flow rate. If $v_0(1)$ is expressed in dm^3 min^{-1} the corresponding 'first guess' value for $1/\tau$, $1/\tau(1)$, is $v_0(1)/130$ min^{-1}. Starting from concentration C_{A0} construct the line $A(1)B(1)$ of slope $-(1/\tau(1))$ (cf. equation (4.37)). This line intersects the reaction rate versus reactant concentration curve at point $B(1)$. From $B(1)$ drop a perpendicular to point $A(2)$. The concentration at this point is that in the fluid leaving reactor 1 and entering reactor 2. From point $A(2)$ construct the line $A(2)B(2)$. The slope of this line will be the same as that of $A(1)B(1)$, being equal to $(-1/\tau)$. $(-1/\tau$ will be the same for all three reactors since their volumes are equal and the volumetric flow rate is taken to be constant for all.) From point $B(2)$ drop a perpendicular to point $A(3)$, which gives the composition of fluid entering the final reactor. Proceeding as before, the concentration of A in the fluid leaving the third and final reactor is given by point $A(4)$. As drawn it is seen that $A(4)$ falls well to the right of the required concentration C_{AF}. The overall conversion achieved was therefore not adequate and a lower flow rate is required. Try different flow rates until finally a rate is found such that the concentration of A in the liquid leaving the third reactor is equal to the desired value C_{AF}. (This is the case with the dotted construction lines in Figure 4.10.) With the data given in the problem, this occurs when

$$\nu_0 = 13\,\text{dm}^3\,\text{min}^{-1}$$
$$= 780\,\text{dm}^3\,\text{h}^{-1}$$

4.6 Comparison of the performance of the various reactors and reactor combinations considered in Sections 4.3 to 4.5

For the purposes of the comparison we will consider a restricted class of reactions, the rates of which can be expressed as a power (n) of the concentration (C_A) of the reactant A only.

$$(-r_A) = k_n C_A^n \tag{4.38}$$

where k_n is an nth-order rate constant and $(-r_A)$ is the rate of consumption of A in unit volume of reaction mixture.

The reaction type (4.38) has been selected for its simplicity rather than its general applicability. Nevertheless, it is applicable directly in some useful situations. If $n = 2$, for example, (4.38) reduces to

$$(-r_A) = k_2 C_A^2 \tag{4.39}$$

This form of kinetics is followed in some decomposition reactions such as the gas phase decomposition of NO and O_2 and also can result from second-order reactions which are first order in each of two reactants if the initial concentrations of the reactants are equal. If, for example, the second-order rate equation

$$(-r_A) = k_2 C_A C_B \tag{4.40}$$

applies and the initial concentrations of A and B are equal, the reaction rate will be given by

$$(-r_A) = k_2 C_A^2 = k_2 C_B^2 \tag{4.41}$$

as in equation (4.45). Equation (4.41) would clearly not be applicable if the initial concentrations of A and B were not equal.

Despite its obvious limitations, equation (4.38) is, because of its simplicity, useful when arriving at generalisations about the performance of one reactor type as compared with another and of the influence of reaction order on these comparisons, as in the following examples.

Combining equation (4.38) with the expression for concentration as a function of conversion given by equation (3.56),

$$-r_A = k_n C_{A0}^n (1 - X_A)^n / (1 + \varepsilon X_A)^n \tag{4.42}$$

ε in equation (4.42) (here termed the 'expansion factor') allows for volume changes during the reaction. C_{A0} is the concentration of A in the feed to the reactor and X_A is the conversion of A in the reactor fluid.

The examples in this section will either be concerned with continuous stirred tank reactors only or with the comparison of these with a plug flow reactor in which the same reaction is taking place. Since continuous stirred tank reactors are usually used for liquid phase reactions only (Section 4.1.2.2), the appropriate values for ε will normally be very small. The denominator in equation (4.42) will accordingly be taken as unity, giving

$$-r_A = k_n C_{A0}^n (1 - X_A)^n \tag{4.43}$$

The volumes (V_r (plug) and V_r (CSTR)) of the plug flow and continuous stirred tank reactors required to produce a conversion X_A of A for specified values of the molar flow rate (F_{A0}) and feed concentration (C_{A0}) of A can now be obtained by substituting ($-r_A$) from (4.40) into the material balance equations (4.8) and (4.24):

$$V_r(\text{plug}) = F_{A0} \int_0^{X_A} (1/(-r_A)) \mathrm{d}X_A = Z \int_0^{X_A} (1/(1 - X_A)^n) \mathrm{d}X_A \tag{4.44}$$

and

$$V_r(\text{CSTR}) = F_{A0} X_A/(-r_A) = Z X_A/(1 - X_A)^n \tag{4.45}$$

where $Z = F_{A0}/(k_n C_{A0}^n)$.

4.6.1 Comparison of volume requirements of ideal CSTRs and PFRs under isothermal conditions

Example 4.7
(a) Compare the volumes of isothermal CSTRs and PFRs required to obtain the same conversion for the same inlet conditions.
(b) Comment on the manner in which the ratio of the two volumes depends on the form of the reaction and on the required conversion.

Solution
From equations (4.44) and (4.45) the ratio of the plug flow to the CSTR volume is

$$V_r(\text{plug})/V_r(\text{CSTR}) = \tau_p/\tau_c = (1 - X_A)^n/X_A \int_0^{X_A} (1/(1 - X_A)^n) \mathrm{d}X_A \tag{4.46}$$

(a) It is a consequence of the form of equation (4.46) that the plug flow reactor volume is lower than that of the CSTR for all positive values of n and that the ratio becomes smaller as n increases.

Considering a conversion of 90%, for example, the ratio for $n = 1$ is $(-\ln(0.1)/9) = 0.25$ while for $n = 2$ it is 0.10, and so on.
For $n = 2$:

$$V_r(\text{plug})/V_r(\text{CSTR}) = (1 - X_A)^2/X_A \int_0^{X_A} \left(1/(1 - X_A)^2\right) dX_A \qquad (4.47)$$

$$= (1 - X_A) = 0.1$$

(b) It is a consequence of the form of equation (4.43) that the ratio of the plug flow volume to that of the CSTR falls progressively towards zero for all n as the required conversion rises towards unity. For very low conversions, however, the two volumes are similar (see equation (4.44) for the second-order case).

The large differences in the volumes of CSTRs and PFRs required to obtain high conversions have already been noted in Example 4.4 in Section 4.3.3. The reason for the difference is that the reactant concentrations during reactions in the CSTRs are those of the final product (Section 4.1.2.2) while in the plug flow type of reactor the reactant concentrations exceed those of the final product, particularly for reaction in the vicinity of the feed point. It follows from the reaction rate laws ((3.14) and (3.15)) that overall reaction rates per unit volume of reactor fluid are higher in the plug flow reactor than in the CSTR. The reactor volume required for a given conversion is therefore lower.

The discrepancy in the volume requirement becomes less pronounced when the plug flow reactor is compared with two CSTRs in series. As shown in Example 4.8 below, the total volume requirement when two CSTRs are used in series is, for a 90% conversion and $n = 2$ in the kinetic equation (4.38), only 30% of that for a single CSTR. It follows from equation (4.47), therefore, that the combined volume of the CSTR pair is 3 times that of the equivalent PFR, whereas the volume requirement of the single CSTR was 10 times that of the PFR. The addition of further CSTRs to the chain of CSTRs would reduce the volume discrepancy further until, in the limit, the volume of the reactor chain approached that of the PFR for a given duty.

Although the volume comparison made in the above example is of interest, it is only one of many factors which contribute to the economic viability of one reactor system as opposed to another. In practice, when all factors are considered, tubular plug flow reactors are usually favoured for homogeneous gas reactions while continuous stirred tank reactors are usually favoured for liquid phase reactions.

4.6.2   Comparison of performance of various arrangements of CSTRs under isothermal conditions

Example 4.8   *Reduction in reactor volume (for given conversion) achieved by addition of extra CSTR in series*

The volume $V(1)$ of a CSTR tank required to obtain 90% conversion of reactant A in a reaction with the kinetics given by equation (4.39), i.e.

$$(-r_A) = k_2 C_A^2 \qquad \text{(i)}$$

is inconveniently large. What reduction in the total volume requirement can be obtained by replacing the single CSTR by two reactors, each of volume V', in series? The volumetric flow rate (ν_0) of the feed is the same in the two cases, as is the feed composition and conversion. Volume changes during the reaction are negligible.

Solution
Combining equation (4.24) with the kinetic equation (i) gives, for the single tank of volume $V(1)$,

$$C_{A0} = C_{AP} + k_2 C_{AP}^2 \tau(1) \qquad \text{(ii)}$$

where C_{A0} = concentration of reactant A in feed
C_{AP} = concentration of A in product

$$\tau(1) = V(1)/\nu_0$$

From data given, $C_{AP}/C_{A0} = 0.1$. Dividing equation (ii) through by C_{AP} and rearranging, gives

$$k_2 C_{AP} \tau(1) = C_{A0}/C_{AP} = 9$$

so, when one tank only is used,

$$V(1) = 9\nu_0/k_2 C_{AP} = 90\nu_0/k_2 C_{A0}. \qquad \text{(iii)}$$

When two tanks each of volume V' are used, equations (4.34b) and (4.34c) give

$$C_{A0} = C_{A1} + k_2 C_{A1}^2 \tau' \qquad \text{(iv)}$$

and

$$C_{A1} = C_{A2} + k_2 C_{A2}^2 \tau' \qquad \text{(v)}$$

where C_{A0} is the concentration of A in the feed to the first tank, C_{A1} is the concentration of A in the product from the first tank and hence in the feed to the second tank, and C_{A2} is the concentration of A in the product stream from tank 2.
τ' = space time for each reactor = V'/ν_0
$(C_{A2} = C_{AP}$ = desired product concentration)
From equations (iv) and (v), on setting $C_{A2} = 0.1 C_{A0}$,

$$k_2 \tau' = 100(C_{A1}/C_{A0}^2) - 10/C_{A0} = (C_{A0}/C_{A1}^2) - (1/C_{A1}) \qquad \text{(vi)}$$

Multiplying through by C_{A0}^2/C_{A1} and rearranging,

$$(C_{A0}/C_{A1})^3 - (C_{A0}/C_{A1})^2 + 10(C_{A0}/C_{A1}) - 100 = 0 \qquad \text{(vii)}$$

and

$$k_2\tau'C_{A0} = 100(C_{A1}/C_{A0}) - 10 \qquad \text{(viii)}$$

From equation (vii), $C_{A0}/C_{A1} = 4.23$ where, from (viii), $k_2\tau'C_{A0} = 13.7$. Hence

$$\tau' = \text{space time for each tank} = 13.7/k_2C_{A0} \qquad \text{(ix)}$$

Space time for the two tanks in series $= 27.4/k_2C_{A0}$ and combined volume of the two tanks $= 27.4v_0/k_2C_{A0}$. Hence

(volume of original single tank) / (combined volume of two tanks in series)

$$= V(1)/2V'$$
$$= 90/27.4$$
$$= 3.3$$

There is thus a three-fold reduction in the overall volume requirement. The volumes of the individual reactors in the modified layout are of course substantially smaller than the original, the volume of each one being only one-seventh that of the original one.

Example 4.9 *Improvement in conversion (for a given throughout) achieved by addition of extra CSTR in series*
The conversion (90%) of reactant A in the CSTR described in Example 4.8 is no longer considered adequate at the designated feed rate. What improvement in conversion can be achieved by placing a second reactor tank of the same volume as the first in series with it?

Solution
The rate of consumption of reactant A per unit volume of reactant/product mixture is given by equation (4.39), as in Example 4.5. The feed rate to the two series reactors is to be the same as to the single one. Since the reactor volumes and volume flow rates are to be the same, the 'space time' τ will be the same for each of the reactors considered.

The material balance equations for the series reactors (here designated 1 and 2) are given by equations (4.34b) and (4.34c) respectively.

Let the overall conversion achieved by the two reactors in series be X_A (overall) and let the concentrations of A in the streams leaving reactors 1 and 2 be C_{A1} and C_{A2} respectively. C_{A0} is the concentration of A in the feed to reactor 1 and the concentration of A in the stream entering reactor 2 is C_{A1}. From the definition of conversion,

$$1 - X_A(\text{overall}) = C_{A2}/C_{A0}$$

But

$$C_{A2}/C_{A0} = (C_{A2}/C_{A1})(C_{A1}/C_{A0}) = (C_{A2}/C_{A1})(1 - X_{A1})$$

where X_{A1} is conversion in reactor 1, which is known to be 0.9. Hence, since

$$C_{A2}/C_{A0} = 0.1(C_{A2}/C_{A1})$$

and

$$X_A(\text{overall}) = 1 - 0.1(C_{A2}/C_{A1}) \qquad \text{(i)}$$

C_{A2}/C_{A1} may be found in terms of the group $k_2 C_{A0}\tau$, from the mass balance equation for the second tank (equation (4.34c)). Dividing this through by C_{A1} and substituting $C_{A1}(C_{A2}/C_{A1})$ for C_{A2} and $0.1\,C_{A0}$ for C_{A1} gives,

$$1 = (C_{A2}/C_{A1}) + 0.1 k_2 C_{A0}\tau(C_{A2}/C_{A1})^2 \qquad \text{(ii)}$$

From Example 4.8 we know that the dimensionless group $k_2\tau C_{A0}$ for a single CSTR with kinetics given by equation (4.34a), and for which the conversion (X_{A1}) is 0.9, is given by

$$k_2\tau C_{A0} = 90 \qquad \text{(iii)}$$

Substituting this value in equation (ii) and solving the quadratic for (C_{A2}/C_{A1}) gives

$$C_{A2}/C_{A1} = 0.282$$

and so, from equation (i)

$$X_A(\text{overall}) = 0.972$$

Hence, addition of a second identical tank in series increases the fractional conversion from 0.9 to 0.972. Addition of further tanks in series would obviously produce progressively smaller absolute increments in conversion as unity was approached.

Example 4.10 *Improvement in throughput for a given conversion achieved by addition of extra CSTR in series: would it be better to place the extra CSTR in parallel?*
In this example we wish to determine the increase in throughput which can be achieved, while holding conversion constant at 90%, by adding an extra continuous stirred tank reactor to the single one at present in use. As in the previous example, the original CSTR and the additional one are to have the same volume and the feed to the reactor system remains unchanged. No appreciable volume change occurs during the reaction and, for illustrative purposes, the reaction kinetics will again be taken to be given by the second-order relationship

$$(-r_A) = k_2 C_A^2$$

Two options will be examined

(a) the second reactor is added in parallel to the first
(b) the second reactor is added in series to the first.

Solution

(a) **When added in parallel,** the second reactor will give the same conversion as the first and, when giving a conversion of 90%, each will process the same amount of feed as the original single reactor.

The throughput for this (or any other) conversion is therefore doubled.

(b) **The two reactors in series.** The solution to this problem is based on the equations developed in Example 4.8. From equation (iii) in that example, the volumetric flow rate (v_0) through a single CSTR of volume $V(1)$ for a conversion of 90% and kinetics given by equation (4.39) is

$$v_0 = k_2 C_{A0} V(1)/90 \qquad \text{(i)}$$

where C_{A0} is the concentration of reactant A in the feed.

The space time τ' for each of two reactors in series giving an overall conversion of 90% and reaction kinetics as above, is given by equation (ix) in Example 4.8 to be

$$\tau' = 13.7/k_2 C_{A0}$$

But, from the definition of space time (equation (4.24b), $v =$ (reactor volume) / τ'.

In the present problem the volume of each reactor is equal to that of the original single one, i.e. (reactor volume) $= V(1)$. It follows, therefore, that the volumetric flow rate v through the two tanks in series is given by

$$v = k_2 C_{A0} V(1)/13.7 \qquad \text{(ii)}$$

The enhancement in flow rate is (from (i) and (ii)),

$$v/v_0 = 90/13.7 = 6.6 \text{ times}$$

This is clearly a much better option than placing the second reactor in parallel.

Although most of the above examples are specific to a single form of reaction kinetics (equation (4.38)) and a single value (90%) for the conversion and apply only to systems in which the fluid density does not change during the reaction, they do illustrate some important principles. These should be borne in mind, particularly when designing reactor systems containing more than one reactor.

4.7 Thermal effects in 'ideal' chemical reactors

4.7.1 Introduction

During the course of a reaction heat will normally either be evolved or absorbed and the design of the reactor will be strongly influenced by the magnitude and sign of this heat of reaction.

The reactor may be operated in one of the following three ways:

1. Closely isothermal operation, which implies the provision of full heat exchange facilities and may be impractical for some very powerfully exothermic or endothermic reactions.
2. Adiabatic operation, in which there is no deliberate heat exchange with the surroundings. This type of reactor is cheap to build but suffers from two disadvantages: (a) the equilibrium conversion deteriorates as the reaction proceeds; (b) if the reaction is endothermic, the reaction rate may fall to a very low value.
3. Non-isothermal operation with some heat exchange, in which there is sufficient heat exchange of the reactor contents with the surroundings to maintain the reactor temperature between reasonable operating limits. These are computed to optimise construction and operating costs.

All three of the above modes of operation will be considered in the illustrations below, which will be for ideal batch, plug flow and continuous stirred tank reactors.

In addition to the material balance equations for these reactor types, the energy and enthalpy balance equations developed in Chapter 2 will also be required.

Example 4.11 *Isomerisation in a batch reactor*
The reaction $A \rightarrow B$ is an irreversible first-order isomerisation which has already been considered briefly in Example 2.8. Both A and B are liquids of very low volatility with a relative molecular mass (or molecular weight) of 250. Both liquids have densities (ρ) of 900 kg m^{-3} and heat capacities of 525 J mol^{-1}K^{-1}. The reaction is to be carried out in the liquid phase in a **batch reactor** and a conversion (X_A) of 0.97 is specified. Calculate the reactor volume required for an annual production of 10^6 kg of B:

(a) if the reactor is operated isothermally at 436 K
(b) if it is operated adiabatically starting at 436 K.

(The plant is operational for 7000 h in the year). In case (a) calculate the maximum heat duty for the batch heat exchanger. For each batch 10 minutes are required to fill the reactor, 12 minutes are required to bring it to temperature and 14 minutes are required to empty the reactor at the end of the reaction period (cf. Example 4.1).

The rate of consumption of reactant A $(-r_A)$ per unit volume of reactant/product mixture is given by the first-order kinetic equation

$$(-r_A) = kC_A \qquad (i)$$

where C_A is the concentration of A and

$$k = A \exp(-E/RT) \qquad (ii)$$

The activation energy E is 121 kJ mol^{-1} and the heat of reaction per mole of A reacting $(\Delta H_{R,A})$ is -87.15 kJ mol^{-1}.

At 436 K, $k = k_{436} = 0.80$ h^{-1}.

Solution

The reaction involves liquids of equal density and it will be assumed that any volume changes during the reaction are of negligible magnitude. In consequence,

$$C_A = C_{A0}(1 - X_A) \qquad (iii)$$

(a) Isothermal operation

Proceeding as in Example 4.1, determine first the required reaction time per batch (t_r) from equation (4.4):

$$t_r = C_{A0} \int_{X_A=0}^{0.97} (1/(-r_A)) \mathrm{d}X_A$$

where C_{A0} is the initial concentration of reactant A. But

$$(-r_A) = kC_A = kC_{A0}(1 - X_A)$$

Therefore,

$$t_r = (1/k) \times \int_{X_A=0}^{0.97} (1 - X_A)^{-1} \mathrm{d}X_A$$
$$= (-\ln(1 - 0.97))(1/k)$$
$$= 4.38 \text{ h}$$

Total time per batch $= 4.38 + (10 + 12 + 14)/60 = 5$ h

No. of batches in 7000 h $= 7000/5 = 1400$

Therefore, mass of A which must be processed per batch is:

$$(1 \times 10^6)/(1400 \times 0.97) = 736 \text{ kg}$$

Volume of reaction mixture which must be contained in the reactor is:

$$736/900 = 0.82 \text{ m}^3 = V_r$$

The required vessel volume would exceed this by about 20%, giving a shell volume of about 1 m^3.

Heat duty. The maximum heat load (Q_{max}) occurs when the reaction rate is fastest, i.e. initially. Therefore,

$$Q_{max} = \Delta H_{R,A}(\text{initial}) \times k C_{A0} V_r$$

where $\Delta H_{R,A}(\text{initial}) = $ heat of reaction per mole of A at 163°C $= -87.15 \text{ kJ mol}^{-1}$

$$C_{A0} = 736/0.82 = 898 \text{ kg m}^{-3}$$
$$= (898/250) \text{ kmol m}^{-3}$$
$$= 3.6 \text{ kmol m}^{-3}$$

Since $k = 0.80 \text{ h}^{-1}$ and $V_r = 0.82 \text{ m}^3$,

$$Q_{max} = -87.15 \times 0.80 \times 3.6 \times 10^3 \times 0.82$$
$$= -206 \text{ MJ h}^{-1}$$

So heat must be removed from the reactor shell at an initial rate of 206 MJ h^{-1} if isothermal conditions are to be maintained.

(b) Adiabatic operation
(1) We first find the reactor temperature T as a function of the conversion (X_A) of A. From equation (2.42), on setting $X_A = X_{AF}$,

$$\int_{T0}^{T} C_p(X_A, T) dT = -n_{A0} \Delta H_{R,A}(T_0) X_A \tag{iv}$$

where $C_p(X_A, T)$ is the total heat capacity of the fluid in an adiabatic reactor when conversion of reactant A is X_A and the temperature is T. $\Delta H_{R,A}(T_0)$ is the enthalpy of reaction (or 'heat of reaction') at the initial temperature T_0 defined as in equation (2.36). n_{A0} is the initial number of moles of A in the (batch) reactor. In the present illustration the only components present in the reactor are A and B and, applying the additivity rule (Section 2.3.3.1),

$$C_p = n_A(X_A)C_{pmA} + n_B(X_A)C_{pmB} \qquad \text{(v)}$$

where $n_A(X_A)$ and $n_B(X_A)$ are the numbers of moles of components A and B present in the reactor when conversion of A is X_A. C_{pmA} and C_{pmB} are the molar heat capacities of A and B. Combining (iv) with (v) gives

$$-\Delta H_{R,A}(T_0) \, n_{A0}X_A = n_A(X_A) \int_{T_0}^{T} C_{pmA}dT + n_B(X_A) \int_{T_0}^{T} C_{pmB}dT$$

where $T_0 = 436\,\text{K}$.

Inserting average values of the heat capacities ($C_{pmA} = C_{pmB} = C_{pm} = 523\,\text{J mol}^{-1}\text{K}^{-1}$), and writing

$$n_A(X_A) = n_{A0}(1 - X_A) \quad \text{and} \quad n_B(X_A) = n_{A0}\,X_A$$

gives

$$-\Delta H_{R,A}X_A = C_{pm}(T - T_0)$$

Inserting $T_0 = 436\,\text{K}$, $\Delta H_{R,A}(T_0) = -87.15\,\text{kJ mol}^{-1}$ and $C_{pm} = 0.525\,\text{kJ mol}^{-1}\text{K}^{-1}$ gives

$$\begin{aligned} T &= 436 + (87.15/0.525)X_A \\ &= 436 + 166X_A \end{aligned} \qquad \text{(vi)}$$

(2) We now evaluate the reaction rate constant as a function of temperature. For this purpose we substitute

$T = 436\,\text{K}$, $k_{436} = 0.80\,\text{h}^{-1}$ and $E = 121 \times 10^3\,\text{J mol}^{-1}$ in equation (ii), giving

$$\begin{aligned} \ln A &= 121\,000/(8.314 \times 436) + \ln\,(0.8) \\ &= 33.380 - 0.223 = 33.157 \end{aligned}$$

and

$$k = 2.511 \times 10^{14} \, \exp(-121\,000/(8.314T)\,)\text{h}^{-1} \qquad \text{(vii)}$$

The rate of consumption of $A(-r_A)$ is given from (i) with (iii) as

$$(-r_A) = kC_{A0}(1 - X_A) \qquad \text{(viii)}$$

where k is given (from (vi) and (vii)) as

$$k = 2.511 \times 10^{14}\exp(-14\,550/(436 + 166X_A)) \qquad \text{(viiia)}$$

The above expression for $(-r_A)$ may now be substituted in the batch reactor design equation (equation (4.4)) to give the reaction time t_r required to obtain the conversion $X_A = 0.97$:

$$t_r = -C_{A0} \int_{X_A=0}^{X_A=0.97} (1/r_A)\, dX_A$$

$$= \int_{X_A=0}^{X_A=0.97} (1/(k(1-X_A)))\, dX_A \tag{ix}$$

where k is given by equation (viiia).

Because of the complexity of the functional dependence of k on X_A, equation (ix) is best solved by a graphical or numerical technique. In either case values of $k(1-X_A)$ are calculated for a set of convenient values of X_A over the range of conversions of interest. $(1/(k(1-X_A)))$ is plotted versus X_A in Figure 4.11 for values of X_A between zero and the required final conversion (0.97).

The required reaction time is given by the area between the curve and the X_A axis. The diagram shows a very sharp initial fall in the parameter $1/k(1-X_A)$ followed by a substantial range in X_A over which the value of this parameter remains 1/100 or less of the starting value. Clearly the evaluation of the required area requires care.

A numerical method, which would clearly be very suitable for computer application, is illustrated in Table 4.1. In this table $(1/[k(1-X_A)]$ (henceforth denoted Z) is shown at each of a set of twenty equal increments (ΔX_A) in the value of X_A ($\Delta X_A = 0.97/20 = 0.0485$). The rate of consumption of isomer A per unit volume is also shown. It is seen that, due to the rise in temperature accompanying the reaction, the reaction rate increases as the reaction proceeds until a conversion of A of about 0.87 is reached. Only then does the rate start to fall off.

The reaction time (t_T) may be obtained from the data shown in Table 4.1 using the trapezoidal rule or other more efficient relationships given by the calculus of differences. Using the trapezoidal rule with twenty increments,

$$t_T = \Delta X_A((Z_0/2) + Z_1 + Z_2 + \ldots + Z_{n-1} + Z_n/2)$$
$$= 0.0485[(1.25/2) + 0.710 + 0.420 + \ldots + 0.003 + (0.005/2)] \tag{x}$$
$$= 0.12\,\text{h}$$

The total time requirement for each batch is given by

$$t_T = 0.12 + [(10 + 12 + 14)/60] = 0.72\,\text{h}$$

Proceeding as in the previous example,

No. of batches/year: $7000/0.72 = 9722$
Mass of A which must be processed in each batch: $1 \times 10^6/9733 = 103\,\text{kg}$
Volume to be held in reactor: $103/900 = 0.11\,\text{m}^3$

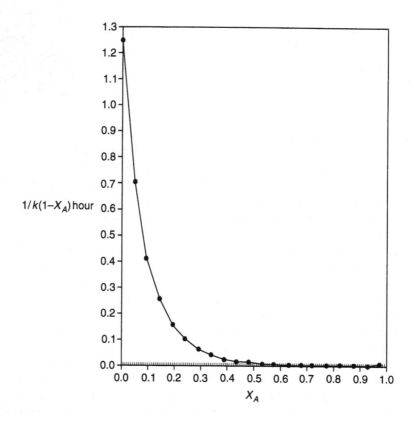

Figure 4.11 Group $(1/[k(1 - X_A)])$ as function of X_A area under curve gives hours required for adiabatic batch reactor to achieve conversion of 0.97 with data given in Example 4.11.

In practice, fewer, larger batches would be employed. The temperature rise of 161°C could be reduced by carrying out the reaction in a solvent.

Example 4.12   *Isomerisation in a CSTR*

The isomerisation reaction $A \rightarrow B$ specified in the previous example is to be carried out in a single continuous stirred tank reactor, the contents of which are assumed to be perfectly mixed.

The contents of the reactor are to be maintained at 436 K using cooling coils but the feed stream consisting of liquid A is to be at 293 K. As previously, the densities of A and B at 436 K are each 900 kg m^{-3} and their heat capacities and relative molecular masses (or molecular weights) are 0.525 kJ mol^{-1} K^{-1}) and 250 respectively. The reaction kinetics are again given by the equation (i)

$$(-r_A) = kC_A \tag{i}$$

in which the rate constant (k_{436}) at 436 K = 0.80 h^{-1}.

Table 4.1 Data used for numerical evaluation of $\int_0^{0.97} \frac{1}{k(1-X_A)} dX_A$ (see text)

n	X_A	T (K)	$k\,(h^{-1})$	$Z = [k(1-X_A)]^{-1}$ (h)	$-r_A$(kmol h^{-1} dm^{-3})
0	0.000	436	0.80	1.25	2.9
1	0.0485	444.1	1.48	0.710	5.1
2	0.0970	452.1	2.65	0.420	8.6
3	0.1455	460	4.60	0.254	14
4	0.1940	468	7.90	0.157	23
5	0.2425	476	13.32	0.099	36
6	0.2910	484	22.08	0.064	56
7	0.3400	492	36.00	0.042	86
8	0.3880	500	57.8	0.0283	127
9	0.4365	508	91.4	0.0194	185
10	0.4850	516	142	0.014	263
11	0.5335	525	231	0.009	388
12	0.5880	533	350	0.007	527
13	0.6805	541	524	0.005	697
14	0.6790	549	775	0.004	896
15	0.7275	557	1135	0.003	1113
16	0.7760	565	1640	0.003	1322
17	0.8245	573	2350	0.002	1484
18	0.8730	581	3340	0.002	1527
19	0.9215	589	4690	0.003	1325
20	0.9700	597	6550	0.005	707

(T is calculated from equation (vi) and k from equation (viiia) in Example 4.11).
Rate of consumption of isomer A per unit volume $(-r_A)$ is also shown.

An annual production rate of isomer B of 10^6 kg is required at a specified conversion of 0.97.

Calculate (a) the liquid hold-up in the tank (V_r) and (b) the heat load on the cooler if the plant is to be operational for 7000 h in the year.

Solution
(a) Inserting the kinetic equation (i) into the CSTR material balance equation (4.24a),

$$C_{A0} = C_{A1} + kC_{A1}\tau$$

where C_{A1} is the concentration of A in stream leaving the reactor, gives

$$\tau = (1/k)((C_{A0}/C_{A1}) - 1)$$
$$= (1/k)(X_A/(1 - X_A))$$

Since $k = 0.8\,h^{-1}$ and $X_A = 0.97$,

$$\tau = 40.4\ h$$

that is,

$$V_r = 40.4\nu_0 \tag{ii}$$

where

$$\nu_0 = \text{volumetric flow rate through reactor}$$

$$= \left(\frac{\text{mass flow rate in kg h}^{-1})}{900}\right) \text{m}^3 \text{ h}^{-1}$$

But

$$\text{mass flow rate} = \left(\frac{\text{mass of } A \text{ to be processed annually}}{7000}\right)$$

$$= 10^6/(0.97 \times 7000) \text{ kg h}^{-1}$$

$$= 147 \text{ kg h}^{-1}$$

Hence $\nu_0 = 0.164 \text{ m}^3 \text{ h}^{-1}$.

Substituting in (ii), the required liquid hold-up in the reactor (V_r) is 6.6 m³.

(b) Applying equation (2.53a) between a point in the cold feed stream at temperature $T_0 = 293$ K and a point in the product stream as it leaves the reactor at temperature $T = 436$ K:

$$Q = \Delta H_{R,A,293}(F_{A0}X_A) + (436 - 293)(F_A \bar{C}_{pmA} + F_B \bar{C}_{pmB}) \qquad \text{(iii)}$$

where F_A = molar flow rate of A in product stream = $F_{A0}(1 - X_A)$
    F_B = molar flow rate of B in product stream = $F_{A0}X_A$

F_{A0} = molar flow rate of A entering reactor

$$= \left(\frac{\text{mass flow rate (g h}^{-1})}{250}\right) = \left(\frac{147 \times 10^3}{250}\right) = 588 \text{ mol h}^{-1}$$

X_A = conversion of A in outlet stream = 0.97
$\bar{C}_{pmA} = \bar{C}_{pmB} = \bar{C}_{pm}$ = average molar heat capacity of both A and B over temperature considered = 523 J mol⁻¹K⁻¹
$\Delta H_{R,A,293}$ = reaction enthalpy per mole of A at 293 K.

Since the molar heat capacities of the reactant and product are in this case equal, $\Delta H_{R,A}$ should be independent of temperature (see equation (2.60a)) and $\Delta H_{R,A,293K} = \Delta H_{R,A,436K} = -87.15$ kJ mol⁻¹.
    Substituting the above values in equation (iii),

$$\dot{Q} = (-87.15 \times 1000 \times 588 \times 0.97) + (143 \times 523 \times 588) \text{ J h}^{-1}$$
$$= (-4.97 \times 10^7) + (4.40 \times 10^7) \text{ J h}^{-1}$$
$$= -5.7 \times 10^6 \text{ J h}^{-1}$$

The heat load on the cooler is thus about 6×10^6 J h^{-1}.

This load is substantially lower than that required to cool the isothermal batch reactor, due to the continuous supply of cold feed in the CSTR case.

Example 4.13 *Heat transfer requirements of a series of three CSTRs operated isothermally*

The isomerisation reaction $A \rightarrow B$ specified in Example 4.12 is in this case carried out in a series of three identical CSTRs rather than in a single one. The feed to the first reactor is at 293 K and the contents of the reactor tanks are maintained at 436 K by heat exchangers. A conversion of 0.97 is again specified and the throughput, the heat of reaction and reaction kinetics and also the heat capacities, densities and molecular weights of A and B are as given in the previous example.

> Calculate: (a) the total reactor volume
> (b) the heat transfer requirements for each reactor.

Solution
The terminology shown in Figure 4.7 will be used.

(a) Combining the material balance equation (4.24a) with the first-order kinetic expression (4.30) gives, for reactor 1

$$C_{A0} = C_{A1}(1 + k\tau) \qquad \text{(i)}$$

where τ (the space time) $= V_r/\nu_0$. V_r is the reactor volume (which is the same for all three reactors) and ν_0 is the volumetric flow rate. Since no volume change is anticipated in the reaction, ν_0 (and hence τ) will be the same for all these reactors. For reactors 2 and 3, therefore,

$$C_{A1} = C_{A2}(1 + k\tau) \qquad \text{(ii)}$$

$$C_{A2} = C_{A3}(1 + k\tau) \qquad \text{(iii)}$$

where C_{A1}, C_{A2} and C_{A3} are the concentrations of A in the streams leaving reactors 1, 2 and 3. From the above equations, we note that

$$C_{A1}/C_{A0} = C_{A2}/C_{A1} = C_{A2}/C_{A0} = 1/(1 + k\tau) \qquad \text{(iv)}$$

Hence

$$C_{A2}/C_{A0} = (C_{A2}/C_{A1})(C_{A1}/C_{A0}) = (1/(1 + k\tau))^2 \qquad \text{(v)}$$

$$C_{A3}/C_{A0} = (C_{A3}/C_{A2})(C_{A2}/C_{A1})(C_{A1}/C_{A0}) = (1/(1+k\tau))^3 \qquad \text{(vi)}$$

The desired overall conversion X_{overall} is given by

$$X_{\text{overall}} = 1 - (C_{A3}/C_{A0}) = 0.97$$

So

$$C_{A3}/C_{A0} = 0.03$$

Substituting these values in (vi) with $k = 0.8\,\text{h}^{-1}$, gives $\tau = 2.78\,\text{h}$
From the previous example $v_0 = 0.164\,\text{m}^3\,\text{h}^{-1}$. Hence $V_r = v_0\tau = 0.456\,\text{m}^3$
and
Total volume of reactor system $= 3 \times 0.456 = 1.37\text{m}^3$

(b) Before the heat load for each reactor is calculated, values are required
   for the conversions X_{A1}, X_{A2} and X_{A3} in the streams leaving reactors 1, 2
   and 3.

Substituting $1/(1 + k\tau) = 1/(1 + 0.8 \times 2.78) = 0.311$ in equations (iv), (v)
and (vi) gives

$$C_{A1}/C_{A0} = 0.310$$
$$C_{A2}/C_{A0} = (0.310)^2 = 0.0961$$
$$C_{A3}/C_{A0} = (0.310)^3 = 0.0298$$

Hence

$$X_{A1} = 1 - (C_{A1}/C_{A0}) = 0.690$$
$$X_{A2} = 1 - (C_{A2}/C_{A0}) = 0.904$$
$$X_{A3} = 1 - (C_{A3}/C_{A0}) = 0.970 \text{ as given}$$

The heat duty Q_1 for the first tank is obtained by applying equation (2.53a)
between a point in the cold feed stream ($T_0 = 293\,\text{K}$) to reactor 1 and a point
in the product stream as it leaves the reactor at $T_1 = 436\,\text{K}$. The conversions
of A in the feed and product streams for reactor 1 are $X_{A0} = 0$ and
$X_{A1} = 0.69$. In this notation, equation (2.53a) gives the required rate of
supply of heat to reactor 1 (Q_1) to be

$$\begin{aligned} Q_1 =& (\Delta H_{R,A,293})(F_{A0}(X_{A1})) \\ & + (T_1 - T_0)[F_{A1}C_{pmA} + F_{B1}C_{pmB}] \end{aligned} \qquad \text{(vii)}$$

F_{A1} and F_{B1} are the molar flow rates of components A and B in the stream
leaving reactor 1.

$$F_{A1} = F_{A0}(1 - 0.69)$$

$$F_{B1} = F_{A0} \times 0.69$$

F_{A0} = molar flow rate of A entering reactor

        = 588 mol h^{-1}, as in previous example

$\Delta H_{R,A,293}$, C_{pmA} and C_{pmB} are also as in the previous example.
   Substituting the above values in equation (vii)

$$Q_1 = (-87.15 \times 1000 \times 588 \times 0.69) + (143 \times 523 \times 588)$$
$$= (-3.54 \times 10^7) + (4.40 \times 10^7)$$
$$= 8.6 \times 10^6 \, \text{J h}^{-1}$$

Heat must therefore be **supplied** to the first tank to maintain isothermal
conditions.
   The heat duty for the second tank is obtained by applying equation (2.53b)
between a point in the feed to this reactor (conversion $X_{A1} = 0.690$ and
temperature 436 K) and a point in the outlet from it (conversion 0.904 and
temperature 436 K).
   In this case there is no sensible heat term and

$$Q_2 = \Delta H_{R,A} F_{A0}(X_{A2} - X_{A1})$$
$$= -87.15 \times 1000 \times 5.88 \times (0.904 - 0.690)$$
$$= -1.10 \times 10^7 \, \text{J h}^{-1}$$

In this case heat must be removed to maintain isothermal conditions. This
can be used to supply heat to the first reactor.
   The heat duty for the third tank is again obtained from equation (2.53b). In
this case it is applied between the feed to reactor 3 (conversion $X_{A2} = 0.904$
and temperature 436 K) and the product stream (conversion $X_{A3} = 0.970$
and temperature 436 K). Again there is no sensible heat term and

$$Q_3 = \Delta H_{R,A} F_{A0}(X_{A3} - X_{A2})$$
$$= -87.15 \times 1000 \times 5.88 \times (0.970 - 0.904)$$
$$= -0.34 \times 10^7 \, \text{J h}^{-1}$$

Thus tanks 2 and 3 together supply 1.44×10^7 J h^{-1} while tank 1 requires
0.86×10^7 J h^{-1} to maintain isothermal conditions.

Example 4.14   *Performance of two continuous stirred tank reactors operated
adiabatically in series*
The reaction considered is again the isomerisation reaction $A \rightarrow B$ with
kinetics given by equations (i) and (v) in Example 4.11. The heat of reaction
and the pure component physical data are as specified in Example 4.11.

Calculate the mass of isomer B which can be produced annually from 1.03×10^6 kg of A using two CSTRs operating adiabatically and in series (the plant is operational for 7000 hours annually). Each reactor has a capacity of 4.25 m³.

Solution

From the above requirements we can readily calculate the feed rate and the reactor space time (τ)

Molar feed rate $= (1.03 \times 10^9)/(7000 \times 250) = 589 \, \text{mol h}^{-1}$
Volumetric feed rate $= (1.03 \times 10^6)/(7000 \times 900) = 0.163 \, \text{m}^3 \, \text{h}^{-1}$
Space time (τ) for each reactor $= (4.25)/(0.163) = 26.1 \, \text{h}$

Overall approach. First obtain the conversion of $A(X_{A1})$ in the outlet stream from the first reactor and the temperature T_1 in this reactor. This involves the simultaneous solution of the mass balance and material balance equations.

The temperature and conversion in the stream leaving reactor 1 are determined first.

Tank 1, material balance. Density changes arising from temperature and other effects are neglected in the present example and the constant density form of material balance equation (equation (4.24a)) will again be used. Combined with the kinetic equation $(-r_A) = kC_A$, this gives, for reactor 1,

$$C_{A0} = C_{A1} + kC_{A1}\tau = C_{A1}(1 + k\tau) \, \text{mol m}^{-3} \quad \text{(i)}$$

and

$$X_{A1} = 1 - (C_{A1}/C_{A0}) = 26.1k/(1 + 26.1k) \quad \text{(ii)}$$

As in Example 4.11, k is given by

$$k = 2.51 \times 10^{14} \exp(-121\,000/(8.314T)) \, \text{h}^{-1} \quad \text{(iii)}$$

C_{A0} and C_{A1} are the concentrations of A in the streams entering and leaving the reactor.

Energy balance. Applying equation (2.53a) between a point in the feed stream to reactor 1 (zero conversion and temperature $T_0 = 293$ K) and a point in the outlet stream from this reactor (conversion X_{A1} and temperature T to be evaluated) gives, for adiabatic operation,

$$F_{A0}X_{A1}(\Delta H_{R,A,293}) + (T - T_0)(F_{A1}\bar{C}_{pmA} + F_{B1}\bar{C}_{pmB}) = 0 \quad \text{(iv)}$$

$\bar{C}_{pmA}$ and $\bar{C}_{pmB}$ are average molar heat capacities over the range of interest and are taken to be as in Example 4.12. F_{A1} and F_{B1} are the molar flow rates of components A and B in the stream leaving the reactor. Substituting $F_{A1} = F_{A0}(1 - X_{A1})$ and $F_{B1} = F_{A0}X_{A1}$ and rearranging,

$$X_{A1} = \bar{C}_{pm}(T - T_0)/ - \Delta H_{R,A,293} \qquad \text{(v)}$$

where $\bar{C}_{pm} = \bar{C}_{pmA} = \bar{C}_{pmB}$ as in Example 4.12, i.e.

$$X_{A1} = 523(T - 293)/(87\,150) = 0.0060(T - 293) \qquad \text{(vi)}$$

X_{A1} given by equation (vi) is now equated with X_{A1} given by equations (ii) and (iii):

$$0.006(T - 293)(1 + 26.1k) = 26.1k \qquad \text{(vii)}$$

with k given by equation (iii)

Solving (vii) by trial and error gives

$$T = 409\,\text{K and}\quad X_{A1} = 0.0060(409 - 293) = 0.69$$

.

The **second reactor** is analysed in the same way. The **material balance** equation for this reactor gives

$$C_{A1} = C_{A2}(1 + k\tau) \qquad \text{(viii)}$$

where C_{A2} = concentration of A in stream leaving the reactor. On substituting

$$C_{A2} = C_{A0}(1 - X_{A2}), \quad C_{A1} = C_{A0}(1 - X_{A1})$$

and rearranging,

$$X_{A2} = 1 - [0.31/(1 + 26.1k)] \qquad \text{(ix)}$$

with k given as a function of temperature by equation (iii).

The energy balance is obtained by applying equation (2.53b) between the inlet stream to the reactor and the outlet stream from it. The result, for adiabatic operation ($\dot{Q} = 0$) is

$$F_{A0}(X_{A2} - X_{A1})\Delta H_{R,A} + (T_2 - T_1)(F_{A2}C_{pmA} + F_{B2}C_{pmB}) = 0$$

where T_2 and X_{A2} are the absolute temperature and conversion of the stream leaving the reactor. F_{A2} and F_{B2} are the molar flow rates of A and B.

Rearranging and substituting physical data as for tank 1,

$$\begin{aligned} X_{A2} - 0.69 &= \bar{C}_p(T_2 - 409)/ - \Delta H_{R,A} \\ &= [523/(87\,150)](T_2 - 409) \end{aligned}$$

i.e.

$$X_{A2} = 0.69 + 0.0060(T_2 - 409) \qquad \text{(x)}$$

Solving equations (ix) and (x) by trial and error (with k given by equation (iii)),

$$T_2 = 460\,\text{K}$$

From equation (x),

$$X_{A2} = 0.69 + 0.0060(40 - 409)$$
$$= 0.996$$

The mass of isomer B which can in theory be produced annually from 1.03×10^6 kg of A is thus $(0.996 \times 1.030) \times 10^6$, i.e. 1.026×10^6 kg. Because of probable small deviations from ideal reactor behaviour, side-reactions and other losses it is unlikely that this degree of conversion will actually be achieved, though it is clear that production rates close to 10^6 kg yr^{-1} of B are a good possibility.

Example 4.15   *Diels–Alder reaction of butadiene with ethylene in ideal homogeneous plug flow reactor*
1,3-Butadiene (A) is known to react with ethylene (B) in the gas phase at temperatures above 400°C via a Diels–Alder reaction to give cyclohexene (C).

$$CH_2 = CH - CH = CH_2 + C_2H_4 \rightarrow C_6H_{10}$$

If an equimolar mixture of 1,3-butadiene and ethylene at 450°C and 1 atm is fed to a plug flow reactor, calculate the space time (τ) required to convert 10% of the butadiene to cyclohexene for (a) isothermal and (b) adiabatic operation.

Data The reaction is second order, the second-order rate constant k_2 being given by

$$k_2 = 10^{7.5}\exp(-115\,500/RT)\,\text{dm}^3\,\text{mol}^{-1}\text{s}^{-1} \tag{i}$$

The heat of reaction per mole of A reacting at 450°C is given by $\Delta H_{R,A,723} = -150\,000\,\text{J}\,\text{mol}^{-1}$.
The reverse reaction may be neglected.
The following average heat capacity values may be assumed over the temperature range of interest:

$$C_{pmA} = 154.6\,\text{J}\,\text{mol}^{-1}\text{K}^{-1}$$
$$C_{pmB} = 84.8\,\text{J}\,\text{mol}^{-1}\text{K}^{-1}$$
$$C_{pmC} = 249.9\,\text{J}\,\text{mol}^{-1}\text{K}^{-1}$$

Solution
The material balance equation. In parts (a) and (b) the material balance equation

$$V_r/v_0 = \tau = C_{A0}\int_0^{X_{AF}} dX_A/(-r_A) \tag{4.10}$$

is used in which $(-r_A) = kC_A C_B$ with k given by equation (i) in the data listing. X_{AF} is the conversion of A in the stream leaving the reactor.

Since $C_{A0} = C_{B0}$ (given) and 1 mole of butadiene (A) reacts with 1 mole of ethylene (B), the concentrations of ethylene and butadiene will remain equal as the reaction proceeds so that

$$(-r_A) = kC_A^2 \qquad \text{(ii)}$$

In order to carry out the integration in equation (4.10) it is necessary to express $(-r_A)$ as a function of the conversion X_A of A. The present example involves a gas phase reaction in which substantial volume changes occur as the reaction proceeds. These arise partly because the number of moles of product produced is only one-half of the number of moles of reactant consumed and (in part 2) also partly because of a change in temperature. Under these conditions the equation

$$C_A = C_{A0}(1 - X_A) \qquad \text{(iii)}$$

used in previous examples involving liquid phase reactions, is no longer applicable. The appropriate equation is now (3.68), which, for a constant pressure, becomes

$$C_A = C_{A0}\left[\left(\frac{1 - X_A}{1 + \varepsilon_A X_A}\right)\frac{T_0}{T}\right] \qquad \text{(iv)}$$

T_0 and T are the initial temperature and the temperature when the conversion is X_A. ε_A is given by equation (3.67) to be the increase in moles per mole of A reacting multiplied by the mole fraction of A in the feed. In the present example, 2 moles of gaseous feed give 1 mole of gaseous product for each mole of A which reacts and the mole fraction of butadiene (A) in the feed is 0.5. Hence

$$\varepsilon_A = -\frac{1}{2}$$

(a) Calculation of τ for isothermal operation
In this case $T_0 = T$ and equation (iv) becomes

$$C_A = C_{A0}[(1 - X_A)/(1 - (X_A/2))] \qquad \text{(v)}$$

Substituting (v) and (ii) in the material balance equation (4.10) with $X_A = 0.1$ gives

$$\tau k C_{A0} = \int_0^{0.1} [(1 - (X_A/2))^2/(1 - X_A)^2]dX_A$$

$$= -\int_1^{0.9} [(1/(2u)) + (1/(4u^2)) + (1/4)]du \quad \text{(where } u = (1 - X_A))$$

$$= 0.1055$$

From equation (i), k at $723.2 \, \text{K} = 0.1437 \, \text{dm}^3 \, \text{mol}^{-1} \, \text{s}^{-1}$.
From inlet data given,

$$C_{A0} = 0.5(P/RT) = 8.43 \times 10^{-3} \, \text{mol} \, \text{dm}^{-3}$$

Hence $\tau = 87 \, \text{s}$.

(b) Calculation of τ for adiabatic operation

In this situation it is necessary to solve the material balance and energy balance equations simultaneously.

According to the material balance equation (4.10), combined with the kinetic equation (ii),

$$\tau = C_{A0} \int_0^{X_{AF}} (1/(kC_A^2)) \, dX_A \qquad \text{(vi)}$$

where X_{AF} is the conversion in the stream leaving the reactor. Since temperature as well as the number of moles change as the reaction proceeds, C_A is now given by

$$C_A = C_{A0}[(1 - X_A)/(1 + \varepsilon X_A)]T_0/T \qquad \text{(vii)}$$

where, as in part (a), $\varepsilon = -\frac{1}{2}$.

Combining (vi) and (vii)

$$C_{A0}\tau = \int_0^{X_{AF}} [T(1 - (X_A/2))/(T_0 k^{1/2}(1 - X_A)]^2 dX_A \qquad \text{(viii)}$$

As in part (a), $C_{A0} = 8.43 \times 10^{-3} \, \text{mol} \, \text{dm}^{-3}$.

Before equation (viii) can be integrated an expression is required giving T and hence (from equation (i)) k as a function of X_A. This expression is obtained by applying the energy balance equation (equation (2.53)) between a point in the feed stream (temperature 723 K and zero conversion) and a point within the plug flow reactor where the conversion is X_A. For adiabatic operation $Q = 0$ and, in terms of average heat capacities over the temperature range of interest,

$$F_{A0}(\Delta H_{R,A,723})X_A + (T - T_0)[F_A C_{pmA} + F_B C_{pmB} + F_C C_{pmc}] \qquad \text{(ix)}$$

where F_A, F_B and F_C are the molar flow rates of components A, B and C at a position in the reactor where the conversion of A is X_A.

$$F_A = F_{A0}(1 - X_A)$$
$$F_B = F_{A0}(1 - X_A)$$
$$F_C = F_{A0}X_A$$

Making these substitutions and inserting the values given for $\Delta H_{R,A,723}$ and the average heat capacities,

$$T - T_0 = -(\Delta H_{R,A,723})X_A[C_{pmA} + C_{pmB} - (C_{pmA} + C_{pmB} - C_{pmC})X_A]^{-1}$$

and

$$T = 723 + (150\,000 X_A)/(239.4 + 10.5 X_A) \qquad \text{(x)}$$

Equation (viii) may now be integrated by a numerical method using a similar technique to that employed in Example 4.11. Following that calculation routine, T is first evaluated at a convenient set of values of X_A between zero and the desired final value and k is evaluated at these temperatures. Using the values of T and k at each selected value of X_A, the quantity

$$\left\{T[1 - (X_A/2)]/[Tk^{1/2}(1 - X_A)]\right\}^2 = Z \qquad \text{(xi)}$$

is then evaluated as outlined in Table 4.2. The integral in equation (viii) is then obtained numerically as in Example 4.11.

Application of Simpson's rule to the initial, final and midpoints in Table 4.2 gives a first approximation value for $C_{A0}\tau$ of 0.399... mol s dm^{-3}, corresponding to $\tau = 47$ s. Greater precision is obtained by using a closer grid of points, though in this case without substantially altering the value for τ from the initial approximate value. This type of calculation is well suited to 'spreadsheet' methods.

4.7.2 Non-isothermal tubular reactors with heat exchange: simultaneous application of differential heat and mass balances

Reactors of the above type will henceforth be described simply as 'non-isothermal'.

The characteristics of 'non-isothermal' tubular reactors fall between those of adiabatic reactors (in which there is no heat exchange between the reservoir fluid and external heat sources or sinks) and isothermal reactors in which any heat evolved or absorbed in the reaction is removed or supplied

Table 4.2 Illustrating the evaluation of $C_{A0}\tau$ from equation (viii) by a numerical routine

X_A	T (equation (x))	k (equation (i))	Z (equation (xi))
0	723	0.144	6.94
0.025	739	0.214	4.99
0.050	754	0.317	3.61
0.075	770	0.460	267
0.100	785	0.658	2.00

A first approximation value is given by Simpson's rule as
$$(0.05/3)(6.94 + 2.0 + 4 \times 3.61) = 0.390 \text{ mol s dm}^{-3}$$

to maintain a constant reaction temperature. In the non-isothermal reactor sufficient heat exchange occurs to limit the magnitude of temperature changes, rather than eliminate them altogether. As in the case of adiabatic reactors both material and energy balance equations are involved in the calculation of reactor size for a given conversion and throughout. However in this case, since neither the temperature change during reaction nor the heat exchange with the surroundings is zero, the integral energy balance equation (2.53) for 'non-isothermal' reactors in general contains three rather than two non-zero terms and one of these (the heat conduction term) involves reactor length and reactor diameter. In this situation the most straightforward approach is probably to express both the heat and the mass balance relationships in differential form leading to two differential equations which can be solved simultaneously by a stepwise integration technique, as in Example 4.16. The same technique can, of course, be applied to adiabatic reactor calculations, though in this intrinsically simpler case other methods may be more convenient.

Before turning to Example 4.16 we first discuss the differential mass and heat balance equations in the form in which they will be used. These equations are obtained by taking mass and energy balances about a length dl of reactor across which the conversion of limiting reactant A increases by dX_A and the temperature increases by dT.

4.7.2.1   The differential mass balance. Considering Figure 4.12, if the cross-sectional area of the reactor is S,

Rate of consumption of A in length dl of (of homogeneous) reactor $= Sdl(-r_A)$

$$(4.48)$$

where $(-r_A) =$ rate of consumption of A per unit volume of reaction mixture.

But

Rate of consumption of $A =$ (moles of A flowing into element dl in unit time) $-$ (moles of A flowing out of element in unit time) $= F_{A0}dX_A$ (4.49)

Equating (4.48) with (4.49) and rearranging,

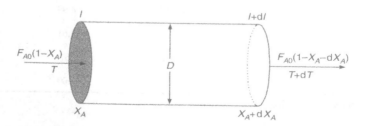

Figure 4.12 The differential mass balance ($F_{A0} =$ molar flow rate of A at reactor inlet, i.e. at $l = 0$).

$$\frac{dl}{dX_A} = \left(\frac{1}{S}\right)\frac{F_{A0}}{(-r_A)} = \frac{4F_{A0}}{\pi D^2(-r_A)} \tag{4.50}$$

where D is the diameter of the reactor tube.

This is the differential mass balance expressed in terms of reactor length and reactor diameter. When $(-r_A)$ has been expressed in terms of the rate constant and conversion, the differential mass balance provides, for an **isothermal reactor**, a useful link between conversion and distance from the feed point. In a non-isothermal reactor, the temperature is required at each point before $(-r_A)$ can be evaluated: only then can this link be established.

Equation (4.50) is, of course, directly equivalent to the alternative form of differential mass balance which is expressed in terms of an increment dV_r in the reactor volume (equation (4.7) in Section 4.1.2.2). This was used in previous examples and reduces directly to (4.50) on writing $dV_r = (\pi D^2/4)dl$. The mass balance in the form given by (4.50) is more convenient in the present context.

In the case of a heterogeneous reactor, identical reasoning leads to the equation

$$\frac{dz}{dX_A} = \frac{4}{\pi D^2}\left(\frac{F_{A0}}{-r'_A}\right) \tag{4.50a}$$

where z is now the depth in the catalyst bed at which the conversion of limiting reactant A is X_A. D is the bed diameter. $(-r'_A)$ is the rate of consumption of limiting reactant A per unit volume of catalyst bed. F_{A0} is the molar flow rate of A at the bed inlet.

4.7.2.2 The differential heat balance Consider an element of reactor of length dl at a distance l from the reactor inlet (Figure 4.13). Over the element, the conversion of limiting reactant A increases from X_A to

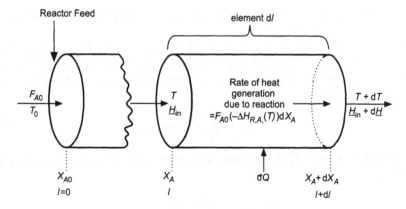

Figure 4.13 Differential heat balance on element dl of reactor. (The use of the symbol d' (rather than d) is a reminder that Q (unlike $\underline{H}$) is not a function of state.)

$X_A + dX_A$ and the stream temperature increases from T to $T + dT$. Let enthalpy entering the element by flow in unit time be $\underline{H}_{in}$ and enthalpy leaving by flow be $\underline{H}_{out}$, where $\underline{H}_{out} = \underline{H}_{in} + d\underline{H}$. Also let the rate at which heat enters the element by conduction through the reactor wall be $đQ$. From equation (2.34) applied over unit time,

$$d\underline{H} = đQ = \text{heat entering in unit time}$$

In the steady state, the enthalpies $\underline{H}_{out}$ and $\underline{H}_{in}$ are associated with equal masses of reactor fluid. As pointed out in Section 2.3.2, the enthalpy of a given mass of fluid undergoing a single reaction in a flow reactor at steady state is a function only of temperature and the conversion of a specified reactant A. $d\underline{H}$ is accordingly given by the two-term expression

$$d\underline{H} = \left(\frac{\partial \underline{H}}{\partial X_A}\right)_T dX_A + \left(\frac{\partial \underline{H}}{\partial T}\right)_{X_A} dT \tag{4.51}$$

where $(\partial \underline{H}/\partial T)_{X_A} = \underline{C}_p$ = total heat capacity of fluid passing through the infinitesimal element dl in unit time. Also, since $n_A = n_{A0}(1 - X_A)$,

$$dX_A = -\frac{dn_A}{n_{A0}}$$

where n_{A0} = moles of A entering reactor in unit time = F_{A0}
$(-dn_A)$ = moles of A consumed in element in unit time
Substituting the above results into (4.51) and setting $d\underline{H} = đQ$,

$$đQ = F_{A0}\left(\frac{-\partial \underline{H}}{\partial n_A}\right)_T dX_A + \underline{C}_p dT$$
$$= F_{A0}\Delta H_{R,A}(T)\, dX_A + \underline{C}_p dT \tag{4.52}$$

where $\Delta H_{R,A}(T)$ is the enthalpy of reaction per mole of A at temperature T as defined by equation (2.36).

A good estimate for $\underline{C}_p$ in equation (4.52) can usually be obtained from the additivity rule (Section 2.3.3.1). In terms of the molar flow rates (F_i) and molar heat capacities (C_{pmi}) of the components flowing through the reactor, this rule states that

$$\underline{C}_p = \sum_i F_i C_{pmi} \tag{4.53}$$

The molar flow rates F_i in the above equation vary with conversion of A as follows:

$$F_i = F_{i0} - \left(\frac{\nu_i}{\nu_A}\right) F_{A0} X_A \tag{4.54}$$

where F_{i0} and F_{A0} are the molar flow rates of components i and A at the entry to the reactor while ν_i and ν_A are the stoichiometric coefficients of i and A. (Equation (4.55) is obtained by applying (2.26a) over unit time interval). Inserting equation (4.53) into (4.52) gives

$$đQ = F_{A0}\Delta H_{R,A}(T)\,dX_A + \left(\sum_i F_i C_{pmi}\right)dT \qquad (4.55)$$

$\Delta H_{R,A}(T)$ is the heat of reaction per mole of A consumed at temperature T. If the combined heat capacity of the reactants is not equal to that of the products formed, $\Delta H_{R,A}(T)$ will be a function of temperature and will vary from point to point in the reactor. If the discrepancy in the heat capacities is substantial, this effect may be significant. It may be allowed for, using equation (2.60a) for the variation of heat of reaction with temperature giving

$$-\Delta H_{R,A}(T) = -\Delta H_{R,A}(T_0) + \sum_i \left(\frac{\nu_i}{\nu_A}\right)\int_{T_0}^{T} C_{pmi}\,dT \qquad (4.56)$$

where $\Delta H_{R,A}(T_0)$ is the heat of reaction at the inlet temperature T_0 to the reactor.

Equation (4.55), with F_i given by (4.54) and $\Delta H_{R,A}(T)$ by (4.56), is the differential heat balance equation. $đQ$ in this equation is the rate of heat input to the element of length dl. This term is zero for an adiabatic reactor.

In an isothermal reactor, dT is zero for all elements along the reactor. Equation (4.55), in conjunction with (4.50), then gives, on integration, the rate of heat input required to maintain isothermal conditions along the reactor length.

When designing a non-isothermal reactor where there is some heat exchange, it is necessary to develop an expression giving $đQ$ as a function of T and dl (and hence dX_A) for the reactor considered. If temperature control is achieved by heat transfer across the reactor wall to an outer jacket, an expression in terms of T and dl can be readily obtained provided the temperature of the circulating fluid in the jacket and the overall heat transfer coefficient over the reactor wall are known. $đQ$ can then be obtained as a function of T and dX_A using the differential mass balance (equation (4.50)). When this expression for $đQ$ has been inserted in the differential heat balance equation, the latter can be integrated to give reactor temperature as a function of conversion. The procedure is illustrated in Example 4.16.

Example 4.16 '*Non-isothermal' tubular reactor calculation*
In order to illustrate the principles involved we will suppose that the homogeneous liquid phase isomerisation reaction $A \rightarrow B$ is to be carried out in a jacketed tubular reactor (plug flow assumed) and that we wish to calculate the reactor volume (V_r), the reactor length (l), the maximum temperature in the reactor and the outlet temperature for an outlet conversion X_{AF} of 0.97.

An internal tube diameter (D) of 4 in (0.1016 m) has been proposed for the reactor.

Data. The circulating heat transfer fluid in the jacket is to be maintained at a temperature (T_m) close to 433 K and the heat transfer coefficient (U) between the jacket fluid and the reactor fluid is $8.36 \times 10^5\,\mathrm{J\,m^{-2}h^{-1}K^{-1}}$.

The reactor feed is at 436 K and consists of isomer A only.

The concentration of A at entry (C_{A0}) is 3600 mol m^{-3} and the molar and volumetric flow rates at entry $(F_{A0}$ and $v_0)$ are 590 mol h^{-1} and 0.164 m^3 h^{-1}.

As in Example 4.11, density changes during the reaction may be taken to be negligible and the rate of consumption of $A(-r_A)$ per unit volume of reactor fluid is given by

$$(-r_A) = kC_A = kC_{A0}(1 - X_A) \tag{i}$$

where

$$k = 2.511 \times 10^{14}\exp(-121\,000/8.314T)\,\mathrm{h^{-1}} \tag{ii}$$

and C_A = concentration of A when the conversion is X_A. The heat of reaction per mole of A reacting at 436 K is given by

$$(\Delta H_{R,A,436}) = -87.15\,\mathrm{kJ\,mol^{-1}}$$

The molar heat capacities of A and B are virtually identical, the average value over the range of interest being 523 J mol^{-1} in each case.

Calculation. The solution below is based on the differential heat balance (equation (4.52) or (4.55)) in conjunction with the differential mass balance (equation (4.50)). On examination of the data it is clear that several simplifications can be made.

1. Since one mole of isomer B is produced for each mole of isomer A consumed, the total molar flow rate is constant through the reactor, being equal to the molar flow of isomer A at entry (F_{A0}).

2. Since the heat capacities of the two isomers are virtually equal (a) the term $\sum F_i C_{pmi}$ is constant through the reactor, being equal to the molar flow of A at entry multiplied by its molar heat capacity, i.e. to $F_{A0}C_{pmA}$, (b) the heat of reaction will be independent of temperature and will, throughout the reactor, take the value -87.15 kJ mol^{-1}, which is the value listed for 436 K. (This follows from equation (2.60a)). Making the above substitutions in the differential mass and heat balance equations (equations (4.50) and (4.55) respectively), we obtain

$$\frac{\mathrm{d}l}{\mathrm{d}X_A} = \frac{4F_{A0}}{kC_{A0}(1 - X_A)\pi D^2} \quad \mathrm{m} \tag{iii}$$

and

$$\mathrm{d}Q = F_{A0}[\Delta H_{R,A}(T)\mathrm{d}X_A + C_{pmA}\mathrm{d}T] \quad \mathrm{J\,h^{-1}} \tag{iv}$$

Inserting numerical values in (iii),

$$\frac{\mathrm{d}l}{\mathrm{d}X_A} = \left(\frac{4 \times 590}{3600\pi \times (0.1016)^2}\right)\left(\frac{1}{k(1 - X_A)}\right) = \frac{20.21}{k(1 - X_A)} \quad \mathrm{m} \tag{v}$$

$\mathrm{d}Q$ in equation (iv) is given by the usual heat transfer equation as the product of the heat transfer coefficient U over the reactor wall with the temperature difference $(433-T)$ across the wall and the heat transfer area $\pi D\mathrm{d}l$, i.e.

$$\mathrm{d}Q = (433 - T)U\pi D\,\mathrm{d}l$$

$$(433 - T)U\pi D\left(\frac{\mathrm{d}l}{\mathrm{d}X_A}\right)\mathrm{d}X_A$$

Inserting $(\mathrm{d}e/\mathrm{d}X_A)$ from equation (4.50) with $-r_A$ given by (i) gives

$$\mathrm{d}Q = \frac{4F_{A0}(433 - T)U}{kC_{A0}(1 - X_A)D} \tag{vi}$$

Equating (iv) with (vi)

$$F_{A0}\left[C_{pmA}\mathrm{d}T + \left(\Delta H_{R,A}(T) - (T - 433)\frac{4U}{kC_{A0}(1 - X_A)D}\right)\mathrm{d}X_A\right] = 0 \tag{vii}$$

Inserting numerical values

$$523\mathrm{d}T - \left[87.15 \times 10^3 + (T - 433)\frac{17.481}{k(1 - X_A)}\right]\mathrm{d}X_A = 0 \tag{viii}$$

Rearranging,

$$\left(\frac{\mathrm{d}T}{\mathrm{d}X_A}\right) = \frac{17.481(433 - T)}{k(1 - X_A)} + 166.6\,\mathrm{K} \tag{ix}$$

where k is given as a function of temperature by equation (ii).

The problem can now be solved provided equations (v) and (ix) can be integrated. Equation (ix) (with (ii)) enables the gradient $(\mathrm{d}T/\mathrm{d}X_A)$ to be calculated at any point in the reactor tube at which temperature and conversion are known. Employing stepwise integration and starting with known conditions at the feed point enables temperature to be determined as a function of conversion over the entire range of conversions considered. In the stepwise integration procedure in its simplest form, the conversion range is split into small increments (ΔX), and $(\mathrm{d}T/\mathrm{d}X_A)$ at the beginning of the first increment (which we will designate [slope(0)]) is evaluated by inserting the feed conditions $X_A(0)$, $T(0)$ into (ix) and (ii). The values of X_A and T at the end of the first increment are then written as

$$X_A(1) = X_A(0) + \Delta X \quad \text{and} \quad T(1) = T(0) + [\text{slope}(0)]\Delta X$$

The values $T(1)$ and $X_A(1)$ can then be used to calculate (dT/dX_A) at the beginning of the second increment, leading to a value for the temperature at the end of this increment and so on over the entire range of conversion of interest. Having established the temperature (and hence the value of k) for each increment, dl/dX_A is given by equation (v), enabling the reactor length to be calculated step by step alongside the reactor temperature. It should be stressed that the above routine is a very simplistic form of stepwise integration and refinements to it would normally be used. The reader should

Table 4.3 Q Basic programme (based on Euler method of stepwise integration) for evaluating the temperature and conversion profiles of the reactor specified in Example 4.16

(The Euler method is overly simplistic and is used here primarily to demonstrate the principles involved though it also has the advantage of leading to a very brief computer program. It has the practical disadvantage of requiring a very large number of integration steps to achieve acceptable accuracy.)

```
DIM Xa (2000)
DIM T(2000), k(2000), SLOPE(2000), LE(2000), GRAD(2000)
Xa(0) = 0
T(0) = 436
k(0) = 2.511E + 14*EXP(−121000/8.314/T(0) )
LE(0) = 0
INPUT "input INCREMENT OF INTEGRATION, INC = ", INC
INPUT "input the final conversion, XAF =", XAF
INPUT Q
N = (XAF − Xa(0))/INC
PRINT "NUMBER OF INTERVALS =", N
PRINT "T = Temperature, k = reaction coefficient, Xa = conversion, L = Length"
FOR i = 1 TO N
Xa(i) = Xa(i − 1)+ INC
k(i) = 2.511E + 14*EXP(−121000/8.314/T(i − 1))
SLOPE(i) = (433 − T(i − 1))/(k(i)*(1 − Xa(i − 1)))*17.481 + 166.63
T(i) = T(i − 1) + (INC*SLOPE(i))
GRAD(i) = 20.21/(k(i)*(1 − Xa(i − 1)))
LE(i) = LE(i − 1) + INC*GRAD(i)
LPRINT ""
LPRINT " T = Temperature, k = reaction coefficient, Xa = conversion, L = Length"
LPRINT " T(K)            k              Xa              L(m)         "
FOR i = N − Q TO N
LPRINT, T(i), k(i), Xa(i), LE(i)
NEXT i
STOP
END
```

Notes:
Q is an integer which should be less than N. (If a complete output is required Q equals N−1.)
'k' is the reaction rate constant given by equation (ii)
'SLOPE' is the gradient dT/dX_A given by equation (ix)
'GRAD' is the gradient dl/dX_A given by equation (v)
'LE' is the reactor length (l in text)
The remaining symbols are self-explanatory.

consult standard mathematical texts (e.g. references [1, 2]) for details of these with special reference to the modified Euler method and the Runge–Kutta method.

The simple form, however, illustrates the principles well and is used in the 'basic' program listed in Table 4.3. The temperature and conversion profiles generated by this program are shown in Figure 4.14. It is seen that temperature rises at first as the reactor is traversed since the heat generated by the

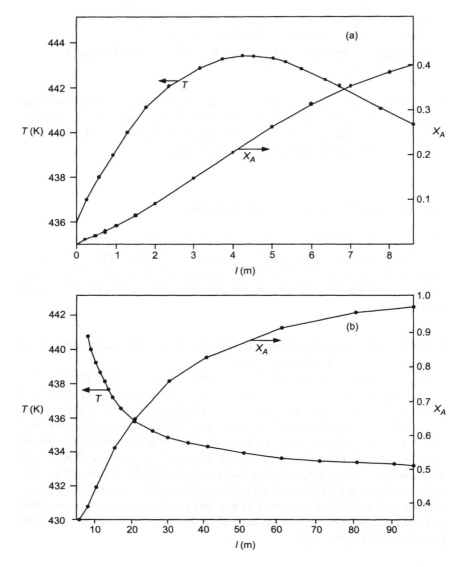

Figure 4.14 Temperature and conversion profiles for reactor detailed in Example 4.16: (a) at inlet end of reactor; (b) at outlet end.

reaction in each element of the reactor then exceeds the heat removed by the cooling jacket. The temperature reaches a maximum at 4.3 m from the inlet. At higher values of l, the reaction rate falls off due to progressive reduction in the reactant concentration. The heat generation by the reaction likewise falls off and the temperature falls.

The output shown in Figure 4.14 is from an 800-step sequence. According to this output

Reactor length $= 93$ m
Reactor volume $= 86 \times \pi \times 0.1016^2/4 = 0.76\,\text{m}^3$
Maximum temperature $= 443.3$ K at $X_A = 0.22$, 4.3 m from inlet
Outlet temperature $= 433.2$ K

The modified Euler method, with the same number of steps, gives the same values for the temperature and position of the hottest point in the reactor, and also for the reactor length. It is probable that preliminary calculations of this type would lead to a redesign of the reactor.

4.8   Heterogeneous tubular reactor calculations

The design equations for heterogeneous tubular reactors are very similar to those for homogeneous ones, with catalyst bed depth z replacing reactor length l and the rate of consumption of reactant per unit volume of bed $(-\underline{r}'_A)$ replacing the rate of consumption per unit volume of reactor mixture $(-r_A)$.

Example 4.17   *A catalysed reaction which is first order with respect to the fluid phase ('pseudo-homogeneous kinetics')*
A process for the catalytic dehydrogenation of benzene to styrene is to be carried out isothermally in a fixed bed of catalyst in the presence of a substantial excess of steam. The suggested mole fraction of ethyl benzene at the inlet to the reactor is 0.11 and the bed diameter is to be 2 m. The total molar flow rate (ethyl benzene + steam) at the inlet is 5000 kmol h^{-1}. At the temperature and pressure at which the reactor is to operate (650°C and approximately 1 bar respectively) the reaction may be taken to be first order with respect to the fluid phase and virtually irreversible. For the catalyst bed which it is proposed to use, the reaction rate $(-\underline{r}'_A)$, expressed in kmol ethyl benzene converted per cubic metre of bed per hour, is, under these conditions, given by

$$(-\underline{r}'_A) = 178 y_A \qquad\qquad (i)$$

where y_A is the mole fraction of ethyl benzene at a point in the reactor where the conversion of ethyl benzene is X_A. If y_{A0}, the mole fraction of ethyl benzene at the inlet to the reactor is 0.11,

(a) Calculate the bed depth required to achieve 80% conversion of the ethyl benzene under the proposed conditions.
(b) State the bed depth which would be required if the size of the catalyst pellets was changed producing a 10% reduction in the value of $(-r'_A)$ for a given value of X_A.
(c) State the volume of catalyst bed required.
(d) What bed depth would be required if the tube diameter was changed to 3 metres?

Solution

Introduction. The dehydrogenation reaction may be written

$$C_8H_{10} \rightarrow C_8H_8 + H_2$$

In this reaction there is an increase of one in the number of moles for every mole of ethyl benzene which reacts.

Accordingly Δn in equation (3.67) $= 1$ and $\varepsilon_A = y_{A0}\Delta n = 0.11$ (conditions are such that the ideal gas behaviour implied in equation (3.67) can be assumed to apply).

Since the reactor is isothermal and at an approximately constant pressure, equation (3.69a) applies and the mole fraction of ethyl benzene (y_A) is given as

$$y_A = y_{A0}(1 - X_A)/(1 + 0.11X_A)$$

and, from equation (i)

$$\begin{aligned} (-r'_A) &= (178 \times 0.11)(1 - X_A)/(1 + 0.11X_A) \\ &= 19.6 \times (1 - X_A)/(1 + 0.11X_A) \end{aligned} \tag{ii}$$

where X_A is the conversion of the ethyl benzene.

Part (a). From the mass balance equation (4.50a),

$$dz/dX_A = (4/\pi D^2)F_{A0}/(-r'_A) \tag{iii}$$

where D is the reactor diameter and F_{A0} is the molar flow rate of reactant A at the feed point.

The required bed depth, z_T, is obtained by combining (iii) with (ii) and integrating to give,

$$z_T = \frac{4F_{A0}}{19.6\pi D^2} \int_0^{0.8} \left(\frac{1 + 0.11X_A}{1 - X_A}\right) dX_A \tag{iv}$$

Rearranging and inserting numerical values,

$$z_T = \int_{X_A=0}^{0.8} \frac{8.93}{9.09} \left(\frac{9.09 + X_A}{1 - X_A}\right) dX_A \, \text{m} \qquad (v)$$

An analytic solution to (v) is readily obtained on setting $(1 - X_A) = u$

$$z_T = \frac{1}{9.09} \int_{u=1}^{u=0.2} \left(1 - \frac{10.09}{u}\right) du$$

$$= \frac{1}{9.09} [(0.2 - 1.0) - 10.09 \ln(0.2)] = 15.2 \, \text{m}$$

Part (b). Inspection of equations (ii) and (iii) shows that at each value of X_A over the integration range, lowering $(-\underline{r}'_A)$ by 10% will raise dz/dX_A by the same percentage. The integral $\int \frac{dz}{dX_A} dX_A$ will likewise be raised by 10%, giving $z_T = 16.7$ m.

Part (c). The reactor volume is $(\pi D^2/4) z_T = 52.4 \, \text{m}^3$. As may be seen from (iv), the product $z_T D^2$ is constant for a given throughput and conversion in the isothermal situation considered here (this is also true for adiabatic beds but not for non-isothermal ones in which some heat exchange with the surroundings takes place).

Part (d). The bed depth required for a diameter of 3 m would be $52.4 \times (2/3)^2 = 23.3$ m.

(The above calculations could equally well have been based throughout on the differential mass balance in the form (4.11), in which case the bed volume would have been obtained directly.)

Example 4.18   *A catalysed reaction under surface film control*
The rate of an irreversible catalysed reaction in a small ideal tubular reactor is known to be limited by the rate of diffusion of a reactant A across the boundary film surrounding each catalyst pellet (see Section 3.6.5). Using a correlation it has been estimated that the film mass transfer coefficient for this component is $0.2 \, \text{m s}^{-1}$ under the operating conditions. The pellet diameter is 6×10^{-3} m, the bed voidage is 0.5 and component A enters the bed at 300 mole per second at a concentration of 5 mol m^{-3}. The reaction takes place at very high and approximately constant temperature (ideal gas conditions prevail) and the volume remains constant during the reaction.

(a) Estimate the catalyst volume required to achieve 70% conversion of A (A is the limiting reactant).
(b) If a bed diameter of 0.6 m has been selected, what bed depth is required?
(c) What factors govern the selection of bed diameter?

Solution

Introduction. Applying the mass balance equation in the form (4.11) (Section 4.1.2.3)

$$\frac{dV_b}{dX_A} = \frac{F_{A0}}{(-r'_A)} \qquad (i)$$

where dV_b is the increment in catalyst bed volume corresponding to an increment dX_A in the conversion of reactant A and F_{A0} is the molar flow rate of A at the inlet. $(-r'_A)$ is the rate of consumption of reactant A per unit bed volume. This is related to the rate of consumption of A per unit volume of catalyst pellet $-r'_{AV}$ by equation (4.17)

$$(-r'_A) = (1 - \varepsilon_b)(-r_{AV}) \qquad (ii)$$

where ε_b is the bed voidage.

When the reaction rate is limited only by the rate at which A can diffuse over the surface film surrounding each pellet, equation (3.157) (Section 3.6.5) applies and

$$(-r_{AV}) = (3/R)k_{CA}C_{AB} \qquad (iii)$$

where R is the effective radius of the catalyst pellets and k_{CA} is the mass transfer coefficient for component A across the surface film. C_{AB} is the concentration of A in the bulk fluid phase.

Since volume changes do not occur during the reaction, the concentration of A in the bulk phase is related to the conversion X_A of this component by equation (3.61)

$$C_{AB} = C_{A0}(1 - X_A) \qquad (iv)$$

Combining the above equations

$$\frac{dV_b}{dX_A} = \frac{RF_{A0}}{3(1 - \varepsilon_b)k_{CA}C_{A0}(1 - X_A)} \qquad (v)$$

Part (a).   The catalyst volume required to achieve 70% conversion of A is obtained by integrating (v) from $X_A = 0$ to $X_A = 0.7$. Inserting numerical values from the data given and arranging in integral form, equation (v) gives

$$V_b = \int_0^{0.7} \left(\frac{3 \times 10^{-3} \times 300}{3 \times 0.5 \times 0.2 \times 5(1 - X_A)} \right) dX_A$$

$$= \int_0^{0.7} \frac{dX_A}{1.67(1 - X_A)}$$

$$= -\ln(0.3)/1.67 = 0.72 \text{ m}^3$$

$$= \text{required bed volume}$$

Part (b).   If a bed diameter of 0.6 m had been selected, the corresponding bed depth would have been

$$0.72/(\text{bed cross-section area}) = (4 \times 0.72)/\pi \times 0.6^2 = 2.5 \, \text{m}$$

Part (c).   In simple isothermal or adiabatic catalyst bed calculations, the catalyst bed volume V_b is determined solely by kinetic data and the specified throughput and conversion. Having calculated V_b, any combination of values of bed depth z_T and bed diameter D which satisfies the expression

$$V_b = z_T \pi D^2/4$$

is then theoretically possible. Possibilities range from tall slender beds to short ones of large diameter. The selection of the best value for D out of the range which is technically feasible, depends on economic conditions. The tall slender bed will incur a higher pressure drop (and hence higher annual compression energy costs) than the short bed while the short large-diameter bed will be more expensive than the tall slim one and will therefore incur a higher annual amortisation charge. Other costs (such as labour cost involved in catalyst replacement) may also be involved and may differ in the two cases.

(When estimating the pressure drop, the Ergun equation described in Section 4.9 should be helpful. Vessel cost (and hence amortisation charges) for pressures up to about 70 bar may be estimated from correlations in reference [3].) The optimum value for D is the one which leads to the lowest overall annual cost, i.e. the one for which compression costs + amortisation costs + other costs is a minimum.

Example 4.19
The reaction $CO + H_2O \rightleftharpoons CO_2 + H_2$ is carried out in an adiabatic fixed bed catalytic reactor. The composition of the gas stream at entry is:

CO   7.9 mol%
CO_2 4.8 mol%
H_2 34.4 mol%
N_2 13.5 mol%
H_2O 39.4 mol%

The total flow rate is 8983 kmol h^{-1} and the diameter of the catalyst bed is 4 m. Tests have shown that diffusion from the bulk gas to the surface of the catalyst pellets does not limit the reaction rate. The reaction rate $(-r'_{CO})$ (expressed in kmol CO converted per m^3 of catalyst bed per hour) can be expressed in the form:

$$(-r'_{CO}) = A(f(T))(y_{CO}y_{H_2O} - y_{CO_2}y_{H_2}/K(T)) \tag{i}$$

where $f(T) = \exp(15.95 - (4900/T))$
$\quad K(T) = \exp(-4.33 + (4578/T))$
$\quad\quad A = 0.170$ kmol per hour per cubic metre of catalyst bed at the proposed operating pressure of 25.3 bar absolute.

$y_{CO}, y_{H_2O}, y_{CO_2}$ and y_{H_2} are the mole fractions of carbon monoxide, water vapour, carbon dioxide and hydrogen respectively in the fluid phase and T is the absolute temperature in Kelvin. The inlet temperature to the reactor is known to be 616.5 K.

The reaction is reversible, the rate of the back reaction being given by the term in $K(T)$ in equation (i) (an infinite value for $K(T)$ would correspond to an irreversible reaction).

(a) If isothermal conditions are maintained in the catalyst bed ($T = 616.5$ K) calculate (dz/dX_{CO}) at points where the conversion of the CO in the bed (X_{CO}) is 0.0, 0.2, 0.4, 0.6 and 0.8 (z is the axial distance from the entry point to the bed).

(b) In the above isothermal situation calculate the bed depth required to achieve 80% conversion of the CO entering the bed. What fraction of the total depth is required to increase the conversion from 70% to 80%?

(c) If the bed is operated adiabatically, evaluate the reactor temperature at points where the CO conversion is 0.2, 0.4, 0.6, 0.7 and 0.8.

(d) In the adiabatic situation estimate the bed depth required to achieve 70% conversion of the CO.

Solution
(a) Calculation of gradients (dz/dX_{CO}) in the isothermal case
(dz/dX_{CO}) is given by the differential mass balance equation (4.50a) to be

$$\frac{dz}{dX_{CO}} = \frac{4F_{CO,0}}{\pi D^2 (-r'_{CO})}$$

where $F_{CO,0}$ is the molar flow rate of CO at the inlet to the catalyst bed. (CO is in this case the limiting reactant A.)

From the data given, $F_{CO,0} = 0.079 \times 8983 = 710$ kmol h^{-1} and $D = 4$ m. Substituting these values gives

$$\frac{dz}{dX_{CO}} = \left[\frac{56.5}{(-r'_{CO})}\right] m \tag{ii}$$

Before (dz/dX_{CO}) can be calculated from (ii), values are required for the rate of consumption of CO per unit volume $(-r'_{CO})$ at each of the specified conversions. $(-r'_{CO})$ is given by equation (i) which in turn requires evaluation of the terms $f(T)$, $K(T)$ as well as the mole fractions of CO, H_2O and CO_2.

In the isothermal situation, $T = 616.5\,\mathrm{K}$ and hence

$$f(T) = 2987$$
$$K(T) = 22.10$$

The mole fractions may conveniently be obtained by considering the molar flow rates F of each component. From the stoichiometry of the reaction, it follows that no change will occur in the total number of moles as the reaction proceeds. In consequence the total molar flow rate (F_T) throughout the reactor remains constant at the inlet value

$$F_T = 8983\,\mathrm{kmol\ h^{-1}}$$

At any point in the reactor, $F_{CO} = F_{CO,0}(1 - X_{CO}), F_{CO_2} = F_{CO_2,0} + F_{CO,0}X_{CO})$ and so on.
So

$$y_{CO} = \left(\frac{F_{CO}}{F_T}\right) = \frac{F_{CO,0}(1 - X_{CO})}{F_T} = y_{CO,0}(1 - X_{CO})$$

where $y_{CO,0}$ = mole fraction of CO at inlet
$\quad\quad y_{CO}$ = mole fraction of CO at a point in the reactor where the conversion of CO is X_{CO}.

Similarly,

$$y_{CO_2} = y_{CO_2,0} + \frac{F_{CO,0}X_{CO}}{F_T} = y_{CO_2,0} + y_{CO,0}X_{CO}$$

$$y_{H_2} = y_{H_2,0} + y_{CO,0}X_{CO}$$

$$y_{N_2} = y_{N_2,0}$$

and

$$y_{H_2O} = y_{H_2O,0} - y_{CO,0}X_{CO}$$

The component mole fractions and the values of $(-r'_{CO})$ and dz/dX_{CO} calculated from the above relationships for each of the specified conversions are listed in Table 4.4.

(b) Calculation of bed depth in the isothermal case

In the longer term the best approach would be to computerise the calculation of bed depth. However, at this stage there is much to be said for adopting a graphical approach since this demonstrates more clearly the basic principles involved. When using this approach, dz/dX_{CO} is evaluated at a series of X_{CO} values (as above), and is then plotted versus X_{CO}; the required depth is obtained as the area between the curve and the X_{CO} axis over the conversion range required, as in Figure 4.15.

Successive application of Simpson's rule to this area indicates a total bed depth of about 7.7 m, about half of which is required to raise the CO conversion from 0.6 to 0.8. The percentage increase in length required to raise conversion from 0.7 to 0.8 is:

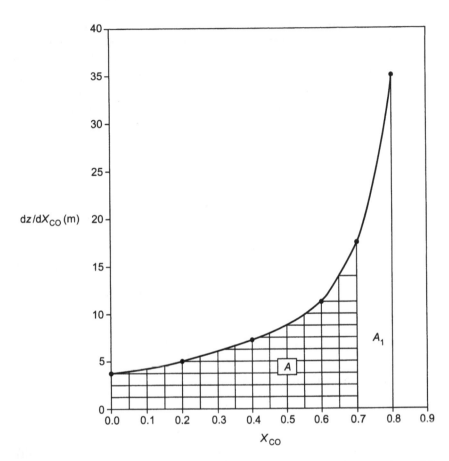

Figure 4.15 Graphical evaluation of catalyst bed depth required for 70% conversion and for 80% conversion of CO in isothermal reactor (see Example 4.19 part (b)). (Required bed depth for 80% conversion = area $(A + A_l)$ = 7.7 m; required bed depth for 70% conversion = area A (shaded) = 5.2 m.)

$$[(7.7 - 5.2)/5.2] \times 100 = 48\%$$

(c) Calculation of temperature at specified conversion in the adiabatic case

The temperature at the specified conversion can be calculated explicitly from the data given by applying the integral energy balance equation (equation (2.53)) under adiabatic conditions. Rearranging this equation with Q=0 gives,

$$X_A = -\left(\frac{1}{F_{A0}\Delta H_{R,A,T_0}}\right) \int_{T_0}^{T} \sum_i F_i C_{pmi}(X_A, T)\mathrm{d}T \qquad \text{(iii)}$$

where $\Delta H_{R,A,T_0}$ is the enthalpy of reaction per mole of reactant A consumed at the reactor inlet temperature T_0 and F_i is the molar flow rate of component i at a point where the conversion of A is X_A. C_{pmi} is the molar heat capacity of the pure component i. F_{A0} is the molar flow rate of the limiting reactant A at the reactor inlet. In the present context, A is carbon monoxide and the molar heat capacities are regarded as sufficiently insensitive to changes in temperature to enable average values over the temperature range of interest to be used. Under these conditions (iii) may be rewritten as

$$X_A = \frac{\sum_i F_i \bar{C}_{pmi}(T - T_0)}{F_{A0}\left(-\Delta H_{R,A,T_0}\right)} \tag{iv}$$

or

$$T = T_0 + \frac{F_{A0}\left(-\Delta H_{R,A,T_0}\right)X_A}{\sum_i F_i \bar{C}_{pmi}} \quad \text{K} \tag{v}$$

$\bar{C}_{pmi}$ is the average molar heat capacity for component i over the temperature of interest. As noted above, the total molar flow rate F_T is constant through the reactor in the present illustration and it is convenient to express equation (v) in terms of the mole fractions (previously evaluated) as follows:

$$T = T_0 + \frac{F_{A0}\left(-\Delta H_{R,A,T_0}\right)X_A}{F_T \sum_i y_i \bar{C}_{pmi}} \tag{vi}$$

Inserting the numerical values for $F_{A0}, (-\Delta H_{R,A,T_0})$ and F_T

$$T = T_0 + \left(\frac{710 \times 38\,000}{8983}\right)\frac{X_a}{\sum_i y_i \bar{C}_{pmi}}$$

$$= T_0 + 3003\frac{X_A}{\sum_i y_i \bar{C}_{pmi}} \tag{vii}$$

The calculation of T from equation (vii) using the given values of $\bar{C}_{pmi}$ and values of y_i from Table 4.4 is detailed in the top part of Table 4.5.

The temperatures listed in Table 4.5 were calculated using the estimated average heat capacities provided. (Such estimates are, of course, based on heat capacity versus temperature data for the relevant substances but their accuracy, when applied over the initially unknown range of temperatures occuring in the reactor, is not known until this temperature range has been established with sufficient accuracy.) The temperatures shown in Table 4.5 would therefore normally be regarded as first approximation values only. Using them, better estimates for the average heat capacities would be obtained from heat capacity versus temperature data for each component using the expression

$$\bar{C}_{pmi} = \frac{\int_{T_0}^{T_1} C_{pmi} \, dT}{T_1 - T_0}$$

Table 4.4 Component mole fractions, reaction rates and the gradient dz/dX_{CO} as functions of CO conversion for the isothermal reactor detailed in Example 4.19 part (a)

	$X_{CO} \rightarrow$					
	0.0	0.2	0.4	0.6	0.7	0.8
y_{CO}	0.079	0.063	0.047	0.032	0.024	0.016
y_{CO_2}	0.048	0.064	0.080	0.095	0.103	0.111
y_{H_2}	0.344	0.360	0.376	0.391	0.399	0.407
y_{N_2}	0.135	0.135	0.135	0.135	0.135	0.135
y_{H_2O}	0.394	0.378	0.362	0.347	0.339	0.331
$y_{CO}y_{H_2O} - y_{CO_2}y_{H_2}/K(T)$	0.0304	0.0258	0.0157	0.0094	0.0063	0.0032
$-r'_{CO}$(kmol h^{-1}m^{-3})	15.43	11.56	7.97	4.77	3.20	1.62
dz/dX_{CO}(m)	3.66	4.88	7.09	11.8	17.7	34.9

Table 4.5 Component heat capacity contributions ($y_i\bar{C}_{pmi}$), reactor temperatures (T), reaction rates ($-r'_{CO}$) and the gradient dz/dX_{CO} as functions of CO conversion for the adiabatic reactor detailed in Example 4.19 part (c)

	$X_{CO} \longrightarrow$					
	0.0	0.2	0.4	0.6	0.7	0.8
$y_{CO}\bar{C}_{pm,CO}$		1.95	1.46	0.99	0.74	0.496
$y_{CO_2}\bar{C}_{pm,CO_2}$		2.75	3.44	4.08	4.43	4.77
$y_{H_2}\bar{C}_{pm,H_2}$		10.44	10.90	11.34	11.57	11.80
$y_{N_2}\bar{C}_{pm,N_2}$		4.19	4.19	4.19	4.19	4.19
$y_{H_2O}\bar{C}_{pm,H_2O}$		13.99	13.39	12.84	12.54	12.25
$\sum y_i\bar{C}_{pmi}$		33.32	33.38	33.43	33.47	33.51
$T - T_0$ (K)	0.0	18.0	36.0	53.9	62.8	71.7
T (K)	616.5	634.5	652.5	670.4	679.3	688.2
$f(T)$	2987	3742	4630	5659	6227	6836
$K(T)$	22.10	17.91	14.67	12.17	11.12	10.20
$[y_{CO}y_{H_2O} - (y_{CO_2}y_{H_2}/K(T))] \times 10^4$	304	225	151	80	44	7.8
$-r'_{CO}$(kmol h^{-1}m^{-3})	15.4	14.3	11.9	7.7	4.7	0.91
dz/dX_{CO}	3.7	3.9	4.7	7.3	12.0	62.0

where T_1 is the first approximation value for T and the integral $\int_{T_0}^{T_1} C_{pmi}dT$ is evaluated from the heat capacity versus temperature data. In the present illustration the initial estimated heat capacities were sufficiently close to the actual values to be used without correction. The total heat capacity of the stream $\sum y_i\bar{C}_{pmi}$ did not change appreciably in this case during the reaction, and for this reason the calculated temperature rise varied nearly linearly with carbon monoxide conversion. This will not always be the case, however.

(d) Evaluation of bed depth in the adiabatic case

The overall approach to the evaluation of bed depth in the adiabatic case is similar to that in the isothermal one, though it is necessary to evaluate the reactor temperature at each of a set of selected conversions, so that the appropriate reaction rate ($-r'_A$) in equation (4.50a) can be evaluated.

Temperatures at appropriate conversions have already been obtained in part (c) and the reaction rates $(-r'_{CO})$ calculated from equation (i) at conversions between 20 and 80% are shown in Table 4.5. It is seen by comparing Tables 4.4 and 4.5 that the reaction rate is higher in the adiabatic case for conversions up to and including 70% but that at 80% it falls substantially below that in the isothermal reactor. The reason for this behaviour is the progressively increasing temperature in the adiabatic reactor. This enhances the rates of the forward and reverse reactions in the adiabatic case and also lowers the equilibrium conversion. At low conversions the effect of the reverse reaction is negligible. The dominant effect of the greater temperature in the adiabatic reactor is then to increase the rate of the forward reaction and hence the net reaction rate. As the equilibrium conversion is approached, however, the influence of the reverse reaction becomes increasingly important, particularly at the higher temperatures experienced in the adiabatic reactor. This factor results, at high conversions, in higher reaction rates for a given conversion in the isothermal reactor than in the adiabatic one.

Values of (dz/dX_{CO}) in the adiabatic reactor calculated from equation (ii) are shown at the foot of Table 4.5 and are plotted as a function of carbon monoxide conversion in Figure 4.15.

4.9   Pressure drops in catalyst beds

In many cases pressure drops in catalyst beds will not be sufficient to have a significant effect on reaction rates and it may not be necessary, therefore, to evaluate a detailed pressure profile through the bed when calculating the bed volume. However, it will be necessary to estimate the total pressure drop through the bed as part of the reactor design (see part (c) of Example 4.18) since pressure drops add to running costs.

The semi-empirical Ergun equation [4] is often used for predicting pressure drops both in catalyst beds and in other packed beds. According to this equation

$$-\frac{dP}{dz} = 150 \left[\frac{(1-\varepsilon_b)^2}{\varepsilon_b^3} \right] \frac{\mu u}{d^2} + 1.75 \left[\frac{(1-\varepsilon_b)}{\varepsilon_b^3} \right] \frac{\rho u^2}{d}$$

where z = axial distance from the entry point to the bed (m)
   ε_b = fractional bed porosity
   μ = fluid viscosity $(N\,s\,m)^{-2}$
   u = superficial velocity (i.e. the 'empty tube' velocity) $(m\,s^{-1})$
   d = average pellet diameter (m)
   ρ = fluid density $(kg\,m^{-3})$
   P = pressure (Pa, i.e. $N\,m^{-2}$)

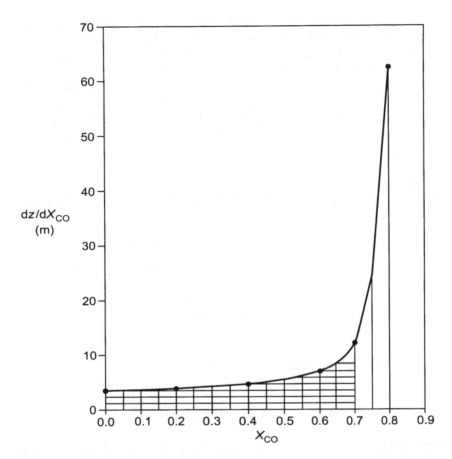

Figure 4.16 Graphical evaluation of catalyst bed depth required for 70% conversion of CO in adiabatic reactor (see Example 4.19 part (d)). The required bed depth for a 70% conversion is given by the shaded area to be 3.7 m as compared with 5.2 m in the isothermal case with the same feed temperature.

Although the initial pressure drop after commissioning a catalyst bed is probably well predicted by the Ergun requation, it has been pointed out that catalyst bed characteristics, and hence pressure drop, tend to change during operation [5] and it is wise to make allowance for the increase which can then occur.

Example 4.20

Estimate the pressure drop over the catalyst bed described in Example 4.19 for the isothermal situation (part (b)), given the following additional information. The catalyst pellets are approximately spherical with an effective diameter d of 6×10^{-3} m. The viscosity μ of the reaction mixture is estimated from pure component data to vary comparatively little from the inlet to the outlet, and an approximate value of $0.021\,\mathrm{mN\,s\,m^{-2}}$ will be used. The procedure should produce little error in the calculated pressure drop since viscosity only appears in the first term of the Ergun equation and, under conditions normally prevailing in reactors, this term is a small one [5]. The bed voidage is 0.4. The fluid density (evaluated from the ideal gas law and the molecular weights) is $7.95\,\mathrm{kg\,m^{-3}}$. (Since there is no change in the total number of moles during the reaction and ideal conditions can be assumed as a good approximation (see Chapter 2), the above figure for the density should apply throughout the bed.)

Solution

Before the Ergun equation is applied, the superficial velocity u_s is required. By definition

$$u_s = (\text{volumetric flow rate})/(\text{cross-section area})$$

From the gas law, the molar volume of the reaction mixture is given by

$$V_m = (22.414 \times 10^{-3}) \times \left(\frac{616.5}{273.2}\right) \times \left(\frac{1.013}{25.3}\right) = 2.03 \times 10^{-3}\,\mathrm{m^3\,mol^{-1}}$$

Hence
volumetric flow rate $= (8983 \times 10^3) \times 2.03 \times 10^{-3}/3600 = 5.06\,\mathrm{m^3 s^{-1}}$
and

$$u_s = 5.06 \times 4/(\pi \times 16) = 0.403\,\mathrm{ms^{-1}}$$

Substituting the above values for ε_b, μ, d and u_s in the Ergun equation gives

$$\left(\frac{-\mathrm{d}P}{\mathrm{d}z}\right) = 150\left(\frac{0.6^2}{0.4^3}\right)\left(\frac{0.021 \times 10^{-3} \times 0.403}{36 \times 10^{-6}}\right) + 1.75\left(\frac{0.6}{0.4^3}\right)\left(\frac{7.95 \times 0.403^2}{6 \times 10^{-3}}\right)\,\mathrm{Pa\,m^{-1}}$$
$$= (198 + 3581)\,\mathrm{Pa\,m^{-1}}$$
$$= 0.037\,\mathrm{bar\,m^{-1}}$$

The bed depth required to give conversion X_A of 0.8 is known to be 7.7 m, giving a pressure drop of 0.29 bar.

References

1. Jenson, V.G. and Jeffreys, G.V. (1977) *Mathematical Methods in Chemical Engineering*, 2nd edition, Academic Press; New York.
2. Lapidus, L. (1962) *Digital Computation for Chemical Engineers*, McGraw-Hill, New York.

3. Coulson, J.M., Richardson, J.F. and Sinnolt, R.K. (1983) *Chemical Engineering, Volume 6, Design*, Pergamon, Oxford.
4. Ergun, S. (1952) *Chem. Eng. Prog.*, **48**, 89.
5. Lywood, W. J. Process design, rating and performance, Chapter 2 in *Catalyst Handbook*, 2nd edition (ed. M.V. Twigg), Manson Publishing Co., London.

5 Non-ideal flow in chemical reactors and the residence time distribution

J. BOURNE

5.1 Introduction

Continuously operated chemical reactors with idealised flow patterns have played a central conceptual role in the development and teaching of chemical reaction engineering. These comprise (a) the plug (or piston) flow reactor and (b) the perfectly mixed stirred tank. In plug flow all fluid elements crossing a given plane are moving at the same speed and in the same direction, i.e. their velocities are all equal. Detailed measurements of the velocities in long straight pipes when the flow is turbulent have shown a very close approximation to plug flow. The radial variation of the axial velocity across the pipe's cross-section and the chaotic motion of eddies in the flow cause an almost negligible departure from plug flow. Perfect mixing means that the temperature, the composition and other properties are spatially uniform. Applied to a continuously operated stirred tank, this signifies that the composition of the tank's contents is homogeneous and that the feed mixes very rapidly with these contents, thereby losing its identity. Obviously this blending needs a finite, albeit short time, during which the feed becomes distributed over the whole of the tank's contents, so that perfect mixing is a state which can be approached, rather than fully attained.

Flow patterns in real processing equipment often depart from the two idealised types presented above. Consider, first, deviations from plug flow in tubular devices. One possible cause is channelling in which the flow is unevenly distributed over the cross-section. This might arise in a packed absorption or distillation tower or a packed catalytic reactor, where variable packing density causes variations in the resistance to flow and hence in the local fluid velocity through the packing. A second cause is a severe obstruction in the flow reducing the local velocity to zero and forming a dead zone. This could happen, for instance, behind a deflector or baffle installed for a different purpose in the equipment. A third cause is the formation of recirculatory or secondary flows imposed on the main flow. This occurs when a high-speed, axisymmetric jet discharges into a pipe and entrains the fluid in its vicinity. The main flow of both streams is along the pipe, but recirculation loops are formed by the action of the jet. Turning now to deviations from perfect mixing in tanks: one cause, namely that the blending time of the feed is not zero, has already been described. Channelling can be a further cause. This can occur, for example, when the feed pipe and the overflow pipe are too closely positioned, allowing short-circuiting of the feed into the overflow. Dead zones can form in regions far removed from the stirrer, e.g. in the bottom corners and near the surface, and this will be more likely in high-viscosity liquids and in those possessing plastic and pseudoplastic rheological characteristics.

Consideration of non-ideal flow leads to two broad questions. How can non-ideal flow be quantitatively described? What are the consequences of deviations from idealised flow, particularly with respect to heat transfer, mass transfer and chemical reaction? This chapter aims at treating these questions at an introductory level. To illustrate the principles involved in as simple a way as possible, volume changes accompanying the reactions considered will be assumed to be negligible.

5.2 Quantitative descriptions of non-ideal flow

Various approaches to the quantitative description of non-ideal flow are in use: they are not mutually exclusive, indeed it is often helpful to proceed simultaneously along different paths. Four of these approaches will be outlined here.

The most fundamental approach requires a knowledge of the complete velocity distribution. In principle, this approach requires solution of the Navier–Stokes equation in three dimensions as a function of time. This procedure would give the time-averaged flow field and all turbulence characteristics, including root-mean-square velocity fluctuations, macro- and micro-scales, and energy dissipation rates. Even with the most powerful computers available today, this approach is not (yet) able to solve the wide range of problems arising in practice. Nevertheless, DNS (direct numerical simulation) is an area of rapid growth, which is attracting significant human and financial resources and which has already produced substantial, ex-perimentally verified results. Chemical engineers would be well advised to follow and preferably to participate in these developments.

The flow field and some turbulence characteristics are also calculated by CFD (computational fluid dynamics), whereby the Navier–Stokes equation is solved with the aid of some modelling assumptions. Whereas the validity of such assumptions and the suitability of the model structure are known for certain classes of turbulent flow, this is not (yet) always the case. Moreover thorough experimental confirmation of CFD predictions is as yet available for only a limited range of problems. The applicability of CFD methods looks certain to increase and, provided they are critically evaluated, such methods will be useful in predicting, for example, mean flow fields and hence in characterising non-ideal flows.

Some non-ideal flow patterns can be decomposed into a scheme of simpler elements, such as plug flow, perfectly mixed, dead zone and short circuit, which are interconnected in series and in parallel. This results in a block diagram of elements, representing certain aspects of the whole flow, such that the performance of the whole, for example as a reactor, can be estimated from the behaviour of its elements. The basis for such a flow scheme and block diagram is information about the flow field, which often is obtained experimentally. Flow visualisation is sometimes relatively simple and can identify parts of the flow which are dead, or in plug flow or are well mixed or channelled and so on. LDV (laser Doppler velocimetry) is non-invasive and can measure the mean flow field as well as some turbulence properties.

Measurement of the RTD (residence time distribution) can sometimes lead to values of the parameters in a flow scheme, as will be explained later. The RTD approach is rather more specific to chemical engineering, whereas the other approaches mentioned here form parts of fluid mechanics. RTD measurement is relatively simple, including application at industrial scale, and knowledge of both its capacities and its limitations is necessary for any study of non-ideal flow in reactors.

The following worked example illustrates the representation of non-ideal flow in a reactor with a simple flow scheme and shows how this can be used to calculate the reactor's performance. Here, as elsewhere in this chapter, fluid density changes will be taken to be negligible.

Example 5.1
A continuously operated stirred tank reactor carries out a reaction having first-order kinetics. It was designed on the basis of perfect mixing to give a conversion of 80%.

It was subsequently found that (a) 10% of the feed is short-circuited, and (b) 20% of the volume is dead. Calculate the conversion **under these non-ideal conditions**

A block diagram of the reactor is shown in Figure 5.1.

The reactant concentration in the feed is C_0 and the volumetric flow rate both in the feed and at the system outlet is ν. The reactant concentration in the well-mixed part of the reactor is C and in the product stream it is $\bar{C}$. A steady-state mass balance on the reactant at point M gives

$$0.9\nu C + 0.1\nu C_0 = \nu\bar{C} \tag{i}$$

A similar mass balance over the well-mixed part of the reactor gives

$$0.9\nu C_0 = 0.9\nu C + 0.8kCV \tag{ii}$$

where k is the first-order rate constant. This is unknown, but can be related to the design conversion (80%) by the following mass balance:

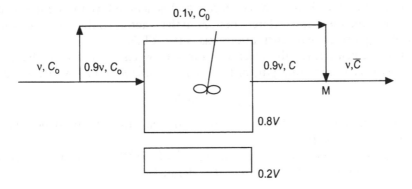

Figure 5.1 Block diagram of the reactor.

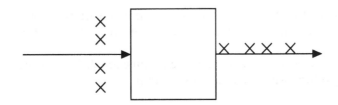

Figure 5.2 The residence time distribution: fluid elements ($\times$), which enter a processing unit simultaneously, leave after various times.

$$\nu C_0 = 0.2\nu C_0 + kV(0.2C_0) \qquad \text{(iii)}$$

From equation (iii), $kV/\nu = 4$, so that from equation (ii), $C = 0.2195C_0$ and, from equation (i), $\bar{C} = 0.2976C_0$. The non-ideal reactor would thus achieve a conversion of 70.2%. Improving the mixing to eliminate the dead zone and the short circuit could raise conversion to the design level of 80%.

5.3 Residence Time Distribution (RTD)

In 1953 P.V. Danckwerts introduced the RTD by defining it and working out some of its properties. This and subsequent developments now form a solid approach to the quantitative treatment of non-ideal flow in all types of processing equipment. Discussion of the residence time distribution and its application to reactors has become a classical part of chemical reaction engineering.

5.3.1 Definitions

Consider n fluid elements which simultaneously enter a continuously operated reactor of any type (Figure 5.2). The residence time of any element is the time spent by the element in the reactor and is denoted by the symbol t. In general, elements spend different periods in the reactor, so that t is a quantity possessing a statistical distribution: this is the RTD.

If dn elements leave the reactor in times between t and $t + dt$, the probability density function (pdf) of the residence time $f(t)$ may be defined by

$$f(t)dt = dn/n \qquad (5.1)$$

This equation gives the fraction of fluid elements with residence times between t and $t + dt$.

By defining the cumulative function $F(t')$ of the RTD, another aspect of this distribution becomes apparent. F is defined by

$$F(t') = \int_0^{t'} f(t)dt \qquad (5.2)$$

The cumulative function expresses the fraction of fluid elements with residence times up to and including t'. It follows that $F(0) = 0$ and $F(\infty) = 1$.

The mean and variance of the residence time are defined in the same way as for any other distributed quantity, so that

$$\bar{t} = \int_0^{\infty} tf(t)dt \qquad (5.3)$$

$$\sigma^2 = \int_0^{\infty} (t - \bar{t})^2 f(t)dt \qquad (5.4)$$

When the range of possible values of t does not extend from zero to infinity, then the lower and upper limits of integration become the shortest and the longest residence times, respectively.

The residence time can be expressed non-dimensionally by dividing it by its mean value

$$\theta = t/\bar{t} \qquad (5.5)$$

The fraction expressed by equation (5.1) may also be written $f(\theta) d\theta$. From equation (5.5), $dt = \bar{t}d\theta$ so that

$$f(\theta) = \bar{t}f(t) \qquad (5.6)$$

This result makes possible the interconversion of the dimensional and dimensionless forms of the RTD.

5.3.2   Measurement of RTD

The procedure for measuring a residence time distribution is sufficiently simple and robust to enable flow patterns in industrial equipment to be investigated by this technique. A small quantity of a material, termed a tracer, is introduced into the inlet stream to the piece of process equipment under test and the tracer concentration is measured as a function of time in the exit stream. To be suitable, the tracer material should be (a) unreactive, (b) not adsorbed within the system, (c) completely soluble so that it follows the same trajectories through the equipment as the main fluid, (d) quantitatively detectable at low concentrations and (e) uniformly introduced over the cross-section of the inlet fluid. Questions of cost, toxicity and disposability are also relevant. Some examples of tracers include (i) concentrated potassium chloride solution added to an aqueous stream, where the KCl concentration at the exit is determined from the electrical conductivity of the solution; (ii) concentrated dyes whose exit concentration is measured photometrically; (iii) helium injected into a gas stream, where the helium concentration in the exit stream is measured using a mass spectrometer or a thermal conductivity detector.

The term **pulse injection** (Figure 5.3a) is used to describe rapid addition of the tracer to the inlet stream in a time which is much smaller than the mean residence time.

Between the times t and $t + dt$ after injection, the mass of tracer leaving the reactor is $\nu C(t) dt$. This mass is also equal to $mf(t) dt$, where m is the injected tracer mass. Hence

$$f(t) = \nu C(t)/m \tag{5.7}$$

This equation enables the probability density function defined in equation (5.1) to be calculated from the volumetric throughput ν, the tracer mass m and the measured exit tracer concentration as a function of time. The quality of the measurements can be assessed by comparing the two sides of the tracer mass balance, equation (5.8)

$$m = \nu \int_0^\infty C(t) dt \tag{5.8}$$

The term **step injection** (Figure 5.3b) is used when the concentration of tracer in the inlet stream is changed from one level (often zero) to a new and constant level (C_0) in a time which is much smaller than $\bar{t}$.

In the exit fluid at any time t, measured from the step change, the fraction of new fluid, having the concentration C_0, is $F(t)$, while the remainder is the fraction $1 - F(t)$ and is unmarked with tracer, i.e. $C = 0$. The measured exit concentration $C(t)$ is then equal to $F(t)C_0$ and the cumulative form of the RTD is given by

$$F(t) = C(t)/C_0 \tag{5.9}$$

As noted earlier, F increases from 0 to 1 as $C(t)$ increases from 0 to C_0.

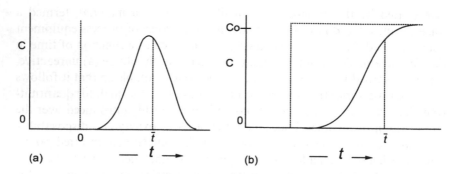

Figure 5.3 Measurement of RTD. (a) Pulse injection: at time zero a quantity of tracer is rapidly introduced into the feed and its concentration in the exit stream is measured as a function of time. (b) Step injection: at time zero the tracer concentration in the feed is suddenly increased from zero to C_0, while its exit concentration is continuously measured. [.... feed stream; —— exit stream]

5.4   The residence time distribution for some selected flow models

In this section it will be seen that when the flow pattern in a device is fully known, the RTD can be calculated from unsteady state mass balances. It is important to recognise, however, that when the RTD has been measured, the parent flow pattern cannot be determined uniquely, although some feasible alternatives can be proposed. Since flow patterns in several industrial devices are not known, this limits what can be learned from an RTD.

5.4.1   The perfectly mixed tank

Perfect mixing means, of course, that the tracer concentration in the tank is uniform and will be denoted by C. This means that wherever the exit pipe is located, the tracer concentration in this stream is also C. Supposing that the capacity of the device is constant and the fluid density does not change, the volumetric inflow and outflow rates must be equal and will be denoted by ν. The unsteady mass balance for the tracer, whose inlet concentration is C_0, may be written

$$\nu C_0 = \nu C + V\mathrm{d}C/\mathrm{d}t \qquad (5.10)$$

For **pulse** injection $C_0 = 0$ and, when $t = 0, C = m/V$. The solution of equation (5.10) is then

$$C = (m/V)\exp(-t/\bar{t}) \qquad (5.11)$$

where the term V/ν has been set equal to the mean residence time, $\bar{t}$. Equation (5.7) then yields

$$f(t) = (1/\bar{t})\exp(-t/\bar{t}) \qquad (5.12)$$

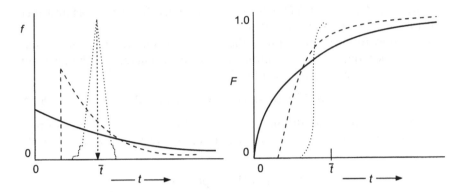

Figure 5.4 Some examples of RTD: (a) differential function f defined in equation (5.1); (b) cumulative function F defined in equation (5.2). [—— well-stirred tank; .... turbulent pipe flow; - - - laminar pipe flow with no Taylor diffusion]

This is the RTD of a perfectly mixed tank.

For **step** injection, when $t = 0$; $C = 0$. The solution of equation (5.10) then becomes

$$(C/C_0) = 1 - \exp(-t/\bar{t}) \qquad (5.13)$$

which, by equation (5.9), is immediately equal to F, the cumulative form of the RTD. From equation (5.2), $f = dF/dt$ and by differentiating equation (5.13), equation (5.12) is again obtained. Whether the RTD of a CSTR is obtained by pulse or step injection, the results are fully equivalent because they express the character of the flow pattern, not the means used to determine it.

Equation (5.12) is an important result, which by equation (5.6) can also be written $f(\theta) = \exp(-\theta)$. Some properties which then follow include the wide range of possible residence times, t, from zero to infinity. From equation (5.4) the standard deviation (σ) of t is also equal to $\bar{t}$. The maximum value of $f(t)$ is $1/\bar{t}$ when $t = 0$: this signifies that the most probable value of t is **zero** and not, for example, $\bar{t}$. When $t = \bar{t}$, equation (5.13) gives $F = 0.632$, so that 63.2% of all the fluid passing through a perfectly mixed tank spends times less than or equal to the mean residence time in the tank.

The residence time and cumulative residence time distribution curves (f and F respectively) for a single well-mixed tank are sketched in Figure 5.4.

5.4.2   Cascade of perfectly mixed tanks

The wide spread of residence times exhibited by a single CSTR can be significantly reduced by using a stirred tank cascade, in which the vessels are connected in series. Usually all tanks have the same volume V. The technique to derive the RTD will first be explained for two tanks and then generalised to N tanks in the cascade.

The inflow to the second tank is the overflow from the first tank having, for pulse injection, the concentration given by equation (5.11). The unsteady state mass balance for tank 2 is therefore

$$(\nu m/V)\exp(-t/\bar{t}) = \nu C_2 + V dC_2/dt \tag{5.14}$$

subject to the initial condition: $t = 0, C_2 = 0$ (C_2 is the concentration of tracer in tank 2 and in its overflow). Equation (5.14) is linear with the solution

$$C_2 = (mt/V\bar{t})\exp(-t/\bar{t}) \tag{5.15}$$

When the cascade consists of N equi-sized well-stirred tanks, solution of the N differential mass balances yields

$$C_n = \frac{t^{N-1}\exp(-t/\bar{t})m}{\bar{t}^{N-1}(N-1)!V} \tag{5.16}$$

which reduces to equations (5.11) and (5.15) when $N = 1$ and 2 respectively. Equation (5.7), which is always applicable to pulse injection, then gives

$$f(t) = \frac{t^{N-1}\exp(-t/\bar{t})}{\bar{t}^N(N-1)!} \tag{5.17}$$

This important result is sometimes written in terms of the mean residence **time in the whole cascade** ($\bar{T}$), rather than the mean value **in a single tank** ($\bar{t}$). Obviously $\bar{T} = N\bar{t}$. This and the important result for the variance of the RTD of an arbitrary cascade can most simply be derived as follows. Suppose a given fluid element spends a time t_1 in tank 1, t_2 in tank 2, etc., then its residence time T in the whole cascade is

$$T = t_1 + t_2 + \ldots + t_N \tag{5.18}$$

We now use the result that the mean value of a sum of random variables is equal to the sum of their individual mean values. In the present application the linear expectation (or averaging) operator of statistics, E, expresses this result in the form

$$E(T) = E(t_1) + E(t_2) + \ldots + E(t_N)$$

so

$$\bar{T} = \bar{t}_1 + \bar{t}_2 + \ldots + \bar{t}_n = N\bar{t} \tag{5.19}$$

The concept of ideal mixing means that a fluid element loses its identity immediately as it merges with the mass in the tank and this further implies that t_2 is **totally independent** of t_1. The variance of a sum of random and mutually independent variables is equal to the sum of their individual variances. Denoting the variance by V, equation (5.18) gives

$$V(T) = V(t_1) + V(t_2) + \ldots + V(t_N)$$

i.e.

$$\sigma^2 = \bar{t}_1^2 + \bar{t}_2^2 + \ldots + \bar{t}_N^2 = N\bar{t}^2 \qquad (5.20)$$

This can also be written

$$\sigma^2 = N(\bar{T}^2/N^2) = \bar{T}^2/N \qquad (5.21)$$

This important result shows that the variance of T (i.e. the residence time in the cascade, according to equation (5.18)) decreases from its highest value when $N = 1$ to zero as N increases without limit, while $\bar{T}$ (i.e. the mean residence time in the cascade) is held constant. Depending upon whether the number of tanks is small or large, the RTD of a cascade is more like that of a CSTR or a plug flow reactor respectively.

For completeness the dimensionless forms of some key results will now be summarised. In this context the general time variable T (or t) is compared to $\bar{T}$, so that

$$\theta = T/\bar{T} \qquad (5.22)$$

The pdf of the residence time T is obtained from (5.17) and (5.22)

$$f(\theta) = \frac{N^N \theta^{N-1} \exp(-N\theta)}{(N-1)!} \qquad (5.23)$$

The mean value of θ is unity and its variance may be written

$$\sigma^2(\theta) = 1/N \qquad (5.24)$$

This last result follows from equations (5.21) and (5.22) since, from statistics, $V(T) = \bar{T}^2 V(\theta)$. It can also be derived by substituting equation (5.17) into equation (5.4) and integrating: this is, however, cumbersome.

This treatment of the RTD of a cascade of ideal stirred tanks contains some techniques and ideas which are helpful for more complex problems. Examples of these include the application of statistical methods to the residence times (especially the use of the operators E and V) and the methods for solving ordinary differential equations which were required for the solution of the N mass balances. Meaningful use of such methods presupposes, however, a clear physical understanding of the concepts involved. The curves shown in Figure 5.3 would typically correspond to the RTD of a cascade made up of a few well-mixed tanks.

Example 5.2
A small quantity of a non-reactive impurity suddenly falls into the first tank of a two-tank CSTR cascade. Each tank has the same volume and an individual mean residence time of 10 min.
(a) When is the concentration of impurity in the second tank a maximum? How large is this?
(b) Calculate the time when the total quantity of impurity in both tanks has fallen to 2% of its initial value.

Solution
(a) Entry of the impurity corresponds to a pulse tracer injection. Equation (5.15) – or equation (5.16) with $N = 2$ – describes the transient concentration C_2 of the impurity in tank 2

$$C_2 = (m/V)xe^{-x}$$

where $x = t/\bar{t}$ and $\bar{t} = 10$ min. The initial quantity of impurity is m, while V is the capacity of **one** tank. Differentiating to find when C_2 is a maximum

$$d(VC_2/m)/dx = e^{-x}(1-x) = 0$$

so $t = \bar{t} = 10$ min. The highest value of C_2 is then given by

$$C_{2(max)} = (m/V)e^{-1} = 0.368m/V$$

(b) Let y be the mass of impurity in both tanks

$$y = V(C_1 + C_2)$$

where C_1 refers to the first tank and is given by equation (5.11) while C_2 is again obtained from equation (5.15). The value of $t = x\bar{t}$ when $y = 0.02\,m$ is defined by

$$0.02m = V\left(\frac{m}{V}e^{-x} + \frac{mx}{V}e^{-x}\right)$$

The following equation may be solved numerically for x:

$$0.02 = e^{-x}(1+x)$$

giving $x = 5.83$ and $t = 58.3$ min.

An alternative solution, using the cumulative form of the RTD, can be formulated as follows. As defined in equation (5.2), $F(t')$ is the fraction of non-reactive solution which has left the system (here the two tanks) at time t^1. In this problem the time when $F = 0.98$ is sought. F is obtained by integration of f (cf. equation (5.2)), where f is given by setting $N = 2$ in equation (5.17). The following expression for F is then obtained:

$$\int_0^t \frac{t\exp(-t/\bar{t})}{\bar{t}^2}\,dt = \int_0^{x^1} xe^{-x}dx = 0.98$$

Integrating and evaluating x' numerically gives the same answer as above.

5.4.3 Laminar flow in a pipe

When a Newtonian fluid flows isothermally and steadily through a straight circular pipe, having a radius R, the velocity $u(r)$ at any radius r is parabolic and given by the following result due to Hagen and Poiseuille (*ca.* 1840). The mean velocity is $\bar{u}$.

$$u(r) = 2\bar{u}\left(1 - \frac{r^2}{R^2}\right) \tag{5.25}$$

The fraction of the total flow ν contained between the radii r and $r + \mathrm{d}r$ is

$$\frac{\mathrm{d}\nu}{\nu} = \frac{2\pi r u(r)\mathrm{d}r}{\pi R^2 \bar{u}}$$

From equation (5.1) the fraction of the flow possessing residence times between t and $t + \mathrm{d}t$ is $f(t)\,\mathrm{d}t$, where here $t = L/u(r)$ and L is the tube length.

$$f(t)\mathrm{d}t = \frac{\mathrm{d}\nu}{\nu} = \frac{2rudr}{R^2\bar{u}} \tag{5.26}$$

In this expression u is a known function of r (refer to equation (5.25)) while t and r are related by

$$t = \frac{L}{u} = \frac{LR^2}{2\bar{u}(R^2 - r^2)}$$

After eliminating u and r from equation (5.26), the pdf of the RTD becomes

$$f(t) = \bar{t}^2/2t^3 \tag{5.27}$$

provided that $t \geq \bar{t}/2$. For shorter times not even the liquid on the pipe axis ($r = 0$) reaches the exit and $f = 0$. In dimensionless form equation (5.27) becomes

$$f(\theta) = 1/2\theta^3$$

provided $\theta \geq 0.5$, while the cumulative RTD is

$$F(\theta) = 1 - (1/4\theta^2)$$

This shows, for example, that 88.9% of the outflow spent times between $0.5\bar{t}$ and $1.5\bar{t}$ in the pipe. When the mean value of the RTD in equation (5.27) is evaluated from

$$\int_{0.5\bar{t}}^{\infty} tf(t)\mathrm{d}t$$

the proper value $\bar{t}$ is obtained, as expected. (The lower and upper integration limits refer to the fluid on the axis $(r = 0)$ and at the pipe wall $(r = R)$ respectively.) The variance of the residence time t can be evaluated from

$$\int_{0.5\bar{t}}^{\infty} (t - \bar{t})^2 f(t) \mathrm{d}t$$

When $f(t)$ in the above expression is obtained from equation (5.27), the result, which the readers can check for themselves, is infinity! This suggests that in addition to the parabolic velocity distribution, equation (5.25), some further phenomenon needs to be included in analysing the transport (or mass transfer) of a solute in a laminar flow.

G.I. Taylor in 1953 showed convincingly that molecular diffusion of the solute is also relevant and, in some circumstances, profoundly modifies the form of the RTD. Obviously no flow occurs in the radial direction, but the radial concentration gradient can produce significant radial diffusion, given sufficient time, invalidating equation (5.27). Solution of Fick's second law, relating to unsteady molecular diffusion, shows that during the mean residence time $\bar{t} = L/\bar{u}$ solute diffuses radially on average over a distance $\sqrt{(2D\bar{t})}$. Provided that this is a significant fraction of the pipe radius, then radial diffusion will modify equation (5.27). Re-expressed, if $D\bar{t}/R^2 \gg 0.1$, radial diffusion will be important. If, however, the inequality is reversed, equation (5.27) will be approximately valid.

Some situations where radial or Taylor diffusion will act include the flow through fine pores or capillaries, e.g. in catalyst particles or fine veins. The RTD then deviates less from plug flow than when only the parabolic velocity distribution is relevant.

When a bend occurs in a pipe, within which the flow is laminar, the flow field is much more complex than that given by equation (5.25). The fluid near the pipe axis is moving faster in the axial direction than that near the wall. The centrifugal force acting as fluid traverses the bend is therefore greater near the axis than near the wall. A radial circulation pattern develops, whereby fluid near the axis moves outwards and then returns inwards near the wall. This secondary flow produces a closer approach to plug flow than is obtained when only the primary flow (with its parabolic velocity distribution) operates.

Figure 5.4 shows sketches of the f and F curves which are obtained when Taylor diffusion may be neglected.

5.4.4   Turbulent flow in a pipe

Although turbulent pipe flow may be well approximated as plug flow, the presence of the pipe wall, where the velocity is zero (no slip condition), produces a boundary layer over the whole cross-section of the flow and hence deviations from strict plug flow. There is a steep velocity gradient near the wall, while on the axis the mean velocity is exceeded by some 20%. The disordered movement of eddies causes radial mixing, so that rather uniform temperatures and compositions are found in the radial direction. Turbulence promotes plug flow in this sense. In the axial direction, however, such fluctuations represent deviations from plug flow and are usually modelled with an eddy diffusivity, D_t. It is assumed that D_t has a single value, irrespective of direction at a given position and irrespective of position in the flow. This assumption of isotropic, homogeneous turbulence is at best acceptable in the core of the flow: near the wall it breaks down. Despite these physical limitations, almost all RTD measurements have been analysed using the simple model according to which $D_t =$ constant. The eddy diffusivity is practically independent of the diffusing solute, unlike the molecular diffusivity, and is strongly influenced by the turbulence. Expressed non-dimensionally by dividing by the internal pipe diameter, d, and the mean velocity, u, the 'diffusivity group' becomes D_t/ud. Experiments have shown that this dimensionless group is roughly 0.2 in fully developed pipe flow.

Detailed development of the influence of eddy diffusion superimposed on a plug flow has shown that, for small levels of dispersion, the variance of the dimensionless residence time θ is given by

$$\sigma^2 = 2D_t/uL \qquad (5.28)$$

When the length (L) to diameter (d) ratio equals 100, $\sigma^2 = 0.004$ and the coefficient of variation of the RTD is 6.3%. With long pipes, deviations from plug flow are often smaller than other uncertainties and errors in typical design calculations and are usually neglected. Various fittings in pipelines (elbows, tees, valves, pumps, etc.) do however increase eddy diffusion and deviations from plug flow relative to straight pipe runs.

Example 5.3
A solution whose kinematic viscosity is 10^{-6} m^2 s^{-1} flows with a mean velocity of 1 m s^{-1} through a straight tubular reactor (diameter = 0.1, m, length = 20 m). Is it reasonable to represent this reactor with a plug flow model?

Solution
The Reynolds number of this flow is 10^5, indicating fully developed turbulence. The turbulent diffusivity is accordingly given approximately by

$$D_t = 0.2ud = 0.02 \text{ m}^2 \text{ s}^{-1}$$

From equation (5.28) the variance of the dimensionless residence time, θ, is equal to 0.002. (The mean residence time, $\bar{t}$, is 20 s and so the standard deviation of the residence time, t, is $20 \times (0.002)^{1/2} = 0.89$ s.) When D_t/uL

(here equal to 0.001) is smaller than about 0.01, the RTD is Gaussian. From tables of the normal distribution, 95% of all fluid elements traverse the tube in times between 18.25 and 21.75 s. This represents a good approximation to plug flow, which would give a single residence time of 20 s. The error in using this model is likely to be small compared to uncertainties in other factors such as the reaction mechanism, rate constant and temperature.

If the flow had been represented by a cascade of N stirred tanks, then from equation (5.24) $N = 1/0.002 = 500$. Once again such a cascade closely approximates plug flow.

Figure 5.4 shows sketches of the f and F curves for turbulent pipe flow.

5.4.5   Packed beds

When the particle Reynolds number in a packed bed indicates well-developed turbulence, the particles cause random deviations from plug flow which are not unlike the eddy movements in pipe flow. The RTD can then be modelled as described in the previous section.

5.4.6   Turbulent tubular reactor with recycle

As seen in Section 5.4.4, turbulent flow in a tube is well approximated by plug flow. When, however, part of the fluid exiting the tube is recycled to the inlet, substantial deviations from plug flow can arise. We will again denote the volumetric flow rate through the whole device as ν. The flow rate in the recycle loop is defined as $R\nu$, where R is the recycle ratio. The volume in the tube is denoted by V and is assumed to be large compared to the capacity of the recycle loop. The mean residence time in the whole device, $\bar{t}$, is equal to V/ν. The flow rate through the tube consists, however, of the feed (ν) and the recycle ($R\nu$), so that the mean residence time there per pass is $\bar{t}/(1 + R)$. Detailed analysis gives the variance of θ as

$$\sigma^2 = R/(1 + R) \qquad (5.29)$$

With no recycle, $\sigma = 0$ and obviously plug flow exists. As R tends to infinity, σ tends to 1, so that the standard deviation of the residence time t tends to $\bar{t}$: the whole device behaves like the well-mixed CSTR considered in Section 5.4.1.

Example 5.4

A turbulent tubular reactor is operated with a recycle loop whose volume is negligible compared to that of the tube. RTD measurements on the whole device gave a mean and a standard deviation of 4 min and 2 min respectively. The conversion of a first-order reaction, whose rate constant $k = 0.5\,\mathrm{min}^{-1}$, is to be calculated.

The result should be compared with the conversions obtainable when the recycle ratio is (a) zero and (b) extremely high.

Solution

To interconvert between dimensional (t) and dimensionless (θ) residence times, a standard result in statistics will be recalled. If x is a random variable whose variance is σ^2, and $y = ax$ is another random variable, where a is a constant, then the variance of y is equal to $a^2\sigma^2$. Here $x = \theta, y = t$ and $a = \bar{t}$, so that the variance of θ equals $(2/4)^2$. From equation (5.29), $0.25 = R/(1 + R)$, giving $R = 1/3$.

Denoting the reagent concentrations in the feed to the plug flow region by C', in the feed to the whole device by C_0 and in the stream leaving both the plug flow region and the whole device by C, the mass balance before the tubular reactor is

$$\nu C_0 + R\nu C = (1 + R)\nu C' \qquad (i)$$

Substituting for R

$$C' = (3C_0 + C)/4 \qquad (i')$$

The mean residence time per pass ($\bar{t}'$) is then

$$\bar{t}' = V/(1 + R)\nu = \bar{t}/(1 + R) = 3\,\text{min} \qquad (ii)$$

For isothermal, first-order reaction in the tube

$$C = C'\exp(-k\bar{t}') \qquad (iii)$$

Here $k\bar{t}' = 1.5$ and substituting from (i') in (iii) for C', it follows that $C = 0.177C_0$ and conversion $= 82.3\%$.

(a) When $R = 0, C' = C_0$ and $\bar{t}' = 4$ min. From (iii), $C = 0.135C_0$ and conversion $= 86.5\%$. The deviation from plug flow caused by the recycle causes a loss of 4.2% in conversion.

(b) When R increases without limit, the reactor behaves like a CSTR, so that the mass balance is

$$\nu C_0 = \nu C + kCV$$

where $V/\nu = 4$ min. Hence, $C = C_0/3$ and conversion $= 66.7\%$.

5.4.7 Dead zones and short circuits

A dead zone (region of no flow) can be directly detected by flow visualisation (see Section 5.2). Dividing the total capacity (V) into active (V_a) and dead (V_d) volumes $V = V_a + V_d$ and the measurable mean residence time ($\bar{t}$) will then be V_a/ν. This is obviously shorter than V/ν, which provides a test for the presence of a dead zone.

A short-circuit refers to a part of the throughput by-passing the equipment. Detailed analysis shows that this does not cause the measured mean retention time to deviate from V/ν, because that fluid which exists too early is compensated by the remainder which leaves later than V/ν. However, when using pulse injection, there will be a sharp peak near $t = 0$, while with step injection a sharp step near $t = 0$ will be recorded.

The six cases described earlier in Sections 5.4.1 to 5.4.6 did not exhibit these two phenomena. They were, however, included in Example 5.1, which shows an appropriate way to calculate conversion when a dead zone or a short-circuit is present in a reactor.

5.5   How can non-ideal flow be quantitatively described?

The short answer is through the RTD, which in many cases means measuring it. When the f curve shows a peak at $t = 0$, short-circuiting is indicated. When the mean residence time $\bar{t}$, evaluated with equation (5.3), is smaller than the nominal value V/ν, a dead zone is indicated. Equation (5.8) should be applied to the measurements to decide whether they satisfy the tracer's mass balance. Application of equation (5.4) to determine the variance of the RTD can encounter problems in the tail ($t \gg \bar{t}$) due to inaccuracy or scatter in the exit concentration measurements.

When no dead zone and no short-circuit are observed, the RTD is bracketed between those for plug flow and perfect mixing. It can then often be represented or modelled by one of the cases considered in Section 5.4.2 (the stirred tank cascade), 5.4.4 (axial turbulent dispersion) or 5.4.6 (plug flow with recycle). Thus a one-parameter flow model, having the same mean retention time and the same variance as the real system, is chosen to represent the real flow. This representation is seldom complete because distributions can differ in detail even when their means and variances are equal. It can, however, be adequate to estimate the performance of the real system and is superior to assuming either plug flow or perfect mixing. If the chosen model is the stirred tank cascade considered in Section 5.4.2, for example, the number of equally sized tanks in the cascade can be found directly by applying equation (5.21). If, however, the model is plug flow with recycle, then equation (5.29) can be solved for R, the recycle ratio. This has already been illustrated in Example 5.4.

5.6   Calculation of performance when flow is non-ideal

When n fluid elements enter simultaneously into a continuously operated unit, they will in general leave at different times, i.e. there exists a distribution of residence times. By definition an element retains its identity in the unit while it undergoes changes due, for example, to heat transfer, mass transfer or chemical reaction. The extent to which such changes occur in an element depends (among other things) upon how long the element spends in the unit, i.e. upon its retention time. The distributions of temperature and humidity among the particles leaving a cement kiln and a dryer, respectively, depend in some way on the RTD in these units. The same applies to the particle size distribution from a crystalliser.

From equation (5.1) the number of fluid elements exiting with retention times between t and $t + dt$ is equal to $nf(t)dt$, while the concentration in such elements, which depends on the retention time, is $C(t)$. The average concentration, $\bar{C}$, in all n elements follows from a mass balance or weighted average

$$n\bar{C} = \int C(t)dn = n \int C(t)f(t)dt$$

Hence

$$\bar{C} = \int C(t)f(t)dt \qquad (5.30)$$

Elements, which retain their identities (Figure 5.2) while they traverse a unit, are said to be totally segregated from their environment and from one another. This applies to gas bubbles and liquid drops when no coalescence or break-up occurs and to solid particles when no aggregation or fracture takes place. Even in miscible fluids elements can be almost segregated. This is the case, for example, when the process occurring in an element is very rapid compared to rates of exchange with the environment and also when the fluid possesses a yield stress and forms gels. Segregated fluid elements are, however, not formed in all circumstances and equation (5.30) is sometimes inapplicable.

The rate of a first-order, irreversible, isothermal reaction is directly proportional to the reagent concentration, $C(t)$. It follows that

$$C(t) = C_0 \exp(-kt) \qquad (5.31)$$

where k is the rate constant. When this equation is substituted into equation (5.30), the mean reagent concentration leaving an isothermally operated reactor having **any** RTD may be calculated. Taking as an example a well-mixed tank, whose RTD is given by equation (5.12), equations (5.30) and (5.31) lead to $\bar{C} = C_0/(1 + k\bar{t})$. This well-known result can also be derived from a simple mass balance which, unlike equation (5.30), ignores segregation and treats the fluid as fully homogeneous. A first-order reaction and indeed a set of such reactions is unaffected by any segregation in a reacting mixture. If, for example, a mixture could be divided into two halves, one free of reagent and the other containing all of it, it is obvious that the reaction rate in the latter half would be exactly double that in the original, homogeneous mixture. While, however, reaction only takes place where reagent is present, the average reaction rate over the two segregated halves would equal that in the original mixture. This compensating effect relies on the linearity between reaction rate and reagent concentration and applies not only to first-order reactions but to other linear rate processes (e.g. conductive and convective heat transfer and mass transfer).

Example 5.5
The results of a RTD measurement, using step injection of tracer, can be represented by

$$F = 0 \qquad\qquad (t < 0.4\,\text{ks})$$
$$F = 1 - \exp[-1.25(t - 0.4)] \quad (t > 0.4\,\text{ks})$$

where t is in kiloseconds.

Calculate (a) the mean residence time in this equipment, and (b) the conversion of a first-order reaction, whose rate constant $k = 2\,\text{ks}^{-1}$.

Solution
(a) From equation (5.3)

$$\bar{t} = \int_{0.4}^{\infty} t f(t)\mathrm{d}t$$

while from equation (5.2),

$$f(t) = \mathrm{d}F/\mathrm{d}t = 1.25\exp[-1.25(t - 0.4)]$$

Substituting and integrating by parts gives $\bar{t} = 1.2\,\text{ks}$.

(b) Since a first-order reaction is not influenced by segregation, conversion can be found by substituting equation (5.31) into equation (5.30) and integrating to give

$$\frac{\bar{C}}{C_0} = \int_{0.4}^{\infty} 1.25\exp[-1.25(t - 0.4)]\exp(-2t) = 0.1728$$

Conversion $= 82.7\%$.

An alternative way is to express the RTD in terms of a flow model and then to calculate its performance for the first-order reaction. The RTD shows plug flow during 0.4 ks and the model would incorporate a plug flow element in series with an ideal macro-mixing element whose mean residence time $= 1.2 - 0.4 = 0.8\,\text{ks}$. The sequence in which the flow encounters these two elements has no influence on a first-order reaction, nor does it matter whether the tank is segregated on the molecular scale, so

$$\frac{\bar{C}}{C_0} = \frac{\exp(-kt_1)}{1 + kt_2}$$

where $k = 2$, $t_1 = 0.4$ and $t_2 = 0.8$, giving $\bar{C} = 0.1782C_0$ as before.

Many chemical reactions exhibit second-order kinetics whose conversions depend not only upon the RTD, but also upon the extent of any segregation present. This statement applies in fact to any reaction exhibiting non-linear kinetics. With equal reagent concentrations and second-order kinetics

$$C(t) = C_0/(1 + kC_0t) \tag{5.32}$$

A stirred tank reactor whose bulk scale mixing is complete and given by equation (5.12), but which is completely segregated on the fine scale, follows equation (5.30). Combining this with equation (5.32) allows the conversion of a second-order reaction in such a reactor to be calculated. This shows a somewhat higher conversion in the segregated reactor, the maximum difference between this conversion and that in a classical, non-segregated CSTR being 7%.

Example 5.6
The length of a tubular reactor (diameter 0.1 m), which is intended to carry out a second-order reaction to 80% conversion, has been calculated on the assumption of plug flow. It has subsequently been found that the Reynolds number is low enough for laminar flow. What is the likely effect on the conversion if the mean residence time is 100 s?

Solution
Equation (5.27) for the RTD is applicable, when insufficient time is available for significant radial molecular diffusion to occur, i.e. when $\bar{t} \ll 0.1R^2/D$. Here R (radius of pipe) $= 0.05$ m and D (diffusivity) is unlikely to exceed $10^{-9} \text{ m}^2 \text{ s}^{-1}$. Hence $0.1R^2/D = 2.5 \times 10^5$ s, whereas $\bar{t} = 100$ s. Taylor diffusion may be neglected and the laminar flow is segregated. Substituting equation (5.27) and, for second-order reaction, equation (5.32) into equation (5.30) and integrating by parts

$$\frac{\bar{C}}{C_0} = \int_{0.5t}^{\infty} \frac{\bar{t}^2 dt}{2t^3(1 + kC_0t)} = 1 - a + \frac{a^2}{2}\ln\left(\frac{2 + a}{a}\right)$$

where $a = kC_0\bar{t}$.

The plug flow design equation (5.27), obtained by directly integrating the second-order rate law gives, for a conversion of 80%,

$$0.2 = (1 + a)^{-1}$$

When $a = 4$, the equation for conversion in laminar pipe flow shows that $\bar{C}/C_0 = 0.24327$, corresponding to 75.6% conversion. The likely effect of laminar flow is to reduce conversion by 4.4%. Insertion of a static (or in-line) mixer would greatly improve radial mixing and increase conversion towards 80%.

To determine the RTD, the transient outlet concentration of a non-reactive tracer is measured. The volume included in a typical measuring cell is seldom less than 1 ml, which corresponds to around 3×10^{22} molecules in an aqueous solution. Even when the RTD, measured from sampling at this scale, follows equation (5.12) and indicates perfect mixing, it is quite possible for much finer samples to be less well mixed and, in particular to be segregated. This effect of the scale at which mixing takes place is brought out by the prefixes 'macro' and 'micro'. Blending at the scale of the whole tank can be judged by the ideality of the RTD, which should agree with equation (5.12) when the tank's contents are homogeneous or macro-mixed. When segregation is absent, so that a random distribution of molecules exists, the mixture is uniform on the molecular scale or micro-mixed.

In addition to states of complete segregation and complete homogeneity, partial segregation or incomplete mixing can also occur. Defining the mean concentration and the local concentration by $\bar{C}$ and C, respectively, the local fluctuation in concentration, C', is given by $C' = C - \bar{C}$. The local rate r of a first-order reaction is then equal to

$$r = kC = k(\bar{C} + C')$$

Averaging over the whole mixture at a given time shows that

$$\bar{r} = k\bar{C}$$

because the average of the fluctuation must be zero. Even when segregation is present $(C^1 \neq 0)$, it also has no effect on the average rate of a first-order reaction, as was noted earlier.

For a second-order reaction the local rate may be written

$$r = kC^2 = k(\bar{C} + C')^2$$

This is again averaged over the whole mixture, giving

$$\bar{r} = k\bar{C}^2 + k\overline{(C')^2}$$

Any segregation present enhances the rate relative to a perfectly micro-mixed system $(C' = 0)$, which was also noted earlier.

The method of concentration fluctuations offers a simple way to decide whether molecular scale segregation will influence a particular reaction and, if so, in which direction. It does not presently offer a prediction of the extent of such an influence.

Does a RTD provide sufficient information about a non-ideal flow to enable the extent of a chemical reaction to be calculated? Unfortunately not. It does enable more accurate calculations to be carried out than is possible using an ideal flow model, but is limited to (a) single feed stream, (b) known mixing sequence, unless the reaction is first order, and (c) complete segregation at the molecular level, unless the reaction is first order.

The single feed stream to a reactor must contain all necessary reagents, initiators and catalyst (if needed). Its entry into the reactor, and especially the way in which it mixes with the reactor's contents, are in many ways analogous to the introduction of a tracer when measuring an RTD. Multiple feed streams, e.g. with one reactant in each, are also employed when feeding highly reactive substances. RTD measurement in this situation gives some information about how the feed stream containing the tracer mixes with the vessel contents, but does not tell us much about how the various feeds mix with each other. The mixing of these reagents is, however, of great import-ance for the reaction.

On its passage through a reactor, fluid often encounters different flow regimes. The sequence of these regimes cannot be deduced from RTD measurements but it can, however, influence the conversion of non-linear reactions. Suppose, for example, that the flow in a reactor can be divided into a zone with plug flow and a zone with perfect mixing. By sketching the overall RTD it is easy to recognise that, whichever zone comes first, the RTD will be the same. The conversion of a first-order reaction is also independent of the sequence. For second-order kinetics the conversion is higher when the plug flow region is encountered first.

Example 5.7
A plug flow reactor having a mean residence time of 100 s and a perfectly mixed tank reactor having a mean residence time of 300 s are connected in series. The direction of flow through this cascade is unspecified. Calculate the conversion to be expected in the following two cases, assuming isother-mal operation:
(a) first-order reaction: $k = 0.005 \text{ s}^{-1}$
(b) second-order reaction: $k = 2.5 \times 10^{-5} \text{ m}^3 \text{ mol}^{-1} \text{ s}^{-1}$ and feed concentra-tion $C_0 = 10^3 \text{ mol m}^{-3}$

Solution
(a) First-order reaction
Using the standard design equations for the two reactor types it can be shown that for **either** flow direction the conversion is given by

$$X = 1 - 0.4 \exp(-0.5) \text{ or } 75.74\%$$

(b) Second-order reaction
Plug flow reactor first: the standard equations show conversions of $1 - 3.5^{-1}$ or 71.43% after the PFR rising to 86.0% after the stirred tank.

Stirred tank first: conversion in the tank, 69.55% rising to 82.7% after the plug flow reactor.

The higher conversion when the tubular reactor is placed first is due to the dilution (mixing) in the tank occurring as late as possible in the cascade. When the tank is first, dilution of the feed occurs immediately upon entry to the cascade, thus lowering the reaction rate to an extent which cannot subsequently be compensated.

5.7   Concluding remarks

The residence time distribution offers a way to learn more about some aspects of non-ideal flows. The experimental technique is easy to implement even on an industrial scale. Applied to simple reactions it allows exact prediction of conversion to be made for a first-order reaction in a non-ideal reactor. Because, however, the RTD does not detect any molecular-scale segregation which might be present, it does not provide a basis for such exact predictions for reactions of second-order or other orders. Nevertheless, simple reactions are rather insensitive to segregation and predicting the conversion of a second-order reaction using the kinetics and the RTD usually does not incur serious error. When multiple reactions occur, the error in predicting product distribution from a non-ideal reactor using the kinetics and the RTD can be more significant. The RTD fails to distinguish the sequence in which the fluid encounters different flow regimes and regards the fluid as totally segregated, which might not be appropriate.

In the last paragraph of Section 5.6, a single feed stream was considered, which contained all the reagents, initiators, catalyst, etc., needed for reaction. In some situations multiple feed streams are employed and the RTD approach is then only of limited value. This is especially the case when the reagents enter in different feed streams and react rapidly with each other. The RTD does not characterise the extent to which various reagents mix and the rate of such mixing: additional information is required. In combustion, for example, the paths through the combuster followed by the fuel and the air respectively are not irrelevant. Information about the rate and extent of mixing between these streams is required before predictions of the burning rate, the production of heat and the rate of formation of products such as CO_2, and NO_x can be made.

An analogous example is the mixing of a stream containing barium ions with a stream of sulphate ions in the precipitation of barium sulphate. The overall flow patterns of the streams need to be known, but the details of the mixing of these two species influence the rate of precipitation and, more importantly, the instantaneous supersaturation, which determines the rates of nucleation and crystal growth and many product properties. The nitration of a reactive aromatic compound provides a third example. In addition to primary nitration forming a mononitro-compound, secondary nitration producing a dinitro-compound frequently occurs too. The product distribution shifts to more secondary – and therefore less primary – nitration as the reactivity increases relative to the attainable rate of mixing the two reagents (the aromatic compound and nitronium ions). Complete analysis of the behaviour of a reactor needs not only information about the RTD, but also about the rate of mixing of the reagents.

Summing up the RTD is necessary, but not always sufficient to characterise reactions in non-ideal reactors.

6 Catalyst design and manufacture

E.M. HOLT, G.J. KELLY and F. KING

6.1 Introduction

Most chemical reactions in the chemical and petrochemical process industries are heterogeneously catalysed. The design of a catalyst for any particular chemical reactor necessitates a compromise of properties in order to obtain the optimal commercial process. The choice of the active catalytic species is usually made by experiment, but this is strongly influenced by experience and the scientific literature. This chapter does not attempt to teach the methods of active species selection but concerns itself with a brief outline of the design and fabrication of heterogeneous catalysts for chemical reactors. For readers who are interested in a greater understanding of the subject, a bibliography of selected texts is given at the end of the chapter. Independent design of catalyst and reactor is highly undesirable since this can result in a more costly design, a lower production rate and a more frequent catalyst replacement. The catalyst properties that have to be considered include activity, selectivity, crush strength, abrasion resistance and resistance to fluid flow. Consequently the design of a catalyst involves both the active phase and the support. Typical forms of industrial catalysts include powder, pellets, rings, spheres, extrudates or granules. A few of the pellet shapes available are shown in Figure 6.1.

Metals can also be used in the form of crystals or woven wire or spun metal. More recent innovations include the monolithic supports in their various forms (see Figure (6.2)) and ceramic foams. The desirable properties of a catalyst can be related to the variable parameters inherent in the method used for manufacture.

6.1.1 Development guidelines

The design and development of a commercial catalyst has to be a compromise of often conflicting requirements. An ideal industrial catalyst should have activity, optimum selectivity, yield strength and abrasion resistance. Moreover, the catalyst should also have a minimum resistance to fluid flow and be relatively cheap to manufacture.

Fulton [1], in the opening article of a series [1–7] on catalyst engineering, proposed a sequence of eight steps necessary to take a catalyst developed in the laboratory to commercial use. These steps can be summarised as follows:

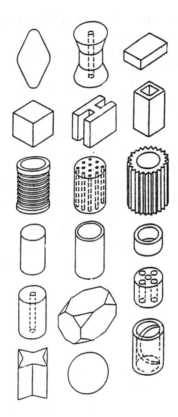

Figure 6.1 Catalyst pellet shapes (reproduced from *Chemical Engineering*, 1986, 97).

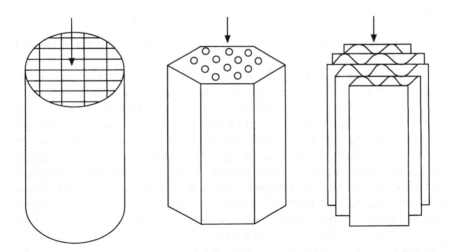

Figure 6.2 Examples of catalytic monoliths (reproduced from *Chemistry and Industry*, 1987, 315).

1. Develop a mathematical model that quantifies the phenomena that influence the performance of the catalyst.
2. Select the catalyst support, the particle size, the particle shape, the mean pore size and the pore size distribution.
3. Establish the method of catalyst manufacture, i.e. decide upon the fabrication, drying and calcining methods.
4. Verify experimentally the performance of the catalyst to determine a reaction rate equation which fits the laboratory results.
5. Evaluate the operating economics of the technically suitable catalysts in order to select the most appropriate candidate.
6. Establish catalyst handling procedures to be used with the process. These include the conditions of charging, discharging and activation.
7. Determine methods of regenerating the catalyst, which may have a significant impact on process economics. Moreover, the problems involved in the disposal of spent catalyst can no longer be considered trivial. Consequently, catalysts with long lives are becoming increasingly desirable.
8. Be prepared to handle reactor breakdowns or maloperation and minimise where possible the damage to the catalyst.

The matters described in this chapter are primarily concerned with guidelines 2 and 3 described above. The development process described by Fulton does, however, emphasise how the design and scale-up of a catalyst fits into the development of a catalyst for a particular process, or the parallel activities that occur concurrent with process flowsheet development for a new catalytic process.

6.1.2   Influence of mass transport on catalyst behaviour

The behaviour of a heterogeneous catalyst in a catalytic reactor is influenced by three transport processes, as shown in Figure 6.3.

As the fluid passes through the reactor it flows around the exterior of the catalyst pellet or particle. If a reaction occurs, a concentration and temperature gradient will be established between the inlet and outlet of the reactor. The transport of reactants through the gas phase to the exterior of the gas film surrounding each pellet is called **intrareactor transport**. The diffusion of the reactants through the stagnant gas film around the pellet is called **interphase transport** and the movement of reactants through the pore system of the catalyst is known as **intrapellet transport**. Distinctions are often drawn between catalysed reactions which are film-diffusion controlled (i.e. whose rate is limited by poor interphase transport) and those which are 'pore-diffusion' controlled (i.e. whose rate is limited by poor intrapellet transport). The thickness of the stagnant gas boundary layer is primarily determined by bulk gas flow but can also be influenced by rough catalyst surfaces which

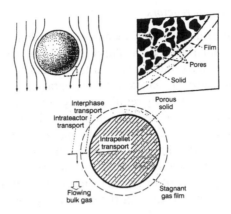

Figure 6.3 Intrareactor, interphase and intrapellet transport (reproduced from *Chemical Engineering*, 1986, 118).

encourage mixing. The internal diffusion rate is dependent upon the nature of the catalyst pore structure and the reactant fluid. If this rate is low, the catalyst material which is furthest from the outer surface of the pellet does not contribute effectively to the reaction process to which it becomes only partially available, as discussed in Section 3.6.1. A measure of the availability of a catalyst pellet's internal active sites is provided by the effectiveness factor.

The size, shape and uniformity of packing of catalysts can have a profound effect on intrareactor transport. Poor fluid flow distribution through, for example, a fixed bed can have a significant impact on processing efficiency and overall chemical yield.

6.1.3 Influence of heat transfer

For reactions which are strongly endo- or exothermic, heat transfer becomes an important part of the reactor design. Typically the surface area of the reactor is increased by packing the catalyst either into or around tubes. Intrareactor transport from the surface to the centre of each bed or section of bed is required to convey heat to or from the catalyst surface. This movement of gas perpendicular to the main fluid flow has a significant pressure drop associated with it and requires the pellets to have an optimum diameter. The ratio of the diameter of the reactor to the diameter of the pellet should typically be greater than 7. Clearly the pressure drop/heat transfer requirements need to be optimised with the other catalyst requirements. As the temperature of reaction increases, the impact of fluid movement on heat transfer is reduced until radiation becomes the dominant mechanism.

6.1.4   Types of catalyst reactors

For the purpose of catalyst design, reactors can be divided into two general classes: fixed bed and fluidised or slurry bed. In a fixed bed reactor the catalyst remains stationary and the reactants are passed through the catalyst. In the fluidised and slurry reactors both the catalyst and reacting medium move. Fixed bed reactors usually require catalysts whose activity and selectivity remain stable for a considerable time, typically in the order of years. Catalysts with a shorter life of, for example, months may be used in a fixed bed but will then probably have to withstand multiple regeneration. Fluid bed reactors are often used with rapidly deactivating catalyst and allow either continuous replacement or regeneration to be performed. A summary of the impact of reactor selection on the design requirements of heterogeneous catalysts is given in Table 6.1.

The catalyst particles used in fixed bed reactors usually have a crush strength of at least 3 kgf per particle (or an axial crush strength of 5–8 $MN\,m^{-2}$ [1]) to avoid the production of fines which could cause an unacceptable increase in pressure drop in the reactor. Fluidisable catalysts generally have a particle size of 30–200 μm [1], the size being dictated by the operating conditions of the appropriate reactor. The selection criteria for the different catalytic reactors are summarised in Table 6.2.

Table 6.1   Catalyst design criteria

Fixed bed reactors	High catalyst strength Porosity Low pressure drop Minimum pore diffusional resistance Thermal/hydrothermal stability
Fluid bed reactors	High attrition resistance Resistance to 'spalling' Tight particle size range Minimum pore diffusional resistance Thermal/hydrothermal stability

Table 6.2   Comparison of reactor catalyst requirements

	Fixed beds	Fluidised beds
Reaction conditions	Favoured for high pressures	Favoured for strongly exo or endothermic reactions
Deactivation behaviour	Catalyst must have a long life	Can utilise rapidly deactivating catalyst
Catalyst strength	Catalyst required high strength	Must have adequate attrition resistance

6.2 Types of catalysts

Catalysts can be made from a wide range of materials including metals, metal oxides, metal sulphides and insulating oxides. Depending on use, the catalytic materials can be prepared either in an unsupported form or supported on a carrier. The actual form of the catalyst is dictated by the intended reactor and the reaction to be catalysed.

6.2.1 Unsupported metals

Few industrial catalysts are used in the form of unsupported metals. Usually a catalytic process requires a large metal surface area to minimise the catalyst requirements. However, there are some processes in which the rate of reaction per unit area of surface is so fast that a small active metal area is sufficient. Unsupported metals can be found in a variety of forms: wires, foils, gauzes, blacks, powders and skeletal (Raney) species.

Wires, foils and gauzes are often termed massive metals since the metal is present in the form of massive metal particles. Typically these catalysts are used for high-temperature mass transport controlled reactions.

A 'black' is a metallic powder which is obtained by the reduction of a metal salt or by condensation of a metal vapour. Blacks and powders are usually composed of relatively large metal particles having a low surface area. These materials are rarely used on a large scale but may be employed in batch slurry reactors in the pharmaceutical industry.

Skeletal metals are produced by the leaching out of one component from an alloy, leaving the active species behind in the form of a porous material having a high surface area. The best-known example of this type of catalyst is Raney nickel, which is prepared from a nickel–aluminium alloy by leaching out most of the aluminium with caustic solution to leave behind a porous nickel catalyst. Raney nickel is an excellent hydrogenation catalyst for use in slurry reactors but is gradually being replaced by supported alternatives.

Three examples of reactions catalysed by unsupported metals are shown in Table 6.3.

Table 6.3 Reactions catalysed by unsupported metals

Reaction	Catalyst and form
(i) Oxidation of ammonia $4NH_3 + 5O_2 \rightarrow 4NO + 6H_2O$	Platinum/rhodium gauzes
(ii) Methyl alcohol oxidation to formaldehyde $CH_3OH + \frac{1}{2}O_2 \rightarrow CH_2O + H_2O$	Silver granules or gauze
(iii) Hydration of acrylonitrile to acrylamide $CH_2CHCN + H_2O \rightarrow CH_2CHCONH_2$	Raney copper

6.2.2   Fused catalysts

The manufacture of catalysts by the fusion of oxides is not extensively used. The product of fusion has negligible surface area and has to acquire this in catalyst activation by, for example, reduction. The most important catalyst produced in this manner is the ammonia synthesis catalyst. This catalyst is manufactured by the fusion of magnetite together with appropriate promoters. Obviously conventional forming techniques are not possible with fused catalysts and the material is cast, crushed and graded to an appropriate dimension range.

6.2.3   Oxide catalysts

Oxide catalysts can be conveniently divided into two general categories: electrical insulators or semiconductors. The insulators are materials where the cationic species has a single valence state. Examples of such insulating oxides include magnesium oxide, aluminium oxide, silicon dioxide and zeolites (aluminosilicates). These materials find most use as solid acids or bases and as supports for other catalytically active species.

Semiconductor oxides are materials in which the metallic species is relatively easily cycled between two valence states. This can be between two different oxidation states as in

$$Fe_2O_3 \leftrightarrow Fe_3O_4$$

or the conversion between the positive ion and the neutral metal, as with the more easily reduced oxides such as ZnO and CdO. The semiconductor oxides are most commonly used in selective oxidation reactions. Examples of reactions which can be catalysed by oxides are shown in Table 6.4.

Table 6.4   Reactions catalysed by oxides

Reaction	Catalyst and form
(i) Oxidation of n-butane to maleic anhydride $n\text{-}C_4H_{10} + 3O_2 \rightarrow C_4H_2O_3 + 3H_2O$	V_2O_5 pellets or spheres
(ii) Methyl alcohol oxidation to formaldehyde $CH_3OH + \frac{1}{2}O_2 \rightarrow CH_2O + H_2O$	Fe–Mo/oxide pellets
(iii) Catalytic cracking of naphthas	Zeolite/SiO_2–Al_2O_3 microspherical particles

6.2.4   Supported catalysts

Catalytic species are often dispersed over support materials. The primary aim of applying a catalytically active component to a support is to improve dispersion significantly and produce a highly active material. The early concept of a support or carrier was as an inert substance that provided a

Table 6.5 Physical and chemical properties of supports

Physical properties	Chemical properties
(i) Mechanical strength	(i) Inert to undesired reactions
(ii) Optimised bulk density	(ii) Stable under reaction and regeneration conditions
(iii) Provide heat sink or source	(iii) React with catalyst to improve specific activity or
(iv) Dilute overactive phase	selectivity
(v) Increase active surface area	(iv) Stabilise the catalyst against sintering
(vi) Optimise catalyst porosity	(v) Minimise catalyst poisoning
(vii) Optimise the metal crystal and particle size	

means of spreading out an expensive catalyst ingredient, such as a precious metal, in order to achieve a greater effective utilisation of the metal compared with a bulk metal system, for example a platinum black. However, with base metal catalysts the use of the support is often primarily aimed at achieving improved catalyst stability. This can be achieved by a suitable interaction of the active component with the support. The choice of support is therefore of crucial importance for the design of a catalyst. The desirable physical and chemical characteristics to be considered in the choice of a support are listed in Table 6.5.

Even a brief perusal of the characteristics summarised in Table 6.5 should indicate that no support could fulfil and optimise all of these requirements. The most common support materials are alumina, silica, zeolites and activated carbon. Table 6.6 summarises the total sales of catalyst support materials for 1990 in Western Europe. Other materials which have some limited use as supports include titania, magnesia, chromia and zirconia. The use of chromia as a catalyst support is in decline because of its toxic nature.

Table 6.6 Consumption catalyst substrates in Western Europe in 1990

Substrate	Tonnes per year
Alumina	47 315
Silica	45 224
Zeolite	24 473
Monolith cordierite	3 600
Activated carbon	825

As has previously been indicated (see Figures 6.1 and 6.2), support materials can be obtained in a variety of forms such as spheres, granules, extrudates, cylinders and powders. These can then be impregnated with a metal salt of the desired active phase. Alternatively, a powdered support may be incorporated into a mixture to be precipitated, or the support may be precipitated from solution in the manufacturing process. Structured supports such as monoliths (see Figure 6.2) are unusual in that the active species is usually dispersed on a high surface area washcoat covering the surface.

Table 6.7 Reactions catalysed by supported catalysts

Reaction	Catalyst and form
(i) Ethyne hydrogenation $C_2H_2 \rightarrow C_2H_4$	Pd/Al$_2$O$_3$ pellets and spheres
(ii) Methyl alcohol synthesis $CO + 2H_2 \rightarrow CH_3OH$	Cu/ZnO/Al$_2$O$_3$ pellets
(iii) NO$_x$ abatement $2H_2 + 2NO \rightarrow 2H_2O + N_2$	Pt impregnated washcoated monoliths

The monolith is typically a single block of ceramic material containing an array of parallel uniform, straight non-connecting channels. This type of support is particularly useful for its very low pressure drop and this advantage is utilised in the car exhaust catalyst to minimise power loss from the engine. Table 6.7 shows some examples of supported catalysts.

6.3 Methods of manufacture

This section will concern itself with the production of supported catalysts because they represent the majority of catalysts in industrial use and offer the catalyst manufacturer many different options in catalyst design.

6.3.1 Incorporation of the active species

Although a variety of techniques has been employed to incorporate an active species onto a support material [8] only two, impregnation and precipitation, are widely employed. Impregnation is achieved by filling the pores of a preformed support with a solution of a metal salt. Impregnation is followed by a thermal treatment in either an inert or an active atmosphere to achieve the dispersion of the active component in either a metallic or oxide form. The theory of the preparation of supported catalysts has been reviewed by Neimark et al. [9] and discussed by Acres et al. [8]. A truly remarkable range of metal placements can be achieved by a suitable choice of co-ingredient. Figure 6.4 illustrates the platinum profiles obtained with chloroplatinic acid and a range of co-ingredients competing for the adsorption sites.

If a mixed metal catalyst is to be produced by impregnation, care has to be taken to ensure that a component in an impregnating solution of salts is not selectively adsorbed. If this occurs an unexpected or undesired concentration profile of each component may result. Physical properties can also be used to influence the location of the active component in a support. Kotter and Riekert [10] demonstrated that the uniformity of the distribution of CuO on gamma alumina could be influenced by the viscosity of the impregnating solution and drying conditions. The advantages of controlling the

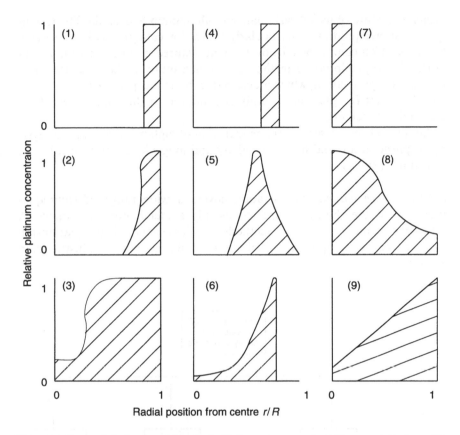

Figure 6.4 Types of platinum profiles obtainable by co-impregnation: (1) outer shell, sharp; (2) outer shell, diffuse; (3) outer shell diffuse to centre; (4) inner shell, sharp; (5) inner shell, diffuse; (6) inner shell diffuse to centre; (7) core, sharp; (8) core, diffuse; (9) linearly increasing from centre. (Reproduced from *Journal of Catalysis*, 1980, **63**, 425.)

surface/bulk distribution of the active component in the support has been summarised by Fulton [3] to be as follows.

1. For a mass transfer controlled reaction the catalyst concentration should be highest near the pellet's external surface.
2. If intrinsic kinetics controls the reaction then the catalyst should have a uniform distribution.
3. If the reactor feed contains a catalyst poison, a shell of catalyst support can be used to guard the active component.

The preparation of supported catalysts by the co-precipitation of the metal ions can produce an intimate mixing of the active component and the support matrix. For instance, an aqueous solution of nickel and aluminium nitrates can be mixed with a solution of sodium carbonate to produce an intimate mixture of nickel carbonate and aluminium hydroxide. The technique of co-precipitation is particularly useful for catalysts with high metal loadings which cannot be effectively manufactured by impregnating a support. The co-precipitation method uses much more of the catalytic metal than does impregnation, which is acceptable for an inexpensive metal such as nickel but not for platinum. Catalysts produced in this manner will have high uniform activity.

Andrew [11] has described the preparation of high metal loading catalysts by co-precipitation and how the activity and longevity is influenced by the preparation stages.

6.3.1.1 Impregnation. Impregnation covers a wide range of techniques used to disperse the active species, usually a metal, onto a support. A schematic summary of the methods available and their relationship with the active component distribution on the support is illustrated in Figure 6.5.

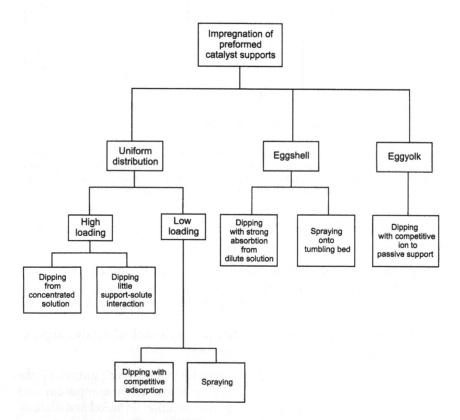

Figure 6.5 Different types of impregnation.

The selection of technique depends on the metal loading, the physical form of the support and the required profile of active species within the support.

Dipping from appropriate solutions can be used to give both high and low metal loadings. The method can also produce various distribution profiles depending on the interaction between the support and the impregnating solution components. Spraying is usually used to produce a low loading of the active component on the support at the surface or slightly subsurface. The latter is particularly important to produce catalysts that are more resistant to large poison molecules and removal of the active component by attrition. In its simplest form, dipping consists of immersing the support in a salt solution until the pores are filled and draining off the excess liquid before transferring to drying and calcining ovens. The anion used determines the fine structure of the deposited metal, and may have a significant effect on catalyst activity. The metal loading is determined by the porosity of the support and the concentration of the active component precursor in the solution. If high metal loadings are required the impregnation process may need to be repeated several times.

A rotating vessel can be used to disperse a predetermined volume of solution over the support. This is then followed by heating and purging to evaporate the solvent in situ and thus deposit a precise loading of active species on the carrier. These vessels can be used with an excess of liquid for high metal loadings, or with a liquid spray when a low metal loading is required at or just under the particle surface. The spraying technique is particularly convenient for the impregnation of powders and fine particles.

Methods of controlling the position of the active species are available to create an egg shell or egg yolk effect (see Figure 6.6 and reference [12]). The former is achieved by creating a strong interaction between the metal and the support, possibly requiring some addition to the support. Impregnation with a dilute solution then results in the metal being chemisorbed onto the outer pores, leaving little to penetrate to the centre of the support. The egg yolk effect requires some passivation of the support exterior by, for example, adsorption of a competing ion to allow full penetration of the metal into the pellet interior.

All these processes are usually followed by drying. This may need to be carefully controlled depending on the desired distribution, as some redistribution of the active species can occur as the liquid fronts move within the pore structure and create concentration gradients (see, for example, reference [10]).

6.3.1.2 Co-precipitation. The process for the production of catalysts by co-precipitation is depicted in Figure 6.7. The technique is particularly useful for the manufacture of catalysts which have a uniform blend of the active

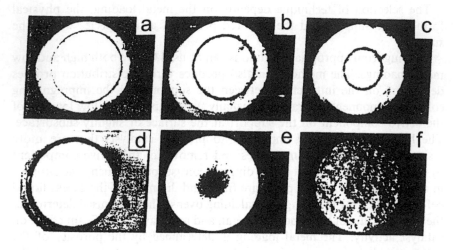

Figure 6.6 Controlled location of a metal: (a) 0.106 M chloroplatinic acid for 15 min followed by 1.0 M citric acid for 15 min; (b) 0.106 M chloroplatinic acid for 15 min followed by 1.0 M citric acid for 60 min; (c) 0.106 M chloroplatinic acid for 15 min followed by 3.0 M citric acid for 60 min; (d) 0.106 M chloroplatinic acid for 15 min; (e) 0.106 M chloroplatinic acid and 1.5 M citric acid for 3 h; (f) 0.106 M chloroplatinic acid and 1.5 M citric acid for 10 h. (Adapted from P. Papageorgiou, MS Thesis, University of Notre Dame, 1984.)

species and support. In practice, in the final activated catalysts the metal particles are coated with smaller oxide particles to prevent metal–metal contacts as this reduces active component sintering. The support must be intimately mixed with the metal, and this is achieved by co-precipitation or precipitation in the presence of finely dispersed oxide particles.

The commercial manufacture of co-precipitated catalysts is usually operated as a continuous process. The metals are introduced as a soluble salt, usually a nitrate or sulphate, and the support precursor as salt or hydroxide. The precipitation is typically achieved by the addition of a solution of sodium carbonate. Other bases can be used to induce precipitation but care must be taken to avoid the formation of ammonium nitrate. Good mixing of the liquid streams is critical to forming a chemically uniform precipitate. In multi-component systems such as the copper/alumina system, consistent co-precipitation of the components only occurs over a narrow pH range (see Figure 6.8).

The precipitates are aged to transform the amorphous material into a gel or crystallite. These are filtered and washed to remove the unwanted ions such as sodium, nitrate and sulphate. This washing process can be critical; as any residual ions may increase the rate of sintering during the subsequent drying and calcination. Similarly, they may increase sintering under the final operating conditions experienced by the catalyst or reduce activity by poisoning the active sites. Spray and flash dryers are both frequently

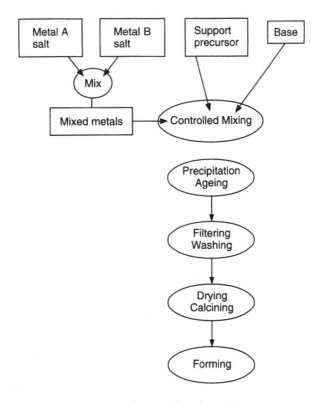

Figure 6.7 Co-precipitated catalysts.

used to remove water and produce powders suitable for the next processing steps.

6.3.1.3   Calcination. The impregnated and co-precipitated catalysts are calcined to convert the salts and hydroxides to oxides. The temperature and duration of calcination can be of critical importance for the final activity and selectivity of the catalyst. The temperature of calcination should not substantially exceed the reduction temperature used to activate the catalyst [11]. Typically the higher the calcination temperature used in preparation the more difficult it will be to activate the catalyst by reduction. The degree of calcination, for co-precipitated catalysts, is controlled within limits so that some residual carbonate is retained, as this makes softer pellet feed.

6.3.1.4   Reduction. Most catalysts require to be activated before use. This usually takes place in the reaction vessel, and its mechanical design may

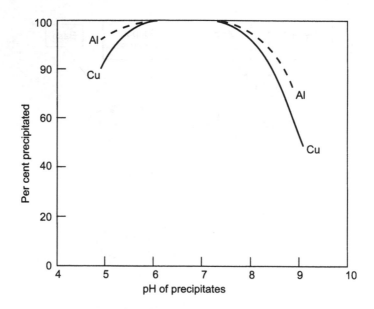

Figure 6.8 Effect of co-precipitation pH on composition.

restrict the process conditions which can be used to activate the catalyst precursor. To give better control of the final active form, and to reduce in-plant reduction times, some catalysts are pre-reduced and stabilised, as part of the manufacturing process. Whether in situ or as part of the manufacturing process, the reduction step and precautions are similar. Some reductions, for example nickel, are thermally neutral, but others, such as copper, are strongly exothermic and need to be limited to protect the catalyst from sintering. This can be achieved by using a dilute reduction gas, for example 1% hydrogen in nitrogen, and careful control of temperature. Another parameter which can have a considerable impact on the metal crystallite size, after reduction, and hence activity of the finished catalysts, is the ratio of water vapour partial pressure to hydrogen partial pressure.

Reduction of precipitated materials leads to an increase in porosity and hence loss of strength. It is in this state that the catalyst must be robust enough to withstand the stresses from start up/shut down and normal operation.

6.3.2 Forming of catalysts

Fixed bed catalysts are used in the form of pellets, rings, worms, matrices or spheres to reduce pressure drop. The processes are similar for making supports prior to impregnation or for making catalysts from co-precipitated precursors and therefore both will be considered together.

6.3.2.1 Pelleting. Pelleting is the preferred method when precise control of dimensions and high strengths are required. It is used to make not only cylindrical shapes but also rings and multi-holed shapes, the ends of which can be flat or domed. The powder is compressed axially in a die to give intimate contact between the particles. To reduce density variations in the pellet a lubricant is required on the die walls, and particularly with high length to diameter ratio catalysts the compaction force is applied from both ends. The pelleting feed needs to have suitable flow properties to fill the die evenly in the short time available on most rotary machines. Precipitated powders are spray dried, granulated or pre-compacted to increase particle size and bulk density. The lubricant, for example graphite or magnesium stearate, can also be added as part of the pellet feed. The punches and dies have a finite life, especially with abrasive materials such as alumina, and tooling costs are significant. The number of pellets required per cubic metre increases as the size falls, so pellets less than 3 mm are seldom made.

6.3.2.2 Extrusion. Extrusion can be used to produce simple cylinders, tubes and more complex shapes with uniform cross-section such as honeycombs. The powders usually require the addition of binders, flow modifiers and water to create a plastic mass. The extrusion aids are usually clays or organic materials and are chosen according to the application, as residues must not be deleterious to the catalyst performance. The powders are kneaded with the liquid to produce a paste, which is forced through the die. Various extruder designs are used to produce worms, e.g. ram, screw or roller with an axial die; and roller or gear with a radial die (see Figure 6.9).

Extruders for complex shapes are restricted to those with an axial die and uniform feed. The extrudate needs to be stiff enough to retain its shape during drying and robust enough to accommodate the shrinkage which occurs as the water is removed. Very small diameters can be produced economically. Consequently, if fixed bed catalyst particles are required to be smaller than 3 mm, extrusion can be the preferred forming process. Larger diameters are also extruded, but these usually have poorly formed ends when compared with a pellet. Honeycombs for car exhaust catalysts are usually cut to length after drying or firing.

6.3.2.3 Granulation. Granulation is a general term used to describe the production of spheres by the agglomeration of powders with a liquid. Many types of equipment can be used from a fluid bed to a pan granulator, but they all have several features in common. Powder particles are held together predominantly by surface tension, powder and water are mixed under conditions which lead to growth; and some energy input is provided to

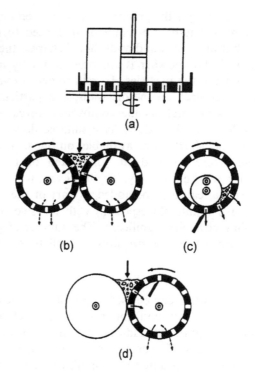

Figure 6.9 Alternative designs of roll type extrusion presses or pellet mills: (a) horizontal perforated plate die; (b) extrusion through side-by-side rollers; (c) outward extrusion through die roll; (d) extrusion through one of side-by-side rollers. (Adapted from C.E. Capes, *Particle Size Enlargement*, Elsevier, 1980, p. 116.)

produce shear within the growing particles leading to rearrangement and densification. The different processes occurring during granulation are illustrated in Figure 6.10. The product is of relatively low density and tends to have a broad size distribution.

6.3.2.4   Spray drying. Spray drying is used to produce spheres directly from a low solids-content feed such as a filter cake. Diameters in the range 5 to 500 μm can be produced, the size being controlled by the atomisation step. This process is used to make catalysts for use in fluidised bed reactors and pellet or extrusion feed materials.

6.4   Characterisation of catalysts

The performance of a catalyst in a reactor system is influenced by both its chemical and physical properties. Moreover, the chemical properties can

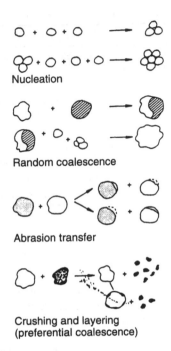

Nucleation

Random coalescence

Abrasion transfer

Crushing and layering
(preferential coalescence)

Figure 6.10 Processes taking place during granulation (adapted from W. Pietsch, *Size Enlargement by Agglomeration*, John Wiley & Sons, 1991, p. 137).

also be split into **bulk** and **surface** properties, the latter being particularly important because the phenomena of catalysis occurs on the surface. The chemical properties of the catalyst are most important in the development of a catalyst, the controlling of catalyst manufacture and in the determination of catalyst deactivation processes. The physical properties of catalysts, such as mechanical strength and physical texture, have more immediate relevance to a process operator. A brief outline of characterisation methods is given in the following sections.

6.4.1   Chemical characterisation

The most commonly used bulk chemical characterisation methods are listed in Table 6.8. The table also gives some indication of the properties that can be measured by the appropriate use of the techniques.

As heterogeneous catalysis is a surface phenomenon, the determination of surface properties plays a major part in catalyst characterisation. A very wide range of surface analysis techniques are now available and many others are in the development stage. Table 6.9 lists some of the more commonly used surface analysis techniques and the properties which they measure.

Table 6.8 Techniques of bulk chemical analysis of catalysts

Technique	Property measured
Elemental analysis (ICP-MS, X-ray fluorescence, atomic absorption, etc.)	Bulk elemental analysis
X-ray diffraction	Crystalline phases present; crystallite sizes
Electron microscopy	Particle shapes and sizes; particle compositions; particle crystal structures
Thermal analysis (DTA, TGA)	Phase changes, weight changes on heating
Temperature programmed reduction (TPR)	Size and temperature range of reduction steps
Infrared and UV-visible spectroscopy	Types of chemical bond present
Mössbauer spectroscopy	Phase identification
Nuclear magnetic resonance	Chemical environment of element

Table 6.9 Techniques for surface analysis of catalysts

Technique	Property measured
Chemisorption of CO, H_2, O_2, N_2O	Surface area of metal components
Chemisorption of acids or bases	Surface concentration of acid/base sites
Temperature programmed desorption (TPD)	Chemical identity of adsorbed surface species
X-ray photoelectron spectroscopy (XPS)	Surface chemical analysis; chemical state of the elements present
Auger electron spectroscopy (AES)	Surface elemental composition
Secondary ion mass spectroscopy (SIMS)	Chemical identity of surface layers
High-resolution electron energy-loss spectroscopy (HREELS)	Type of chemical bond present
Extended X-ray absorption fine-structure analysis (EXAFS)	Atomic structure of surfaces and adsorbates
Work function determination	Surface ionisation

6.4.2 Physical characterisation

The measurement of the mechanical strength of a catalyst pellet is essential to ensure that the catalyst is sufficiently robust to withstand transport, vessel charging and the forces experienced during operation. A number of comparative tests are used to simulate some of the physical stresses placed on a catalyst pellet. These are shown in Table 6.10.

The performance of a catalyst is often dependent upon its textural properties (porosity, surface area and pore size distribution). The two main techniques

Table 6.10 Techniques for measuring the physical strength of catalyst pellets

Technique	Property measured
Compression testing	Pellet strength
Pellet tumbling	Attrition loss
Controlled packing	Bulk density and packing characteristics
Bed crushing strength	Resistance of catalyst bed to breakage

Table 6.11 Techniques for measuring catalyst texture

Technique	Property measured
N_2 adsorption (BET methods)	Surface area (i.e. total accessible area) Pore volume Pore size distribution (as a function of pore size)
Mercury porosimetry	Pore volume Pore size distribution Complexity of pore structure (connectivity/tortuosity)

used to measure the texture of catalysts are nitrogen adsorption and mercury porosimetry (see Table 6.11). By using a combination of these techniques some important physical characteristics of a catalyst can be measured.

Pore sizes in catalysts are often grouped into three classifications: macropores (> 50 nm), micropores (< 2 nm) and mesopores (intermediate size). The textural features of a catalyst have a major influence on properties such as pressure drop across the catalyst bed and the diffusion of reactants into the catalyst pellet.

6.4.3 Characterisation of used catalysts

Investigating spent catalyst samples is an important feature of catalyst research. When discharging catalysts from large reactors, however, it can often be very difficult to obtain representative samples of the spent catalyst. For instance, pyrophoric catalytic materials may be discharged under a water wash or stabilised before discharge with air or carbon dioxide. Both of these methods will obviously alter the condition of the catalyst sample. The most useful spent catalyst samples are those which are extracted from known positions in the catalyst bed. With these samples a profile of the catalyst condition down the catalyst bed can be built up. Although many of the previously mentioned chemical and physical techniques can be used to analyse spent samples, the techniques listed in Table 6.12 are particularly useful.

Table 6.12 Techniques for analysing spent samples

Technique	Information derived
Elemental analysis	Loss of active phase; catalyst poisons
X-ray diffraction	Appearance/loss of crystalline phases (sintering)
N_2 adsorption	Loss of surface area (sintering); changes to pore size/distribution
Temperature programmed oxidation (TPO)	Carbon fouling of catalyst; types of carbon species present; heavies burn-off
X-ray photoelectron spectroscopy (XPS)	Surface chemical analysis
Compression testing	Loss of pellet strength during use
Solvent extraction	Removal of absorbed species for subsequent analysis

An informed use of the techniques listed in this table allows a scientist and engineer to determine why a catalyst has deteriorated in use and this, in turn, can lead to optimisation of process conditions and improved catalyst design.

In concluding this section it should be emphasised that, although catalyst characterisation is very important for the understanding, design and troubleshooting of catalysed industrial processes, there is no universal recipe governing the selection of the most expedient methods to employ. The techniques selected above have been chosen to show the wide range of methods available and the information they can provide.

6.5   Industrial catalyst examples

6.5.1   Methyl alcohol synthesis

6.5.1.1   Active component selection. Methyl alcohol is produced from synthesis gas which is a mixture of carbon dioxide, carbon monoxide and hydrogen.

$$CO + 2H_2 \rightarrow CH_3OH \quad \Delta H = -90.64 \, \text{kJ} \, \text{mol}^{-1} \qquad (6.1)$$

$$CO_2 + 3H_2 \rightarrow CH_3OH \quad \Delta H = -49.47 \, \text{kJ} \, \text{mol}^{-1} \qquad (6.2)$$

The catalyst used in the process developed by BASF in the 1920s was zinc oxide and chromia. In 1966 ICI introduced a low-pressure process which used a new, more active and selective catalyst composed of copper/zinc oxide/alumina (see reference [13]). Copper was identified in the 1920s as the most effective active component, but was severely affected by the catalytic poisons present in synthesis gas produced from coal. Fortunately, zinc oxide is remarkably resistant to poisons such as sulphur or arsenic and thus was selected as the active component for the original catalyst. With the advent of much cleaner synthesis gas produced by steam reforming in the 1960s, the way lay open to use the more active copper catalyst. During the development of the copper catalyst it was discovered that, unlike zinc oxide, the reaction proceeded by the hydrogenation of carbon dioxide, which had significant economical benefits.

6.5.1.2   Support and manufacturing route selection. The synthesis of methyl alcohol is a relatively slow reaction and in the industrial process the reactor is under kinetic control except possibly during the very early stages of the catalyst life. The activity of the catalyst is directly related to the surface area of copper present, and this is illustrated in Fig. 6.11.

Attempts to produce a high copper area catalyst by the impregnation of a support have been unsuccessful. The copper is prone to sintering under the reaction conditions and quickly sinters. As the reaction is under kinetic

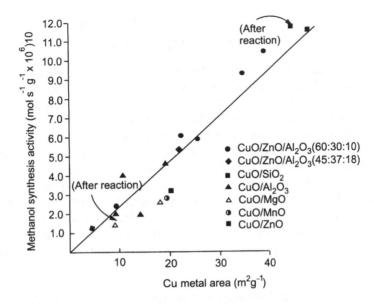

Figure 6.11 Relationship between methyl alcohol synthesis activity and copper metal area (reproduced from *Applied Catalysis*, 1986, **25**, 101–7, *The Activity and State of the Copper Surface in Methanol Synthesis, Catalysts* by G.C. Chinchen, K.C. Waugh and D.A. Whan, with kind permission of Elsevier Science – NL., Sara Burgerhartstraat 25, 1055 KV Amsterdam, The Netherlands.

control, the active phase does not have to be concentrated near the surface of the catalyst. Preparation methods for the catalyst, other than impregnation, can therefore be considered. The problem of overcoming the propensity of copper to sinter has been achieved by the development of a co-precipitated manufacturing route. This technique allows the production of a stable high copper area catalyst which lasts for many years under normal operating conditions. The role of zinc oxide and alumina is primarily to support and separate the copper crystals during preparation. The zinc oxide and alumina, however, carry out additional roles during the life of the catalyst in the plant. For instance, the support prevents the agglomeration of the copper particles during use (which may be due to some epitaxy between copper metal and zinc oxide) and in the early levels of the reactor can act as an absorbent for poisons such as sulphur and chlorine. Alumina cannot be used as the support on its own because it encourages the unwanted side-reaction to produce methyl ether. The alumina present in the catalyst is locked within the zinc oxide lattice.

6.5.2 Selective ethyne removal from ethene

6.5.2.1 Active component selection. The production of polymer grade ethene from pyrolised feedstocks incorporates the selective hydrogenation of the trace by-product ethyne, using a fixed bed catalyst.

$$C_2H_2 + H_2 \rightarrow C_2H_4 \quad \Delta H = -172\,kJ\,mol^{-1} \tag{6.3}$$

$$C_2H_4 \rightarrow C_2H_6 \quad \Delta H = -137\,kJ\,mol^{-1} \tag{6.4}$$

The active component selected for this transformation is palladium metal, which has the highest intrinsic selectivity. This intrinsic selectivity arises from the high strength of chemisorption of alkynes to palladium. The difference in free energies of adsorption of ethene and ethyne is sufficiently large to eliminate any ethene coverage in the presence of ethyne. At zero or low concentrations of ethyne the catalysts still do not hydrogenate ethene because carbon monoxide, from hydrocarbon steam reforming, is present in sufficient quantities to compete for sites with ethene. Carbon monoxide has a very similar adsorption strength to alkynes. The performance of the industrial catalyst has been found to be profoundly influenced by the design of the catalyst. The relationship between the design and performance of trace ethyne hydrogenation catalysts has been reported [14] and is described below.

6.5.2.2 Effect of palladium loading. The amount of palladium present on the catalyst has a significant effect on the selectivity of the catalyst (see Figure 6.12). At very low palladium loadings the catalysts have too little volumetric activity for commercial use. However, at relatively high palladium loadings there is a major loss of selectivity. Higher loadings increase the sites available and it becomes more difficult to control location. Both experimental and theoretical work have shown that for consecutive reaction an outer 'egg shell' location of the active metal improves performance and

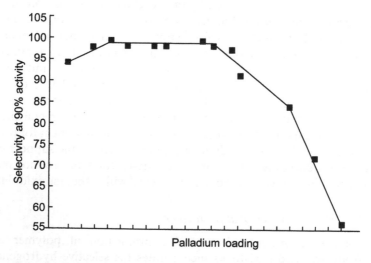

Figure 6.12 Effect of palladium loading on ethyne hydrogenation selectivity.

the typical loading of palladium in the industrial catalyst is 250–350 parts per million.

6.5.2.3 Support selection. The choice of support material for this reaction is far from simple. The support must be stable to water, as a steam/air mixture is used to regenerate the catalyst. In addition, the catalyst must not contain acid sites which encourage undesirable side-reactions and, in practice, the support is alkali treated. The usual support employed is alumina which in present-day catalysts has a surface area of $< 20\,m^2\,g^{-1}$. The average pore radius of the support affects the catalyst selectivity, and this is shown in Figure 6.13.

Increasing the average pore radius of the catalyst reduces the probability of the ethene adsorbing inside the catalyst and being hydrogenated. Even though increasing the average pore diameter will obviously reduce the area of support available for the active component, the overall result is a more stable and more selective catalyst.

Finally, in addition to the parameters described above, the industrial catalyst still has to have an adequate pressure drop, crush strength and attrition resistance. The latter is obviously particularly important for this catalyst where the active component is located at or near a support surface.

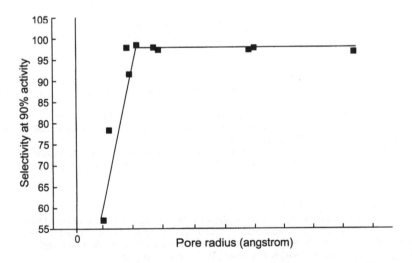

Figure 6.13 Effect of average pore radius on ethyne hydrogenation selectivity.

References

1. Fulton, J.W. (1986) *Chemical Engineering*, 17 February, 118.
2. Fulton, J.W. (1986) *Chemical Engineering*, 12 May, 97.

3. Fulton, J.W. (1986) *Chemical Engineering*, 7 July, 59.
4. Fulton, J.W. (1986) *Chemical Engineering*, 13 October, 71.
5. Fulton, J.W. (1987) *Chemical Engineering*, 19 January, 107.
6. Fulton, J.W. (1987) *Chemical Engineering*, 11 May, 59.
7. Fulton, J.W. (1987) *Chemical Engineering*, 14 September 99.
8. Acres, G.J.K., Bird, A.J., Jenkins, J.W. and King, F. (1981) *Catalysis*, Volume 4 (eds C. Kemball, and D. Dowden), The Royal Society of Chemistry.
9. Neimark, A.V., Kheifez, L.I. and Fenelonov, V.B. (1981) *Ind. Eng. Chem. Prod. Res. Dev*, **20**, 439.
10. Kotter, M. and Riekert L. (1978) *Proceedings of Second International Symposium, Scientific Bases for the Preparation of Heterogeneous Catalysts*, Louvain-la-Neuve, Belgium.
11. Andrew, S.P.S. (1981) *Chemical Engineering Science*, **36** (9), 143.
12. Papageorgiou, P. (1984) Preparation of Pt– Al_2O_3 pellets with internal step-distribution of catalyst, M.S. Thesis, University of Notre Dame.
13. Chinchen, G.C., Mansfield, K. and Spencer, M.S. (1990) *Chemtech*, November, 692.
14. Hancock, F.E. and Smith, J.K. (1992) A.I.Ch.E. Spring Meeting.

Bibliography

1.   Catalyst design and manufacture

Becker E.R. and Pereira, C.J. (eds) (1993) *Computer-Aided Design of Catalysts*, Marcel Dekker.
Le Page, J.F. (1987) *Applied Heterogeneous Catalysis*, Editions Technip.
Stiles, A.B. and Koch, T.A. (1995) *Catalyst Manufacture*, Marcel Dekker.

2.   Catalyst forming

Pietsch, W. (1991) *Size Enlargement by Agglomeration*, John Wiley & Sons.
Stanley-Wood N.G. (ed.) (1983) *Enlargement and Compaction of Particulate Solids*, Butterworths.

3.   Catalyst characterisation

Anderson, J.R. and Pratt, K.C. (1985) *Introduction to Characterisation and Testing of Catalysts*, Academic Press.
Bell, A.T. and Pines, A. (eds) (1994) *NMR Techniques in Catalysis*, Marcel Dekker.
Bradley, S.A. Gattuso, M.J. and Bertolacini, R.J. (eds) (1988) *Characterization and Catalyst Development*, American Chemical Society.
Gregg, S.J. and Sing, K.S.W. (1969) *Adsorption Surface Area and Porosity*, Academic Press.
Thomas, J.M. and Lambert, R.M. (eds) (1980) *Characterisation of Catalysts*, John Wiley & Sons.

4.   General catalyst texts

Augustine, R.L. (1996) *Heterogeneous Catalysis for the Synthetic Chemist*, Marcel Dekker.
Bond, G.C. (1974) *Heterogeneous Catalysis Principles and Applications*, Clarendon Press.
Satterfield, C.N. (1991) *Heterogeneous Catalysis in Industrial Practice*, McGraw-Hill.
Twigg, M.V. (ed.) (1989) *Catalyst Handbook*, Wolfe Publishing Ltd.

7 Overview of catalytic reactor design
R. CRANE

The chemical engineering required to develop and design a reactor ranges from the trivial to the complex and on any scale of difficulty the design of a heterogeneous catalytic reactor would come close to or at the top. The problems start with understanding the chemistry of the catalysed reaction, then uniting this with the physics of diffusion and hence developing a creditable model of the system which will allow options to be explored and which will finally lead to an optimised design.

7.1 The catalyst

The definition of catalysis as taking a whip to a sluggish horse understates the advantages of many catalysed reactions. The effect which catalysis can have on reaction rate is strikingly illustrated by comparing the reaction of hydrogen and nitrogen to give ammonia with and without a catalyst. It has been calculated that, if the entire volume of the solar system between the sun and the planet Pluto were filled with hydrogen and nitrogen at 500°C and 200 bar, the rate of ammonia production would equal that of a catalysed volume of 0.07 ml. The sluggish horse would have to be accelerated to several times the speed of light to match this enhancement!

The key properties required by a catalyst are:

- High activity
- High selectivity
- Long life.

Looking at these in turn.

7.1.1 High activity

The simplest model of a catalyst is that of a porous solid with a large internal surface. For example, a typical 100 m^3 reactor would have a surface area of about 5000 km^2, i.e. the area, say, of Oxfordshire and Cambridgeshire together. The shape of the catalyst pellets would probably be cylindrical with dimensions of 0.5 to 1 cm in a bed 2 to 5 metres in diameter and 2 to 5 metres deep.

The reactant gases diffuse into these porous pellets and are adsorbed onto the surfaces at so-called active sites thereby becoming more amenable to reaction. The design and development of a catalyst is concerned with increasing the number of active sites and changing the structure to minimise the resistance of the pore system to diffusion.

Taking the activity of the iron-based ammonia synthesis catalyst as an example, the present performance is such that, based on activation energy calculations, only one in a million sites would seem to be active. There is, therefore, still a long way to go despite the fact that it is nearly 100 years since the catalyst was first developed.

7.1.2   High selectivity

In terms of importance, this feature of a catalyst perhaps should be placed first since being able to catalyse one reaction in preference to a less attractive alternative is more difficult to bring about than is 'straightforward' rate enhancement.

Air oxidation processes are the most obvious examples of this aspect of catalysis. Three important examples are ethylene to ethylene oxide, ammonia to NO_x (and hence to nitric acid) and methanol to formaldehyde. All these compounds left to themselves will react with air to give carbon dioxide and water. The purpose of the catalyst is to promote the useful step and suppress the reaction that proceeds without catalysis. To do this it may be necessary to sacrifice activity for selectivity. For example, in the oxidation of methanol to give formaldehyde, platinum will catalyse all possible reactions. In order to be selective the less active silver or iron/molybdenum catalysts are used since these promote only the useful reaction with an efficiency of 95%.

A second variation on this problem of selective oxidation is exemplified by the removal of NO_x compounds from nitric acid tail gases. Two practical options exist for this case where the oxidant is the unwanted component. The first is to react the NO_x with hydrogen or methane over a precious metal catalyst reducing it to nitrogen. Unfortunately the tail gas has a residual of 2% oxygen which is also reduced, using considerably more fuel than that required by the NO_x. Alternatively, the second process can be used, the so-called selective ammonia reduction process, where ammonia is added to the tail gas, the mixture is then heated to a lower temperature of 250°C compared to the 400–600°C of the first option and then reacted over vanadium pentoxide catalyst, of much larger volume, where the ammonia reacts selectively with the NO_x to give nitrogen and water. The process adopted depends upon the price of the various reducing fuels and the overall plant flow sheet.

One final variation on this theme of selectivity is in the lightening of a refinery feedstock, in the so-called 'Cat Cracker'. Here heavy feedstocks are reacted over a zeolite catalyst. Since heavy feedstocks have a higher carbon to hydrogen ratio than the product, the cracking process inevitably leads to the formation of carbon which is laid down as 'coke' on the catalyst. The successful catalyst is one which does not produce 'Smoke and Coke', i.e. it splits the heavy molecules once near the middle of the molecule and does not generate large quantities of short-chain hydrocarbons with the inevitable production of large quantities of carbon.

7.1.3 Long life

The life of a catalyst can vary upwards from a few seconds to an indefinitely long period of time before regeneration or replacement is needed. The loss of activity is principally due to changes in morphology. Mechanical attrition is one obvious cause and part of the design of a reactor is concerned with minimising movement within the bed by keeping the velocities at a suitable level, fitting holding-down grids and designing the catalyst for high strength even at the expense of activity. However, the principal cause in the case of non-precious catalysts is the steady growth of the small crystals making up the catalyst which affects both the number of active sites and the porosity of the catalyst. This, together with solid deposition by unwanted reactions, eventually forces replacement. Delaying the onset of this growth and deposition rank equally with activity and selectivity in catalyst development. Two outstanding examples in the 1950s and 1960s were the development of catalysts to allow naphtha to be reacted with steam to give hydrogen without depositing carbon, and extending the life of copper/zinc catalysts to allow them to be used in the synthesis of methanol at a lower temperature and pressure with fewer by-products.

The sudden death of catalysts can be brought about inadvertently by several means, including heating them rapidly when they contain moisture thus blowing them apart, or allowing air to enter a bed of reduced catalyst to give a fused monolith of oxide. The most common cause of death is to allow poisons to enter the feed stream. These poisons attach themselves to the active sites thus preventing further reaction. Examples of poisons are sulphur on nickel catalysts, oxides of carbon on the iron in the catalyst used for ammonia synthesis and iron on the platinum catalyst used in nitric acid and hydrogen cyanide manufacture.

7.2 The kinetic model

The development of a model giving the rate of a catalysed reaction involves a large amount of testing, to give at the end something to be treated with caution and scepticism. It will have its share of adjustable parameters and how far it can be extrapolated can only be found out the hard way, i.e. by trying it out. For catalysts designed for a simple duty, say for removing contaminants from an effluent gas stream, this uncertainty is not of import-

304   REACTOR DESIGN FOR CHEMICAL ENGINEERS

ance since the process conditions at which the catalyst is tested will generally be close to those in the unit being designed and will be clearly defined. On the other hand, for a catalyst for, say, the synthesis of methanol it will be necessary to use rate data for a wide range of possible conditions to arrive at the most efficient and cost-effective design. The availabily of these data is of paramount importance almost matching that of the actual manufacturing process of the catalyst itself. The model representing these data is also important. To be useful the model should be able to

(a) generate 'intrinsic' reaction rates as functions of temperature, reactant concentrations (or partial pressures) and (for reversible reactions) the concentrations of the products;
(b) account for changes in catalyst size.

In some, though not all, cases the Langmuir–Hinshelwood equations discussed in Chapter 3 may be helpful for the first purpose, though it will be necessary to fit the parameters which they contain to available data. There are several other types of equation. The Temkin equation [1] has been found very useful for the ammonia synthesis reaction [2]. According to this equation the rate of reaction r is given by

$$r = k_2\left\{\underline{K_p}p_{N_2}(p_{H_2}^3/p_{NH_3}^2)^\alpha - (p_{NH_3}^2/p_{H_2}^3)^{1-\alpha}\right\}$$

where

$$k_2 = k_{2(0)}\exp - \{(\Delta E_{k_2}/R)[(1/T) - (1/T_0)]\}$$

K_p is the equilibrium constant for the synthesis reaction and ΔE_{k_2} is about $150\,kJ\,mol^{-1}$. The constant α is usually taken as 0.5 [1], though any value between 0 and 1 is possible.

The effect of catalyst size (Table 7.1) is largely due to poor pore diffusion as discussed in Chapter 3. However, a contributory factor is that the outer regions of the larger particles tend to be more exposed to the sintering action of water during the reduction process [2].

Table 7.1 Influence of catalyst particle size on reaction rate in the ammonia synthesis reaction

Average particle size (mm)	Rate (kmol N_2 h^{-1}per m^3 catalyst)
0.9	300
3.8	112
7.5	61

7.3 Methodology of catalytic reactor design

The methodology listed in Table 7.2 gives an exhaustive list of issues which need addressing in the development of a new catalyst/reactor combination.

Table 7.2 Methodology for development of new catalyst / reactor combination

Factor to be considered	Likely option for solution
Data needed to determine viable reactor operating temperature and pressure	
• Reaction equilibria	Calculation
• Reaction rate	
– Main reaction	Literature, lab testing
– Side reactions	Lab testing
• Heat of reaction	Calculation
Check reaction at plant conditions	
• Conversion	Kinetic model, Lab testing at pressure
• Selectivity/product purity	Kinetic model, Lab testing at pressure
• With real feed rather than pure reactants	Sidestream/pilot plant tests
• With recycles to check on impurity build-up	Final plant tests
• With pelletted/supported catalyst rather than powdered catalyst	Semi-tech rigs
• Margin to avoid runaway	Computer model
– At what temperature can more exothermic reactions kick in?	Lab tests
– Effect of impurities in catalyst	Lab tests
– Control requirements	Computer modelling
Supported catalyst design	
• Catalyst type (pellet/extrudate/monolith/ceramic foam)	Consultation
• Supported or unsupported	Consultation
• Catalyst size	Economic trade-off
• Catalyst shape	Economic trade-off
• Catalyst density	Semi-tech testing/Economic trade-off
Catalyst operation/life	
• Sintering	Semi-tech testing, sidestream
• Poisoning	Sidestream/pilot
• Carbon laydown	Semi-tech testing or sidestream or pilot
• Shrinkage	Lab tests
• Durability	
– Strength	Lab tests
– Vaporisation of catalyst or support	Lab tests
– Catalyst/support interactions	Lab tests
• Pore diffusion	Specialist lab tests
• Catalyst disposal/recycle route	Lab tests
• Design life	Economic trade-off
Reactor feed condition	
• Liquid feed or vapour feed 10 degC above dew point	
• Control of feed condition	ELD review, computer modelling
• Well-mixed feed	CFD modelling
Reactor design	
• Choice of Reactor	Experience
– Axial flow/radial flow	Consultation
– Adiabatic/tubular	Consultation
• Volume determination	Economic trade-off
• Aspect ratio	Economic trade-off
• Inlet velocity	Calculation
• Flow distribution	CFD modelling
• Crush/heave of catalyst	Specialist lab testing
• Catalyst charging	Consultation, lab testing
• Catalyst discharging	Consultation, lab testing

Table 7.2 (*contd*)

Factor to be considered	Likely option for solution
Heat transfer	
• Overall heat transfer	Computer modelling
• Max temperature	Computer modelling
Transient procedures	
• *Start-up/shut-down*	
– Temperature gradients in catalyst bed	Computer modelling
– Temperature gradients in reactor shell	Computer modelling
– Condensation on cold catalyst or cold reactor wall	Calculation
Conditions during reduction / activation / regeneration operations	
• Flow rate	
• Reactant concentrations	
• Flow distribution	Computer modelling/Semi-tech/Pilot
(This is critical for regeneration and must be considered carefully)	plant study
• Exotherm control	Computer modelling
Metallurgy	Lab tests
Reactor Monitoring (*needed to determine performance and predict catalyst change point*)	
• Pressure drop	
• Temperature profiles	
• Conversion	
• Selectivity	
Temperature operating strategy	On-line or off-line computer optimisation

7.3.1   Design of a catalytic converter using ammonia synthesis as an example

Since ammonia manufacture is second only to sulphuric acid in volume and first in terms of value, uses seven different catalysts and involves a worldwide investment of 20 billion dollars to give 100 million tonnes of product per annum, this would seem to be a useful starting point from which to look at the design of catalytic reactors. The process in terms of catalysis is shown in Table 7.3.

The design of the ammonia synthesis reactor will be used as example.

As is often the case the balance is between energy efficiency and mechanical complexity which, in turn, can be expressed in terms of capital cost and reliability.

The formation of ammonia is favoured by high pressure and low temperature and in the preliminary design stages equilibrium conversion and reaction rate over the selected catalyst will have been established as functions of temperature, pressure and other relevant variables. The exothermicity of the reaction will also be known.

The choices open and the questions to be answered as the design proceeds beyond the preliminary stages are as follows:

• What is the optimum design temperature and pressure?

Table 7.3 Catalysts used in ammonia manufacture

Step	Catalyst	Reaction	Conditions	Change
Primary reformer	Nickel	Endothermic	800°C 30 bar	$CH_4 + H_2O$ $= H_2 + CO + CO_2$
Secondary reformer	Nickel	Exothermic	1200°C 30 bar	$CH_4 + O_2 + N_2$ $= H_2 + CO_2 + N_2$
High-temperature shift	Iron	Exothermic	380°C 30 bar	$H_2O + CO$ $= H_2 + CO_2$
Low-temperature shift	Zinc copper	Exothermic	250°C 30 bar	$H_2O + CO$ $= H_2 + CO_2$
Methanation	Nickel	Exothermic	300°C 30 bar	$CO + CO_2 + H_2$ $= CH_4 + H_2O$
Ammonia synthesis	Iron	Exothermic	470°C 100–300 bar	$H_2 + N_2 = NH_3$

- What is the best means of cooling the reaction mixture as it passes through the catalyst beds?
  - Tube-cooled or adiabatic beds with interbed cooling?
  - Is the best form of interbed cooling cold shot or cooling by heat exchange?
- How many beds of catalyst?
- What is the bed aspect ratio?
- Radial or axial flow?

The questions cannot necessarily be answered in the order given since there will be interplay between the decision made in each case and also with economic considerations. We will, however, look at them in turn as listed above.

7.3.1.1 Reactor design temperature. The temperature has lower and upper bounds. The lower bound is set by the so-called 'strike temperature' of the catalyst, i.e. the temperature at which the catalyst starts to catalyse the reaction. In the case of ammonia synthesis this is around 300°C. The upper bound is dictated by the effect of temperature on the rate of catalyst sintering, which accelerates at temperatures around 550°C, and also by the fall-off in strength and increased corrosion rate of the containing stainless steel cartridge. Between these bounds it is a game for any number of players.

7.3.1.2 Reactor design pressure. The pressure chosen over the last 60 years has varied from as high as 350 bar to as low as 90 bar. The reasons for these changes are a mixture of fashion, costs and technical necessity. The choice is influenced by the factors listed below.

7.3.1.3   Method of compression and pressure vessel technology. The early plants used simple slow-speed compressors which could be designed for virtually any pressure. Also the pressure vessels then in use were forged for strength which placed a severe limitation on the maximum diameter. This led to high-pressure plants with small converters. In the 1960s a surge in demand stimulated a move to large plants and the development of centrifugal compressors. This, together with the capability to manufacture larger, non-forged, pressure vessels, resulted in a drop in pressure to 150 bar. A period of improvements in centrifugal compressor design caused a drift up in design pressure in the 1970s. Finally, in the 1980s the pressure fell again to lower than 100 bar because of changes in the upstream flow sheet and the realisation that lower pressures increased the number of machine and vessel manufacturers willing to bid for their supply.

7.3.1.4   Best way of cooling the reaction mixture. At the inlet to the first bed the temperature is set at a level which guarantees a high reaction rate, say 50°C above the 'strike' temperature. As the gases react, the bed temperature rises and cooling is required to prevent the temperature rising to unacceptable levels. The limits to the acceptable temperature rise are set by (a) the need to remain within the operating limits for the catalyst and (b) the need to retain an acceptably high level of ammonia conversion. Cooling may be either by heat exchange tubes inserted in the beds of catalyst or by interbed cooling. In the latter case the space required for the cooling equipment and peripherals can be up to a quarter of the original bed size. The result of this is that although the volume of catalyst theoretically is at a minimum with an infinite number of catalyst beds, the practical design is limited to three to four.

(a) **Tube-cooled beds.**
One way of avoiding interbed cooling is to install tubes within the catalyst bed and inject the feed gas into the reactor via these tubes thereby heating the feed gas at the same time as cooling the reacting gases. This design was popular for small plants but has fallen into disfavour with increasing plant capacity because of mechanical complexity and the lack of control over the life of the catalyst. Interesting features of the design are the technique of fitting core tubes to enhance the inside tube heat transfer coefficient and the calculation of the required surface area for heat transfer. This has to be calculated very carefully. If the area is too large or too small an ineffective reactor is obtained.

(b) **Interbed cold shot or heat exchange?**
An obvious disadvantage in cold shot cooling (Chapter 2) is that it reduces the ammonia concentration at the point of injection and the injected material does not contact all the catalyst. Interbed cooling by heat

exchange is better from this angle but involves expensive heat transfer equipment.

Reactor suppliers, in the past, have offered this feature. In practice the design has not been popular because of the high capital cost and unreliability. The most modern designs which tend to operate at lower pressure and give a lower final ammonia concentration than earlier designs, compromise by providing heat exchange before the last bed where it is of most advantage.

7.3.1.5 Bed aspect ratio. To save on vessel costs the diameter should be kept to a minimum. On the other hand, small-diameter beds are taller for a given bed volume and hence give high pressure drops which leads to excessive power consumption. The optimisation generally leads to an aspect ratio of one to one for a single bed reactor for pressures in the 30 bar region but as operating pressures increase, the optimum length to diameter ratio increases steadily. One advantage of a deep bed is an improvement in gas distribution. The gases can be presented to the bed in a non-uniform fashion and the catalyst quickly redistributes the flow to a uniform loading.

For shallow beds, this evening out of the flow pattern does not occur. For example, in nitric acid plants, where a precious metal catalyst is used, a typical bed diameter can be several metres but the depth is only a few centimetres. In this case presenting the feed gases to the platinum in a uniform manner is the key to a successful design.

7.3.1.6 Radial flow or axial flow?. The obvious arrangement of a catalyst bed is to place the catalyst inside a vertical right cylinder with gas flow down the bed keeping the catalyst in the vessel with a holding-down grid to prevent attrition of the top layers. As plant capacities increase the reactor wall thickness, therefore cost increase with the larger diameter if the pressure drop across the converter is to be kept constant. To get round this problem there has been a move to radial flow designs where the gas is marshalled to a central tube in the bed from whence it flows, radially, to a collecting manifold at the outside diameter of the bed. The design is mechanically more difficult, requiring sealing at the top of the bed to prevent bypassing, a variable velocity through the bed with high, possibly damaging, velocities at the beginning and a lot space taken up getting the reacting gases in and out of the beds. Despite this, the prize of a smaller diameter and, therefore, cheaper containing pressure vessel has led to an almost universal adoption.

7.4 Start-up operation and the need for care

The action required to bring a catalyst on line varies from virtually nothing, in the case of precious metals, to the careful three-day reduction programme required, for example, for the ammonia synthesis catalyst. The reason for this difference is that precious metal catalysts are delivered with the metal surfaces ready for use whereas non-precious metal catalysts are supplied as the oxides which have to be reduced to usually the metal or at least a lower level of oxide before they will act as catalysts. The degree of difficulty in carrying out this reduction process depends upon the heat and the rate of reaction.

For example, although the heat of reaction is about the same for the reduction of nickel based catalysts as it is for copper-based ones, the rate of reaction is much lower. In consequence nickel catalyst can be reduced using high partial pressures of hydrogen without any excessive temperatures being generated, whereas in the case of copper oxide, a nitrogen circulation system is required into which hydrogen, at low concentrations, is introduced to prevent the catalyst melting because of the rapid rate of reaction. In preparing ammonia synthesis catalyst an additional problem is that water vapour, generated by the reduction process, is a poison and the rate of reduction has to be controlled to limit the concentration of water vapour in the circulating gas to between 3000 and 10 000 ppm to give the maximum activity. Once the catalyst has been prepared for use the importance of keeping air out and hence preventing reoxidation varies from desirable in the case of nickel based catalysts to absolutely essential in the case of copper and iron. Nickel-based catalysts warm when exposed to air as the metal surface takes on a thin layer of protecting oxide which inhibits further oxidation whereas with copper- and iron-based catalysts there is a rapid temperature rise leading to fusion.

7.4.1   Pre-reduced catalysts

The delay in preparing catalysts for duty has led to the development of pre-reduced catalysts. Although more expensive they have advantages. In the case of ammonia synthesis catalyst, for example, they save three days' reduction time and with nickel-based steam-reforming catalysts they give a higher activity. The catalysts are prepared in the normal manner, reduced, then carefully reoxidised in a gas stream with a low oxygen content. The aim is to produce a strong monolayer of oxide which will act as a barrier to further oxidation. Obviously these catalysts have to be handled with care since there is a risk of overheating during charging. Despite this their use is on the increase.

7.4.2   Discharging and disposal

With precious metal catalysts, recycling of the spent catalyst has always been an essential part of the economics of the process. With increasing environmental pressures this is now also true for non-precious metal catalysts. After discharge into inert gas blanketed metal containers the catalyst is left to oxidise slowly. It is then recycled either as catalyst or for use by the metal industries.

References

1. Temkin, M.I. and Pyzhev, V. (1940) *Acta PhysioChim. (USSR)*, **12**, 327.
2. Jennings, S.A. and Ward, S.A. (1996) Chapter 8 in *Catalyst Handbook*, 2nd edition (ed. V. Twigg), Manson Publishing, London.

8 Fluidised bed reactors

R.M. NEDDERMAN

8.1 Introduction

Chemical reactions between a granular solid and a fluid are complicated by the difficulties of solids handling. If the solid serves as a catalyst which degrades only slowly, a packed bed can be used, but if the solid itself reacts or if the catalyst needs replacing often, the problems of emptying and filling the reactor become dominant. In these circumstances a fluidised bed can be used with advantage. In a fluidised bed the granular material is subjected to an upflow of fluid sufficient to support its weight. The solid then behaves as a fluid and can readily be transferred to and from the reactor. The classic example of a fluidised bed reactor is the catalytic cracker which first gained importance in the 1930s. Here the objective is to split the long-chain hydrocarbons in the heavier ends of crude oil into light species and this can be done by means of an alumina catalyst, often in the form of spherical particles of diameter about 150 μm. The nature of the reaction is such that carbon is deposited on the catalyst, rapidly deactivating it. By performing the reaction in a fluidised bed, catalyst can be continuously removed and supplied to a second fluidised bed in which the carbon is burnt off by fluidising with air. More recently there has been considerable interest in the use of fluidised beds for the combustion of coal, since, if powdered limestone is fed to the bed, much of the sulphur in the coal gets converted into solid $CaSO_4$, reducing atmospheric pollution. Fluidised beds are also advantageous for exothermic reactions since the efficient mixing within the bed not only helps to eliminate the formation of hot-spots typical of such reactions in fixed beds but also facilitates heat transfer to the walls of the bed.

8.2 Fluid mechanics of fluidised beds

When a fluid is passed upwards through a packed bed of solid particles at sufficiently low Reynolds numbers, the frictional pressure gradient (-dp/dl) is given by the Carman–Kozeny equation [1],

$$-\frac{dp}{dl} = \frac{180\mu U(1-\varepsilon)^2}{d^2\varepsilon^3} \tag{8.1}$$

where, in consistent units, p is the pressure, l is the distance from the bed inlet, μ is the fluid viscosity, U is the superficial velocity of the fluid, ε is the bed voidage and d is the equivalent diameter of the particles. Thus the pressure gradient $(-\mathrm{d}p/\mathrm{d}l)$ increases with increasing flow rate and eventually becomes sufficient to support the buoyant weight of the bed, $(1 - \varepsilon)(\rho_s - \rho_f)g$. When this happens the particles separate and begin to behave in a fluid-like manner. The superficial velocity at which this occurs, U_{mf}, is known as the minimum fluidising velocity. If the Reynolds number at which this occurs is sufficiently low for equation (8.1) to be obeyed, U_{mf} will be given by

$$(1 - \varepsilon_{mf})(\rho_s - \rho_f)g = \frac{180\mu U_{mf}(1 - \varepsilon_{mf})^2}{d^2\varepsilon_{mf}^3} \tag{8.2}$$

This will be true for very small particles, but if the Reynolds number is not small at the onset of fluidisation, it may be preferable to use the Ergun equation (see Chapter 4) giving,

$$(1 - \varepsilon_{mf})(\rho_s - \rho_f)g = \frac{150\mu U_{mf}(1 - \varepsilon_{mf})^2}{d^2\varepsilon_{mf}^3} + \frac{1.75\rho_f U_{mf}^2(1 - \varepsilon_{mf})}{d\varepsilon_{mf}^3} \tag{8.3}$$

Alternatively Leva's correlation [2],

$$\frac{U_{mf}\mu}{gd^2(\rho_s - \rho_f)} = 0.0007\left(\frac{U_{mf}\rho_f d}{\mu}\right)^{-0.063} \tag{8.4}$$

may be used to predict the minimum fluidising velocity, though caution should be exercised as correlations of this type do not generally cope adequately in the transitional flow regime. Since prediction of the value of ε_{mf} is difficult, it is always wise to confirm the predicted values of U_{mf} experimentally, particularly when the bed contains non-spherical particles or a mixture of particle sizes.

The pressure gradient does not increase when the superficial velocity is increased beyond the minimum fluidising value. Instead the bed expands and the pressure drop remains equal to the buoyant weight of the bed. The expansion can take place in two ways. If the fluidising medium is a liquid, the bed expands uniformly and the voidage ε becomes a function of the fluidisation rate. This type of fluidisation is known as **particulate fluidisation** and is of minor interest as far as this chapter is concerned. Much more commonly, the fluidising medium is a gas and the bed expands by forming voids within it, known as bubbles, since in many respects they behave similarly to gas bubbles in a liquid. In the two-phase theory of Davidson and Harrison [3] it is assumed that the bed between the bubbles is in a state of minimum fluidisation, taking a quantity of gas equal to that at minimum fluidisation, i.e. a volumetric flow rate of U_{mf} per unit area. The rest of the gas, i.e. the quantity $U - U_{mf}$, is assumed to pass through the bed in the form of bubbles. Though the two-phase theory is only a model for the behaviour of a fluidised bed and cannot be deduced from theory, it none-theless provides a good approximation to real bed behaviour and forms the basis of all the analyses presented below.

8.2.1 Fluidisation phenomena in gas-fluidised systems

X-radiographs of the bubbles show them to be of a shape similar to the spherical cap bubbles found in low-viscosity liquids and rise relative to the particles with the same velocity U_B given by [4]

$$U_B = 0.71\sqrt{(gD_E)} \tag{8.5}$$

where D_E is the equivalent spherical diameter, i.e. the diameter of the sphere of the same volume. If, however, the bubbles are comparable in size with the diameter of the bed, they form slugs, just as in inviscid liquids, which rise with velocity [4]

$$U_B = 0.35\sqrt{(gD_T)} \tag{8.6}$$

where D_T is the diameter of the bed. In general, bubbles rise with the lesser of the two velocities predicted from equations (8.5) and (8.6).

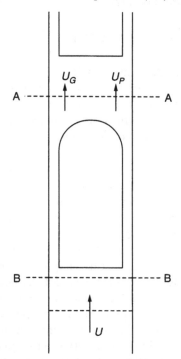

Figure 8.1 Mass balance on section of slugging bed.

These values of U_B are, however, the velocity of the bubble relative to the particles ahead of it, and in general these particles may be moving. Considering a slugging bed as shown in Figure 8.1, we can perform a mass balance from the base of the bed up to the section A–A which passes through the bed between slugs. Let the velocity of the particles at this section be U_P. Recalling that the two-phase theory postulates that the bed between the slug is in a state of minimum fluidisation, we can note that the difference between the particle and gas velocities U_P and U_G must be the same as that at minimum fluidisation. At minimum fluidisation, the particles are stationary and the interstitial gas velocity is the superficial velocity divided by the void fraction, i.e. U_{mf}/ε_{mf}. Thus, in the slugging bed, the gas velocity U_G is given by,

$$U_G = U_P + \frac{U_{mf}}{\varepsilon_{mf}} \qquad (8.7)$$

The total volumetric flow rate across the section A–A is the sum of the gas and solid flow rates and is given by

$$Q = A\varepsilon_{mf}U_G + A(1 - \varepsilon_{mf})U_P \qquad (8.8)$$

which from equation (8.7) becomes

$$Q = AU_{mf} + AU_P \qquad (8.9)$$

But the total volumetric flow rate, assuming incompressibility, should be equal to that of the gas entering the bed at plane B–B and equal to AU. Therefore,

$$AU = AU_P + AU_{mf} \qquad (8.10)$$

(U is the average superficial gas velocity in the bed. It will be constant in the steady state, though the instantaneous superficial velocity across a given plane will fluctuate.)

The velocity U_P of the particles ahead of a slug is given from equation (8.10) as

$$U_P = U - U_{mf} \qquad (8.11)$$

The absolute velocity of the slug U_A is the sum of the velocity of the particles ahead of the slug plus the relative velocity U_B. Thus,

$$U_A = U - U_{mf} + U_B = U - U_{mf} + 0.35\sqrt{(gD_T)} \qquad (8.12)$$

The analysis of the rise velocity of bubbles in a bubbling bed is not so straightforward, since it is not possible to draw a horizontal control surface through the bed which does not pass through any bubbles. Nonetheless a result similar to equation (8.12) is often used, i.e.

$$U_A = U - U_{mf} + 0.71\sqrt{(gD_E)} \qquad (8.13)$$

Example 8.1

Determine the minimum fluidisation velocity for 150 μm spherical catalyst particles of density 1750 kg/m^3 using air of viscosity 1.8×10^{-5} Ns/m^2 and density 1.2 kg/m^3. Take $\varepsilon_{mf} = 0.38$.

Solution

Since we do not yet know the interstitial Reynolds number it is safer to start with equation (8.3), giving

$$(1750 - 1.2) \times 9.81 \times (1 - 0.38) = \frac{150 \times 1.8 \times 10^{-5} \times U_{mf} \times (1 - 0.38)^2}{(150 \times 10^{-6})^2 \times 0.38^3}$$
$$+ \frac{1.75 \times 1.2 \times U_{mf}^2}{150 \times 10^{-6} \times 0.38^3}$$
$$17\,156 = 1\,356\,000 U_{mf} + 255\,100 U_{mf}^2$$
$$U_{mf} = 0.0126 \, \text{m/s}$$

Hence

$$Re = \frac{0.0126 \times 150 \times 10^{-6} \times 1.2}{1.8 \times 10^{-5}} = 0.126$$

This is small enough for equation (8.1) to be valid and, using that equation, we find $U_{mf} = 0.0105$m/s. The slight difference between these two estimates results from the different coefficients, 180 and 150 in equations (8.1) and (8.3). This difference merely reflects the precision of these two equations. Similarly, Leva's correlation, equation (8.4) gives, $U_{mf} = 0.0168$ m/s. Of these three estimates, that coming from the Ergun equation (0.0126 m/s) is probably the most reliable.

8.2.1.1 Bed expansion. As the gas flow rate is increased, the bubbles increase in both number and size and, as a consequence, the bed expands. The height of the bed H will therefore be greater than the height at minimum fluidisation H_{mf} and we can define a quantity ε_{bubble}, known as the bubble voidage which is the volume fraction of the bed occupied by bubbles. Since there are no bubbles at minimum fluidisation the volume of particulate phase, i.e. particles plus the gas between them, is AH_{mf} and hence,

$$AH_{mf} = AH(1 - \varepsilon_{bubble}) \tag{8.14}$$

or

$$\varepsilon_{bubble} = \frac{H - H_{mf}}{H} \tag{8.15}$$

Considering now the flow rate of the gas, we can say from the two-phase theory that all the gas in excess of that required for minimum fluidisation passes through the bed in the form of bubbles, i.e. at velocity U_A, but only through a fraction ε_{bubble} of the bed. Thus,

$$A(U - U_{mf}) = U_A A \varepsilon_{\text{bubble}} \tag{8.16}$$

Hence, from equation (8.15),

$$\varepsilon_{\text{bubble}} = \frac{H - H_{mf}}{H} = \frac{U - U_{mf}}{U_A} \tag{8.17}$$

where U_A is given by equations (8.12) or (8.13) as appropriate.

For a slugging bed there is some uncertainty about the precise meaning of the bed height H since the top surface of the bed moves up and down as the slugs arrive at the top. It can, however, be shown that the appropriate value of H is the maximum bed height, that is, the bed height immediately before a slug breaks through the top surface. The problem in a bubbling bed is less severe in that many bubbles will be breaking simultaneously.

Substituting from equation (8.12) into equation (8.17) gives

$$\frac{H - H_{mf}}{H} = \frac{U - U_{mf}}{U - U_{mf} + U_B} \tag{8.18}$$

which can be rearranged to give

$$\frac{H - H_{mf}}{H_{mf}} = \frac{U - U_{mf}}{U_B} \tag{8.19}$$

The reader is cautioned about the deceptive similarity of equations (8.17) and (8.19) which often causes confusion. Comparing these equations gives

$$\frac{H_{mf}}{U_B} = \frac{H}{U_A} = T \tag{8.20}$$

where T is the time it takes for a bubble to pass through the bed.

Example 8.2
The particles of Example 8.1 are fluidised at $2U_{mf}$ in a bed with a depth at minimum fluidisation of 0.5 m. The mean bubble diameter is 6.0 cm. Evaluate the bubble rise velocity, the bed depth and the bubble residence time.

Solution
From equation (8.5) the rise velocity of the bubble relative to the surrounding material is $U_B = 0.71 \times \sqrt{(9.81 \times 0.06)} = 0.545\,\text{m/s}$. The absolute velocity from equation (8.7) is

$$U_A = U - U_{mf} + U_B = U_{mf} + U_B = 0.0126 + 0.545 = 0.557\,\text{m/s}.$$

Equation (8.17) gives the bubble voidage,

$$\varepsilon_B = \frac{U - U_{mf}}{U_A} = \frac{U_{mf}}{U_A} = \frac{0.0126}{0.557} = 0.0226$$

From equation (8.17), $\varepsilon_B = (H - H_{mf})/H$. Hence

$$H = \frac{0.5}{1 - 0.0226} = 0.512\,\text{m}$$

The residence time is $H/U_A = 0.512/0.557 = 0.919$ s, which is also equal to $H_{mf}/U_B = 0.5/0.545 = 0.917$ s; the difference being solely the result of round-off errors.

Example 8.3
A fluidised bed has diameter 100 mm and depth at minimum fluidisation of 1.0 m. The minimum fluidising velocity is 0.91 m/s. Find the bubble rise velocity (assuming slug flow) and hence the maximum bed height and the fraction of the bed occupied by bubbles when the superficial velocity is 2.0 m/s.

Solution
From equation (8.6) we have, for a slugging bed,

$$U_B = 0.35 \times \sqrt{(9.81 \times 0.1)} = 0.347\,\text{m/s}$$

and from equation (8.12),

$$U_A = 2.0 - 0.91 + 0.347 = 1.437\,\text{m/s}$$

The bubble void fraction is given from equation (8.15),

$$\varepsilon_{\text{bubble}} = \frac{2.0 - 0.91}{1.437} = 0.759$$

and from equation (8.15), $(H - 1.0)/H = 0.759$, from which we obtain $H = 4.14$ m.

8.2.1.2 Bubble growth. In the derivations of Section 8.2.1.1, we implicitly assumed that the bubbles remain at constant size as they pass through the bed. In fact bubbles coalesce and their diameter increases with height. This phenomenon is not fully understood, but Whitehead [5] predicts that the bubble diameter increases with height y above the distributor according to the relationship

$$D = 0.49(3.37\sqrt{a_0} + y)^{4/5} \frac{(U - U_{mf})^{2/5}}{g^{1/5}} \tag{8.21}$$

where a_0 is the area of the distributor divided by the number of holes. For a sintered plate, a_0 is close to zero.

This theoretical expression is of similar form to Rowe's correlation [6],

$$D = \frac{\sqrt{(U - U_{mf})}}{g^{1/4}}(y + H_0)^{3/4} \tag{8.22}$$

where H_0 is an empirical function of the nature of the distributor. A number of studies of scale-up [7] have shown that bubble growth is a very complex process which can vary markedly with the mean size of the particles in the bed. As a simplification and despite the predicted and observed increase in bubble size with height, all the analyses of this chapter assume a constant bubble diameter, which should be taken as some mean of the bubble size. In view of the difficulties in the prediction of bubble growth the optimum effective bubble size is best determined by direct experiment.

Example 8.4
The distributor of a fluidised bed consists of a square array of holes at 5 mm centre-to-centre spacing. The superficial velocity is 0.18 m/s and $U_{mf} = 0.1 \text{m/s}$. Determine the bubble diameter immediately above the distributor and at a height of 20 cm.

The area per hole is the square of the hole spacing and hence $\sqrt{a_0} = 0.005 \text{m}$. Hence, from equation (8.21) the bubble diameter immediately above the distributor is

$$0.49 \times (3.37 \times 0.005)^{0.8} \times (0.18 - 0.1)^{0.4} \times 9.81^{-0.2} = 4.3 \text{mm}$$

At the height of 20 cm the corresponding value is

$$0.49 \times (3.37 \times 0.005 + 0.2)^{0.8} \times (0.18 - 0.1)^{0.4} \times 9.81^{-0.2} = 33.2 \text{mm}$$

showing considerable growth over this distance.

8.2.1.3 Gas flow through a bubble. Unlike bubbles in liquids, in which the gas is trapped within its bubble, the porous nature of the continuous phase in a fluidised bed enables gas to percolate through the particulate phase and to flow into and out of the bubble. The percolation of the gas can be analysed by potential flow theory since the gas velocity is proportional to the pressure gradient and hence the pressure satisfies Laplace's equation, $\nabla^2 p = 0$. From this it can be shown that the gas flow Q through a spherical void is three times that through an equivalent area of particulate phase [2]. However, bubbles are not spherical, but in the absence of better information it is usually assumed that

$$Q = \frac{3\pi}{4} D_E^2 U_{mf} \tag{8.23}$$

where D_E is the effective diameter of the bubble.

The same analysis does not apply in slugging beds where it is usually assumed that the through-flow can be evaluated by assuming that the bubble occupies the whole cross-section of the tube and that the gas enters the base of the bubble with the minimum fluidising velocity. Thus,

$$Q = \frac{\pi}{4} D_T^2 U_{mf} \tag{8.24}$$

where D_T is the bed diameter.

It is of course the flow through the bubble which provides an upward force on the particles immediately above the bubble and hence prevents the roof falling in.

8.3 Reactions in gas-fluidised beds

8.3.1 Reaction on a spherical particle

In a fluidised bed of non-porous particles, the reaction takes place on the particle surface, whether the particle is behaving as a catalyst or reacting itself. Thus the reactants have to diffuse to the particle surface, where reaction takes place and to diffuse back into the gas. For dilute concentrations of reactant, we can define a surface film mass transfer coefficient k_G as the ratio of the flux of reactant to the concentration difference between the particle surface and the bulk gas in the particulate phase. Thus the flux of reactant at the surface F is given by,

$$F = k_G(C_P - C_S) \tag{8.25}$$

where C_P is the reactant concentration in the gas in the bulk of the particulate phase and C_S is the concentration on the surface of the particle. If the particles are small enough for the transport to be purely diffusive, and if there is no change in the number of moles on reaction, the mass transfer coefficient will be given by

$$k_G = \frac{2D}{d} \tag{8.26}$$

where D is the diffusivity and d the particle diameter. At higher Reynolds numbers the mass transfer coefficient should be evaluated from some correlation such as [8]

$$Sh = 2 + 0.37 Re^{0.6} Sc^{0.4} \tag{8.27}$$

where the Sherwood, Reynolds and Schmidt numbers are defined by

$$Sh = \frac{k_G d}{D} \tag{8.28}$$

$$Re = \frac{\rho_G U_{mf} d}{\varepsilon_{mf} \mu_g} \tag{8.29}$$

$$Sc = \frac{\mu_g}{\rho_g D} \tag{8.30}$$

8.3.1.2 Calculation of the reaction rate. The flux of reactant, in moles per unit area of particle outer surface, equals the reaction rate per unit area, which, if the reaction is first order, will be proportional to the surface concentration

$$F = k'C_S \qquad (8.31)$$

If the particles are non-porous as assumed above, k' is simply the surface reaction velocity constant. Commonly, however, particles are porous and reaction takes place both on the outer surface and within the particle. Nonetheless, an effective surface velocity constant k' can be defined by equation (8.31) and is the product of the effective volumetric reaction velocity constant, the effectiveness factor and the volume per unit surface area.

Eliminating C_S between equations (8.25) and (8.31) gives

$$F = \left(\frac{1}{k_G} + \frac{1}{k'}\right)^{-1} C_P = \frac{k_G k'}{k_G + k'} C_P \qquad (8.32)$$

The rate at which a reactant A is being used up by the reaction per particle will be $\pi d^2 F$. If there are n^* particles per unit volume, the rate of reaction per unit volume will be $n^* \pi d^2 F$. We can relate n^* to the void fraction ε since $n^* \pi d^3/6$ is the solid content per unit volume, i.e. $1 - \varepsilon$. Thus, the number of particles per unit volume of the particulate phase is given by

$$n^* = \frac{6(1 - \varepsilon)}{\pi d^3} \qquad (8.33)$$

and hence the rate of reaction per unit volume of particulate phase $(-r_A)$ is given by,

$$(-r_A) = \frac{6(1 - \varepsilon)}{d} \frac{k_G k'}{k_G + k'} C_P \qquad (8.34)$$

which can more conveniently be written as,

$$(-r_A) = k_V C_P \qquad (8.35)$$

where k_V is the apparent volumetric reaction velocity constant.

More complicated kinetics and diffusion through rich mixtures require more detailed analysis. For example, if we were to consider the reaction $2A \rightarrow B$ catalysed by the particles and assume that the reaction is second order in A and that mass transfer is diffusion controlled with pure A at large distances from the particle, we would find that the flux of A to the surface is given by,

$$F = 4CD \ln\left(\frac{2C}{2C - C_S}\right) = k_2 C_S^2 \qquad (8.36)$$

where C is the total concentration, k_2 is a second-order reaction velocity constant and D is the diffusivity of A in the gas. The overall reaction is clearly of complex order and no satisfactory apparent reaction velocity constant can be defined.

8.3.2 First-order chemical reaction in fixed beds

The conversion in a bed operated below the minimum fluidising velocity or in a bed in which expansion is prevented, can be obtained by a simple mass balance on an element,

$$-U\delta C_P = k_v C_P\,\delta y \qquad (8.37)$$

or

$$C_P = C_0 \exp\left(-\frac{k_v y}{U}\right) \qquad (8.38)$$

where C_0 is the inlet concentration.

8.3.3 First-order chemical reaction in a well-mixed fluidised bed

The simplest case of a fluidised bed reactor is when the apparent reaction within the particulate phase is first order and when the particulate phase is well mixed. The assumption of a well-mixed particulate phase is often reasonable due to the stirring action of the bubbles. Well-mixed beds are commonly found when the superficial velocity is considerably in excess of the minimum fluidising velocity and when the bed diameter is large.

The two-phase theory of fluidisation tells us that a quantity of gas equal to U_{mf} per unit area passes through the particulate phase in unit time and that the rest goes through the bed in the form of bubbles. However, as we have seen in Section 8.2.3, there is interchange of gas between a bubble and the particulate phase at a volumetric flow rate Q. If we denote the concentration of reactant in the bubble by C_B and that in the particulate phase by C_P, we can perform a mass balance on the reactant in the bubble as follows:

$$Q(C_B - C_P) = -V\frac{dC_B}{dt} \qquad (8.39)$$

where V is the volume of the bubble. Assuming that the particulate phase is well mixed, we can conclude that C_P is a constant and we can therefore integrate equation (8.39) to give

$$\frac{C_B - C_P}{C_0 - C_P} = \exp\left[-\left(\frac{Qt}{V}\right)\right] \qquad (8.40)$$

where C_0 is the value of the concentration in the bubble at the base of the bed and C_B is the concentration at time t after formation. We are particularly interested in the concentration in the bubbles at the top of the bed, C_{BH}, and this is given by

$$\frac{C_{BH} - C_P}{C_0 - C_P} = \exp\left[-\left(\frac{QT}{V}\right)\right] \tag{8.41}$$

where T is the residence time of the bubble in the bed. Combining equation (8.41) with (8.20) gives

$$\frac{C_{BH} - C_P}{C_0 - C_P} = \exp(-X) \tag{8.42}$$

where

$$X = \frac{QT}{V} = \frac{QH_{mf}}{VU_B} = \frac{QH}{VU_A} \tag{8.43}$$

The parameter X is known as the cross-flow factor and is the ratio of the volume of gas passing through the bubble, QT, to the volume of the bubble V. This is sometimes expressed as the number of times the bubble has been washed through.

We have assumed that there has been no reaction within the bubble since the reaction is either between the fluidising gas and the particles, or is catalysed by the particles. Thus all the reaction is assumed to take place within the particulate phase which has a volume H_{mf} per unit area. Thus the rate of reaction is $H_{mf}k_vC_P$ and this can be equated to the overall change in concentration, thus,

$$H_{mf}k_vC_P = U(C_0 - C_H) \tag{8.44}$$

where C_H is the concentration of the reactant in the exit gases. The quantity C_H should not be confused with the concentration in the bubbles at the top of the bed, C_{BH}, and these two quantities can be related by a mass balance on the reactant at the top of the bed,

$$U_{mf}C_P + (U - U_{mf})C_{BH} = UC_H \tag{8.45}$$

Equations (8.42), (8.43) and (8.45) are sufficient for solution and can be rearranged to give

$$\frac{C_H}{C_0} = \frac{1 + (k-1)\beta \exp(-X)}{(k+1) - \beta \exp(-X)} \tag{8.46}$$

$$\frac{C_P}{C_0} = \frac{1 - \beta \exp(-X)}{(k+1) - \beta \exp(-X)} \tag{8.47}$$

$$\frac{C_{BH}}{C_0} = \frac{1 + (k-\beta) \exp(-X)}{(k+1) - \beta \exp(-X)} \tag{8.48}$$

where,

$$\beta = \frac{U - U_{mf}}{U} \tag{8.49}$$

and

$$k = \frac{k_v H_{mf}}{U} \tag{8.50}$$

The parameter β represents the fraction of the gas passing through the bed in the form of bubbles and k is a dimensionless reaction velocity constant.

Various special cases of these results are worthy of note.

1. If $k = 0$, i.e. if there is no reaction, $C_H = C_{BH} = C_P = C_0$.

2. As $k \to \infty$, $C_P \to 0$, $C_{BH} \to C_0 e^{-X}$ and $C_H \to C_0 \beta e^{-X}$. Thus with a very fast reaction the concentration in the particulate phase tends to zero as expected, but the concentration in the bubbles at the top of the bed and in the output streams are non-zero due to by-passing of the bed via the bubbles. To get a large conversion it is necessary either to operate with a very low value of β, i.e. to have U only marginally greater than U_{mf}, or to operate with a large cross-flow factor X, implying either small bubbles or a deep bed. Since the former is not a variable under our control, deep beds become a necessity for high conversion.

3. If we have a very deep bed, i.e. $X \to \infty$, we have

$$C_H \to C_{BH} \to C_P \to \frac{C_0}{1+k}$$

showing that the bubble and particulate phases have equilibrated and that the bed is behaving as a well-mixed reactor.

We can plot the concentration profiles as in Figure 8.2, which shows a constant value of C_P and an exponential decay in C_B from C_0 towards C_P.

Example 8.5
A fluidised bed with $H_{mf} = 0.5$ m is operated at $3U_{mf}$ where $U_{mf} = 0.11$ m/s. The mean bubble diameter is 80 mm. Find the bubble rise velocity U_B, the absolute bubble rise velocity U_A, the bed depth H, the residence time of the bubbles within the bed T, the bubble volume V, the through-flow Q and the cross-flow factor X.

Solution
From equation (8.5),

$$U_B = 0.71\sqrt{(9.81 \times 0.08)} = 0.629 \, \text{m/s}$$

From equation (8.13),

$$U_A = 0.629 + 2 \times 0.11 = 0.849 \, \text{m/s}$$

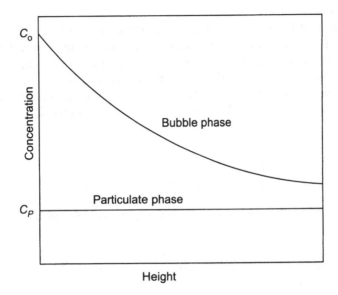

Figure 8.2 Concentration profiles of reactant in the bubble and particulate phases of well-mixed bed (first-order reaction).

From equation (8.18),

$$\frac{(H - 0.5)}{H} = \frac{0.22}{0.849}.$$

Hence

$$H = 0.675 \, \text{m}$$

From equation (8.20),

$$T = \frac{0.675}{0.849} = 0.795$$

The bubble volume is

$$V = \frac{\pi}{6} D^3 = \frac{\pi}{6} 0.08^3 = 2.68 \times 10^{-4} \, \text{m}^3$$

and the through-flow is given by equation (8.23) as

$$Q = \frac{\pi}{4} \times 0.08^2 \times 0.11 = 1.66 \times 10^{-3} \, \text{m}^3/\text{s}.$$

Hence from equation (8.43),

$$X = \frac{1.66 \times 10^{-3} \times 0.795}{2.68 \times 10^{-4}} = 4.92$$

Example 8.6

A fluidised bed is operated with $U = 0.2\,\mathrm{m/s}$, $U_{mf} = 0.05\,\mathrm{m/s}$, $H_{mf} = 1.0\,\mathrm{m}$ and $X = 1.5$. The inlet gas contains a reactant at concentration 0.003 kmol/m^3 which decomposes in the particulate phase with an apparent first-order constant k_v of $1.6\,\mathrm{s^{-1}}$. Determine the concentrations in the particulate phase, in the bubble phase at the top of the bed and in the mixed outlet gases. It should be assumed that the particulate phase is well mixed.

Solution

An overall balance, equation (8.44), gives

$$0.2 \times (0.003 - C_H) = 1 \times 1.6C_P \qquad \text{(i)}$$

and a mass balance at the top of the bed, equation (8.45) gives

$$0.2C_H = 0.15C_{BH} + 0.05C_P \qquad \text{(ii)}$$

The concentration in the bubble phase at the top of the bed is given by equation (8.42),

$$\frac{C_{HB} - C_P}{0.003 - C_P} = e^{-X} = e^{-1.5} = 0.223 \qquad \text{(iii)}$$

Equations (i), (ii) and (iii) can be solved simultaneously to give, $C_P = 2.83 \times 10^{-4}\,\mathrm{kmol/m^3}$, $C_H = 7.38 \times 10^{-4}\,\mathrm{kmol/m^3}$ and $C_{BH} = 8.90 \times 10^{-4}\,\mathrm{kmol/m^3}$.

8.3.4   Second- and higher-order reactions in a well-mixed fluidised bed

The analysis of the preceding section assumed first-order kinetics in the particulate phase but can be readily adapted for any other order of reaction. If the reaction is nth order in the reactant, equations (8.42) and (8.45) will still apply and equation (8.44) can be modified to

$$H_{mf}k_n C_P^n = U(C_0 - C_H) \qquad \text{(8.51)}$$

In particular, for a second-order reaction we have

$$\frac{C_{BH} - C_P}{C_0 - C_P} = e^{-X} \qquad \text{(8.42)}$$

$$H_{mf}k_2 C_P^2 = U(C_0 - C_H) \qquad \text{(8.52)}$$

and

$$U_{mf}C_P + (U - U_{mf})C_{BH} = UC_H \qquad \text{(8.45)}$$

Equations (8.42), (8.52) and (8.45) can be solved for C_{BH}, C_P and C_H. Caution must, however, be exercised while solving these equations as the quadratic form of the equations give two sets of solutions. The appropriate set is immediately apparent if all three concentrations are evaluated. One set will contain a negative root for C_P, which is clearly inadmissible and arises from second-order kinetics which apparently allows a positive rate of reaction with a negative concentration. This root appears to give more conversion than can be achieved in practice and therefore the larger value of C_H is the correct root.

Example 8.7
Rework Example 8.6 for the case when the reaction in the particulate phase is second order in C_P with reaction velocity constant $k_2 = 2000\,\text{m}^3/\text{kmol s}$.
  Equations (ii) and (iii) of Example 8.6 still apply, i.e.

$$0.2C_H = 0.15C_{BH} + 0.05C_P \tag{ii}$$

and

$$C_{HB} - C_P = 0.223(0.003 - C_P) \tag{iii}$$

Equation (i) however now takes the form,

$$0.2 \times (0.003 - C_H) = 1 \times 2000C_P^2 \tag{i}$$

Elimination of C_P and C_{BH} yields the quadratic

$$C_H^2 - 9.35 \times 10^{-4}C_H + 4.39 \times 10^{-8} = 0$$

which has the roots $C_H = 8.85 \times 10^{-4}\,\text{kmol/m}^3$ and $4.96 \times 10^{-5}\,\text{kmol/m}^3$. The corresponding values of C_P are 4.60×10^{-4} and $-5.43 \times 10^{-4}\,\text{kmol/m}^3$. Since negative values of concentrations are inadmissible, the latter root is unrealistic, giving $C_H = 8.85 \times 10^{-4}\,\text{kmol/m}^3$, $C_p = 4.60 \times 10^{-4}\,\text{kmol/m}^3$, and, from equation (ii), $C_{BH} = 1.03 \times 10^{-3}\,\text{kmol/m}^3$.

8.3.5   Combustion of gases

Fluidised beds are sometimes used for the combustion of hydrocarbon gases. This most frequently occurs during the fluidised combustion of coal where the high temperature releases volatile components which burn in the presence of air. However, the combustion of hydrocarbons usually proceeds by a chain reaction involving free radicals. Such radicals are deactivated on contact with a solid and hence the combustion reactions are effectively suppressed within the narrow passages of the particulate phase. However, in the bubbles, the absence of particles allows the combustion to take place rapidly. In these circumstance we have an inversion of the situation analysed above; the reaction rate in the particulate phase is zero and all the reactant entering the bubbles reacts rapidly.
  As we saw in Section 8.3.3 (equation (8.43)), the volume of gas passing through a bubble during its passage through the bed is XV, where V is the

bubble volume. Thus the rate of reaction per bubble is XVC_P since the gas enters with concentration C_P and, in the present context, we can assume that the reaction goes to completion. The number of bubbles produced per second per unit area of distributor is the volumetric flow rate of bubbles gas divided by the bubble volume, $(U - U_{mf})/V$. Multiplying these two quantities gives the total reaction rate, $(U - U_{mf})XC_p$. However, some of the reactant leaves the top of the particulate phase unreacted at a rate $U_{mf}C_P$. Thus the fraction of the gas leaving the bed unconverted is

$$\frac{U_{mf}}{U_{mf} + (U - U_{mf})X}$$

The unburnt gases normally burn rapidly in the freeboard.

Alternatively, if the gases are released within the bed at a rate R per unit volume, we can evaluate the concentration in the particulate phase since the production rate is the sum of the combustion rate and the effluent rate,

$$H_{mf}R = (U - U_{mf})XC_P + U_{mf}C_P \qquad (8.53)$$

8.3.6 Plug flow in the particulate phase

In the previous sections we assumed that the particulate phase was well mixed. This is an idealisation but may be an adequate approximation for a vigorously bubbling bed of large diameter. If the particulate phase is not well mixed, a knowledge of the concentration profile within it is required and usually this can only be obtained by experiment. However, the opposite extreme, plug flow within the particulate phase, can readily be analysed. This will be most appropriate for slugging beds and for narrow beds with few bubbles. The case of a first-order reaction is considered below.

If we assume plug flow within the particulate phase, the concentration C_P will be a function of position and while equation (8.39) in Section 8.3.3 remains valid, it can no longer be integrated to give equation (8.42). Equation (8.39) can be put into a more convenient form by expressing the time derivative as a spatial derivative, i.e. by noting that

$$\frac{dC_B}{dt} = U_A \frac{dC_B}{dx} \qquad (8.54)$$

giving

$$Q(C_B - C_P) = -VU_A \frac{dC_B}{dy} \qquad (8.55)$$

where U_A is the absolute bubble velocity.

Since the particulate phase gas concentration is now a function of position, we must evaluate the reaction rate over a small element of height δy.

Recalling that the reaction takes place only in the particulate phase, i.e. in a fraction $1 - \varepsilon_{\text{bubble}}$ of the bed, a mass balance over the distance δy gives

$$-[(U - U_{mf})\, \delta C_B + U_{mf}\, \delta C_P] = k_v C_P (1 - \varepsilon_{\text{bubble}})\, \delta y \tag{8.56}$$

It is more convenient to express equations (8.55) and (8.56) in terms of the dimensionless height z defined by $z = y/H$; z representing the fractional distance up the bed. In this terminology these equations become

$$\frac{dC_B}{dz} = -\frac{QH}{VU_A}(C_B - C_P) \tag{8.57}$$

i.e.

$$C_P = C_B + \frac{1}{X}\frac{dC_B}{dz} \tag{8.58}$$

and

$$\beta \frac{dC_B}{dz} + (1 - \beta)\frac{dC_P}{dz} = -\frac{k_v H(1 - \varepsilon_{\text{bubble}})}{U} C_P = -kC_P \tag{8.59}$$

where k, β and X are as defined in Section 8.3.3 (equations (8.50), (8.49) and (8.43)).

Eliminating C_P between these equations, gives,

$$(1 - \beta)\frac{d^2 C_B}{dz^2} + (X + k)\frac{dC_B}{dz} + kXC_B = 0 \tag{8.60}$$

Since this is a second-order linear differential equation with constant coefficients, its solution takes the form

$$C_B = Ae^{m_1 z} + Be^{m_2 z} \tag{8.61}$$

where A and B are arbitrary constants and m_1 and m_2 are the roots of the auxiliary equation

$$(1 - \beta)m^2 + (X + k)m + kX = 0 \tag{8.62}$$

The discriminant of the auxiliary equation

$$b^2 - 4ac = (X + k)^2 - 4(1 - \beta)\,kX = (X - k)^2 + 4\beta kX \tag{8.63}$$

is inevitably positive, showing that the auxiliary equation has two real roots. These are both negative since all the coefficients in the auxiliary equation have the same sign. The arbitrary constants A and B have to be determined from the boundary conditions. These can be obtained by noting that, at the base of the bed, the gas entering both the particulate phase and the bubbles is at the inlet concentration C_0, i.e.

$$z = 0, \qquad C_B = C_0, \qquad C_P = C_0 \tag{8.64}$$

and from equation (8.58) it follows that

$$z = 0, \qquad \frac{dC_B}{dz} = 0 \tag{8.65}$$

Thus,

$$A + B = C_0 \tag{8.66}$$

and

$$m_1 A + m_2 B = 0 \tag{8.67}$$

giving

$$A = \frac{C_0 m_2}{m_2 - m_1} \tag{8.68}$$

$$B = -\frac{C_0 m_1}{m_2 - m_1} \tag{8.69}$$

The concentration in the particulate phase can be obtained from C_B by substitution into equation (8.58), giving

$$C_P = A\left(1 + \frac{m_1}{X}\right)e^{m_1 z} + B\left(1 + \frac{m_2}{X}\right)e^{m_2 z} \tag{8.70}$$

Figure 8.3 shows the reactant concentrations in the bubble and particulate phases as functions of bed height, and it can be seen that the particulate phase concentration tends asymptotically to zero, with the bubble phase concentration lagging behind as the reactant is slowly convected out of the bubbles.

The concentrations at the top of the bed are obtained by putting $z = 1$ into equations (8.60) and (8.69) and the use of equation (8.45) leads, after some manipulation, to

$$\frac{C_H}{C_0} = \frac{m_2 e^{m_1} - m_1 e^{m_2}}{m_2 - m_1} + \frac{m_1 m_2 (1 - \beta)}{X(m_2 - m_1)}\left(e^{m_1} - e^{m_2}\right) \tag{8.71}$$

If a fluidised bed is operated well above minimum fluidisation, as is commonly the case, a large fraction of the gas will pass through the bed in the form of bubbles and β will be close to 1. In this case $(1 - \beta)$ will be very much less than both X and k, and the values of m are given approximately by

$$m_1 \approx -\frac{kX}{X + k} \tag{8.72}$$

$$m_2 \approx -\frac{X + k}{1 - \beta} \tag{8.73}$$

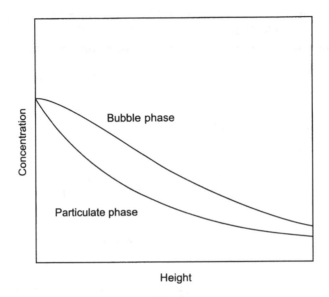

Figure 8.3 Concentration profiles of reactant in the bubble and particulate phases when particulate phase is in plug flow (first-order reaction).

It is seen that m_2 is inevitably large and negative and the term $e^{m_2 z}$ dies away rapidly with distance above the distributor. Thus to a first approximation,

$$C_B = C_0 e^{m_1 z} \tag{8.74}$$

and

$$C_P = \frac{C_0 X}{X + k} e^{m_1 z} \tag{8.75}$$

The apparent violation of the boundary condition $C_P = C_0$ at $z = 0$, arises because we have neglected the term in $e^{m_2 z}$. This term is not insignificant as $z \to 0$, though it rapidly becomes so as z increases above zero..

Example 8.8

Evaluate the concentrations at the top of the bed and also one-tenth of the way up for the fluidised bed of Example 8.6, but on the assumption that plug flow occurs in the particulate phase.

Solution

From equations (8.49) and (8.50) the parameters β and k are given by 0.75 and $(1.6 \times 1.0)/0.2 = 8.0$ respectively. From equation (8.60) the bubble phase concentration is given by

$$0.75 \frac{d^2 C_B}{dz^2} + (1.5 + 8) \frac{dC_B}{dz} + 8 \times 1.5 C_B = 0$$

which has the auxiliary equation $0.75\,m^2 + 9.5\,m + 12 = 0$, giving $m = -1.308$ and -36.69. Thus $C_B = A\,e^{-1.308z} + B\,e^{-36.69z}$.

The boundary conditions $C_B = 0.003\ \text{kmol/m}^3$ and $dC_B/dz = 0$ on $z = 0$ give

$$A + B = 0.003 \quad \text{and} \quad 1.308A + 36.69B = 0$$

Hence

$$A = 3.111 \times 10^{-3}\ \text{kmol/m}^3 \quad \text{and} \quad B = -1.109 \times 10^{-4}\ \text{kmol/m}^3$$

From equation (8.70)

$$C_P = A\left(1 - \frac{1.308}{1.5}\right)e^{-1.308z} + B\left(1 - \frac{36.69}{1.5}\right)e^{-36.69z}.$$

i.e.

$$C_P = 4.043 \times 10^{-4}\,e^{-1.308z} + 2.606 \times 10^{-3}\,e^{-36.69z}.$$

Hence we find,

$z = 0$:	$C_P = 30.0 \times 10^{-4}\ \text{kmol/m}^3$	$C_B = 30.0 \times 10^{-4}\ \text{kmol/m}^3$
$z = 0.1$:	$C_P = 4.21 \times 10^{-4}\ \text{kmol/m}^3$	$C_B = 27.3 \times 10^{-4}\ \text{kmol/m}^3$
$z = 1.0$:	$C_P = 1.09 \times 10^{-4}\ \text{kmol/m}^3$	$C_B = 8.41 \times 10^{-4}\ \text{kmol/m}^3$

It is seen that initially C_P decreases rapidly, undergoing 89% of its overall change in the first 10% of the height, whereas the decrease in C_B is much more even (12.5% change in 10% of the height). The outlet concentration is given by equation (8.45) as

$$C_H = 0.75 \times 8.41 \times 10^{-4} + 0.25 \times 1.09 \times 10^{-4} = 6.58 \times 10^{-4}\ \text{kmol/m}^3$$

It should be noted that the conversion is greater than that in Example 8.6, as is typical for plug flow reactors.

8.3.7   Combustion of solid particles

The fluidised combustor offers considerable advantages over other types of solid fuel combustion, partly because of the ease of pollution control and also because the combustion gases can be used in a gas turbine as part of an efficient integrated power generation system. Not only can sulphur emissions be reduced by adding limestone to the bed but nitrogen oxide formation can also be reduced by running the lower part of the bed with insufficient air for complete combustion and adding further air part way up the bed. In fluidised combustion one of the main topics of importance is the changing size of the particles during their residence in the bed. This changing particle size, however, has no significant effect on the fluidisation characteristics since the bed in a fluidised combustor is predominantly composed of inert particles such as sand or ash. A bed composed entirely of coal particles would not only use up all the oxygen in the air within the first few millimetres but would also turn into a sticky mess as a result of tar formation and very rapid sintering.

We see from equations (8.28) and (8.32) that the flux of oxygen at the surface of the particle is given by

$$F = \left(\frac{d}{D \, Sh} + \frac{1}{k'} \right)^{-1} C_P \qquad (8.76)$$

from which it can be seen that the rate of reaction will vary as the particle size changes. In the limit as the particle size reduces to zero the Sherwood number will tend to 2 and F will tend to $k'C_P$. The value of k' varies from coal to coal, but Field gives the following expression as typical,

$$k' = 595000T \, e^{-149\,227/RT} \text{ m/s} \qquad (8.77)$$

where, in this equation, R is the universal gas constant and T is the absolute temperature. For a typical burning temperature of 1040°C this gives $k' = 0.903$ m/s, which is comparable with the value of k_g for a 0.5 mm sphere, $2 \times 2.08 \times 10^{-4}/0.5 \times 10^{-3} = 0.83$ m/s. Thus it cannot be assumed with confidence that the rate of reaction is limited by diffusion. However, for clarity, the analyses below are developed first for diffusion control and the effects of kinetic limitation are presented later.

In the steady state the consumption of oxygen per unit bed area equals the molar supply rate of coal per unit area, m, since most beds are operated with excess air in order to limit CO production. Thus,

$$m = U(C_0 - C_H) \qquad (8.78)$$

which, from equations (8.42) and (8.45), gives

$$m = (C_0 - C_P)(U - (U - U_{mf}) \, e^{-X}) \qquad (8.79)$$

for the case of a well-mixed bed. Since the oxygen concentration in the particulate phase cannot be negative, equation (8.79) imposes an upper limit on the supply of coal. For lesser flow rates this equation can be used to evaluate C_P.

Assuming diffusion control, the molar flow of oxygen $\dot{N}$ to the surface of the particle is given by equation (8.74),

$$\dot{N} = \pi d^2 F = \pi d^2 \frac{ShD}{d} C_P \qquad (8.80)$$

and thus must equal the burning rate $-dM_P/dt$, where M_P is the number of moles in the particle, given by $\pi d^3 \rho_m/6$, where ρ_m is the molar density of the solid. Thus,

$$\pi d^2 \frac{Sh\,D}{d} C_P = -\frac{\pi d^2 \rho_m}{2} \frac{dd}{dt} \qquad (8.81)$$

or

$$\frac{ds}{dt} = -\frac{2Sh\,DC_P}{\rho_m s} = -\frac{\alpha}{s} \qquad (8.82)$$

where the parameter α is defined by

$$\alpha = \frac{2\,Sh\,DC_p}{\rho_m} \qquad (8.83)$$

and we have temporarily replaced d by s to avoid the confusing group, dd/dt. Assuming that the Sherwood number remains constant, as it will for small particles, we can integrate equation (8.82) to give the burnout time t_B for a particle of initial diameter d_0,

$$t_B = -\frac{1}{\alpha}\int_{d_0}^{0} s\,ds = \frac{d_0^2}{2\alpha} = \frac{d_0^2 \rho_m}{4DSh\,C_P} \qquad (8.84)$$

In order to evaluate the mass of coal within the bed we need to know the particle size distribution $p(s)$ where $p(s)\delta s$ is the fraction by number of the particles within the bed having sizes between s and $s + \delta s$. Considering first the case when all the particles are supplied to the bed with the same initial diameter d_0, we can calculate the number of particles supplied to the bed per second as

$$N = 6m/\pi d_0^3 \rho_m \qquad (8.85)$$

The total number of particles within the bed, n, is given by

$$n = Nt_B \qquad (8.86)$$

Every particle starts with diameter d_0 and decreases in size to zero and disappears. Thus the rate at which particles decrease through any size s is a constant equal to N. This is sometimes referred to as the size cascade.

In time δt, a particle will decrease in size by an amount $-\delta s$. Thus all particles in the size range s to $s + \delta s$ will decrease through size s in time δt. There are $np(s)\delta s$ such particles and hence the rate at which particles decrease through size s is $-np(s)(\,ds/dt)$. Thus,

$$N = -np(s)\frac{ds}{dt} \qquad (8.87)$$

hence, from equations (8.82), (8.86) and (8.84),

$$p(s) = \frac{s}{\alpha t_B} = \frac{2s}{d_0^2} \qquad (8.88)$$

We can confirm that this distribution is correctly normalised by noting that

$$\int_0^{d_0} p(s)\mathrm{d}s = \int_0^{d_0} \frac{2s}{d_0^2}\mathrm{d}s = 1 \qquad (8.89)$$

as it must.

Hence we can find the total number of moles of carbon within the bed from

$$M = n\int_0^{d_0} \frac{\pi s^3 \rho_m}{6} p(s)\mathrm{d}s = \frac{n\pi}{15}\rho_m d_0^3 \qquad (8.90)$$

and substitution from equations (8.84), (8.85) and (8.87) gives

$$M = \frac{2}{5}mt_B \qquad (8.91)$$

Thus the molar content of the bed is two-fifths of the product of the molar flow rate and the burnout time.

In this analysis we assumed that the process was diffusion controlled and that the Sherwood number was constant. While the latter is reasonable for the small particles under consideration, we should always allow for kinetic effects since, from equation (8.76), these will become dominant as the particle size tends to zero. By analogy with equation (8.81), we see that the rate of change of particle size is given by

$$\left(\frac{s}{DSh} + \frac{1}{k'}\right)^{-1} C_P = -\frac{\rho_m}{2}\frac{\mathrm{d}s}{\mathrm{d}t} \qquad (8.92)$$

The rest of the analysis follows similar lines to that presented above and gives rise to the following results:

Burnout time,

$$t_B = \frac{\rho_m}{2C_P}\left(\frac{d_0^2}{2DSh}\right) + \frac{d_0}{k'} \qquad (8.93)$$

Number of particles in the bed,

$$N = \frac{n\rho_m}{2C_P}\left(\frac{d_0^2}{2D\,Sh} + \frac{d_0}{k'}\right) \qquad (8.94)$$

Particle size distribution,

$$p(s) = \frac{2(k's + D\,Sh)}{k'd_0^2 + 2D\,Sh\,d_0} \qquad (8.95)$$

Mass of carbon within the bed,

$$M = mt_B \frac{4k'd_0 + 5D\,Sh}{20(k'd_0 + 2D\,Sh)} \qquad (8.96)$$

It is seen that these results reduce to those given above when $k' \gg DSh$. The case of kinetic control can obtained by putting $k' \ll DSh$.

Example 8.9
Coal, which may be assumed to be pure carbon, is added at the rate of 4.0×10^{-3} kg/s per unit bed area to a fluidised bed operating at 900°C and 1.0 bar. The coal is in the form of spherical particles of diameter 1.0 mm. Assuming diffusion control and perfect mixing in the particulate phase find:

(1) the concentrations of oxygen in the particulate phase and the flue gas
(2) the particle size distribution within the bed
(3) the burnout time, and
(4) the mass of carbon within the bed.

Take $U = 0.3$ m/s, $U_{mf} = 0.05$ m/s, $X = 1, \rho_s = 720$ kg/m^3, $D = 2.0 \times 10^{-4}$ m^2/s and $Sh = 3.0$.

Solution
The inlet oxygen content

$$C_0 = \frac{0.21 \times 10^5}{8314 \times 1173} = 2.15 \times 10^{-3} \text{ kmol/m}^3$$

and an overall mass balance gives the outlet oxygen concentration

$$0.3 \times (2.15 \times 10^{-3} - C_H) = 4.0 \times 10^{-3}/12.$$

Hence $C_H = 1.04 \times 10^{-3}$ kmol/m^3.
From equation (8.45),

$$0.05C_P + 0.25C_{BH} = 0.3 \times 1.04 \times 10^{-3}$$

and from equation (8.42),

$$2.15 \times 10^{-3} - C_P = \exp(C_{BH} - C_P).$$

Hence, $C_P = 0.547 \times 10^{-3}$ kmol/m^3 and $C_{BH} = 1.14 \times 10^{-3}$ kmol/m^3.
From equation (8.88),

$$p(s) = 2s/(1.0 \times 10^{-3})^2 = 2 \times 10^6 \, s$$

The burnout time is given by equation (8.84),

$$t_B = \frac{(1 \times 10^{-3})^2 \times 720/12}{4 \times 2.0 \times 10^{-4} \times 3.0 \times 0.547 \times 10^{-3}} = 45.7 \, s$$

and from equation (8.91) the mass of carbon in the bed is given by

$$M = \frac{2}{5} \times 4 \times 10^{-3} \times 45.7 = 0.0731 \text{ kg/m}^3$$

8.3.7.1 Fluidised combustion calculations when the feed particles are not of uniform size. In all the above results we have assumed that the feed is in the form of uniform spheres of initial diameter d_0. Commonly the feed will contain a distribution of sizes $q(s)ds$. In these circumstances we must rework the size cascade analysis leading to equation (8.88) and we will illustrate this by considering the case of a feed with a uniform size distribution up to diameter d_0, i.e.

$$q(s) = \frac{2s}{d_0^2} \qquad (8.97)$$

The analysis can readily be adapted for any other size distribution in the feed.

The rate r at which particles decrease through size s is given by equation (8.87) as

$$r = -np(s)\frac{ds}{dt} \qquad (8.98)$$

where r is no longer the rate N at which particles are fed to the bed. Instead it is the feed rate of particles of size greater than s, i.e.

$$r = \int_s^{d_0} Nq(s)\,ds = N\frac{d_0^2 - s^2}{d_0^2} \qquad (8.99)$$

If the reaction is diffusion controlled, we can determine ds/dt from equation (8.82), giving

$$N\frac{d_0^2 - s^2}{d_0^2} = np(s)\frac{\alpha}{s} \qquad (8.100)$$

and

$$p(s) = \frac{N}{\alpha d_0^2 n}(d_0^2 s - s^3) \qquad (8.101)$$

Recalling that $\int_0^{d_0} p(s)ds = 1$, we have

$$\frac{N}{\alpha d_0^2 n}d_0^4\left(\frac{1}{2} - \frac{1}{4}\right) = 1 \qquad (8.102)$$

or

$$n = \frac{Nd_0^2}{4\alpha} \qquad (8.103)$$

Alternatively, paying attention to the size range between s and $s + \delta s$, we see from equation (8.98) that particles leave this range at a rate $r = -np(s)$ (ds/dt). Similarly particles enter this range from above at a rate $r+(dr/ds)\delta s$ and enter from the feed at a rate $Nq(s)\delta s$. Thus a balance gives

$$r = r + \frac{dr}{ds}\delta s + Nq(s)\delta s \tag{8.104}$$

or

$$Nq(s) = -\frac{dr}{ds} = n\frac{d}{ds}\left(p(s)\frac{ds}{dt}\right) \tag{8.105}$$

Substituting from equations (8.82) and (8.97) gives

$$\frac{2Ns}{d_0^2} = -n\alpha\frac{d}{ds}\frac{p(s)}{s} \tag{8.106}$$

and hence,

$$p(s) = \frac{N}{d_0^2 n\alpha}(As - s^3) \tag{8.107}$$

where the arbitrary constant A can be evaluated from the boundary condition that $p(d_0)$ is zero since no particles of size greater than d_0 are fed to the bed.

8.3.7.2 Elutriation. In the course of a combustion process the particles will inevitably become of very small size, and fine particles can also occur in a bed due to attrition or may enter the bed in the feed. If a particle is fine enough it may be carried away in the gas stream leaving the bed, a process known as **elutriation**. Particles with terminal velocities less than the fluidising velocity are not elutriated immediately from a bed. They can only be elutriated from the top surface and the rate-limiting step may be the transfer of the particle from the bulk of the bed to the top surface. Transfer of particles within the bed occurs due to the gross circulation caused by the bubbles and, in addition, bubbles carry material to the top of the bed in their wakes. When a bubble bursts through the top surface, particles of a wide range of sizes are ejected into the 'freeboard' (the space between the particulate phase surface and the gas exit port). Many of these are heavy enough to fall back into the bed and even those with terminal velocities less than the gas velocity may hit the wall and fall back into the bed as a curtain of falling particles close to the wall. Thus the flux of upward-moving particles decreases with height above the top surface. This decrease is at first rapid but eventually the flux of particles settles down to a more or less constant value above a height which is known as the transport disengagement height. The actual elutriation rate cannot be predicted from first principles and correlations such as that given by Geldart [9],

$$K = \frac{\mu(U - U_t)^2}{gd^2}(0.0015Re_t^{0.6} + 0.01Re_t^{1.2}) \tag{8.108}$$

are often used. Here d is the diameter of the particles being elutriated, U_t is their terminal velocity and Re_t is the Reynolds number based on d and U_t.

The elutriation constant K has the units of kg/m^2s and is effectively a first-order velocity constant for elutriation, the mass flow rate of the elutriate being given by

$$E = KAx \qquad (8.109)$$

where A is the area of the top surface of the bed and x is the mass fraction of the fines in the bed. For a batch system without production of fines, the elutriation rate is the rate of decrease of the fines content with time, i.e. the mass of the bed multiplied by $-dx/dt$. Thus,

$$E = -AH_{mf}\rho_B \frac{dx}{dt} \qquad (8.110)$$

from which we find that

$$x = x_0 \exp\left(-\frac{Kt}{H_{mf}\rho_B}\right) \qquad (8.111)$$

where x_0 is the initial mass fraction of fines. If, on the other hand, fines are produced at a rate f within the bed, the steady-state mass fraction will be given by

$$x = \frac{f}{KA} \qquad (8.112)$$

For a bed operating in the steady state with inflow and outflow of solid, as shown in Figure 8.4, we can perform a mass balance on the fines,

$$W_0 x_0 = W_1 x_1 + E = (W_1 + AK)x_1 \qquad (8.113)$$

where W_0 is the feed rate, W_1 is the overflow rate and x_0 and x_1 are the mass fraction of fines in the feed and overflow respectively. This latter quantity is also the mass fraction of fines within the bed since we can consider the bed to be well mixed.

An overall mass balance gives

$$W_0 = W_1 + E \qquad (8.114)$$

and equations (8.113) and (8.114) are sufficient to solve for E, W_1 and x_1.

Commonly the bed will contain fines of a range of sizes. If this distribution can be discretised so that we can say that there is a finite number of species of fines, i, j, etc., with mole fractions x_i, x_j, etc., we can perform a mass balance on each species by analogy with equation (8.113), giving

$$W_0 x_{0i} = (W_1 + AK_i)x_{1i} \qquad (8.115)$$

where K_i is the elutriation constant for species i. It is necessary to include the coarse material among these species even though its value of K may be close to zero. However, the mass fractions x_{1i} must add up to 1, and therefore,

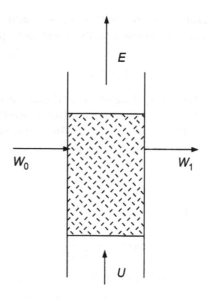

Figure 8.4 Illustrating elutriation.

$$\sum \frac{W_0 x_{0i}}{W_1 + AK_i} = 1 \qquad (8.116)$$

This equation can be solved to give W_1 and hence the x_{1i} can be found from equation (8.115) and the elutriation rate of each species, E_i, from $AK_i x_{1i}$.

Example 8.10
A stream of mass composition $x_{01} = 0.5, x_{02} = 0.25, x_{03} = 0.25$ is fed to a fluidised bed of surface area 10m^2 at a flow rate of 50 kg/s. The elutriation constants are $K_1 = 0.1$ kg/m^{2}s, $K_2 = 1.0$ kg/m^{2}s and $K_3 = 2.0$ kg/m^{2}s. Find the mass flow rates and the compositions of the elutriate and overflow.

Solution
From equation (8.116),

$$\frac{50 \times 0.5}{W_1 + 10 \times 0.1} + \frac{50 \times 0.25}{W_1 + 10 \times 1} + \frac{50 \times 0.25}{W_1 + 10 \times 2.0} = 1$$

This is a cubic in W_1 but can readily be solved numerically to give $W_1 = 43.1$kg/s. Though a cubic, this equation has only one real root. This can be seen immediately since every term on the left-hand side, and hence the left-hand side itself, is a monotonically decreasing function of W_1. The individual terms in this equation are the values of x_{1i}. Hence

$$x_{11} = \frac{50 \times 0.5}{43.1 + 10 \times 0.1} = 0.567$$

and similarly $x_{12} = 0.235$ and $x_{13} = 0.198$. We can check that $\Sigma x_{1i} = 1$ as it must. The flow of each species in the elutriate E_i is $AK_i x_{1i}$. Hence

$$E_1 = 10 \times 0.1 \times 0.567 = 0.567 \text{ kg/s}$$

and similarly $E_2 = 2.35 \text{ kg/s}$ and $E_3 = 3.96$. $E = \Sigma E_i = 6.877$, which accords perfectly with the overall mass balance $E = 50 - 43.1 = 6.9$ kg/s, at least to the precision with which we solved the cubic equation in W_1. The mass fractions in the elutriate x_{ei} are given by E_i/E. Thus

$$x_{e1} = 0.567/6.877 = 0.082$$

and similarly $x_{e2} = 0.342$ and $x_{e3} = 0.576$. Again we can check that $\Sigma x_{ei} = 1$ as it must.

Sometimes the size distribution cannot be discretised conveniently and continuous distribution functions $p_0(s)$ and $p_1(s)$ in the feed and overflow must be considered. In this case x_{0i} must be replaced by $p_0(s) \, ds$ and equation (8.116) becomes

$$\int \frac{W_0 p_0(s)}{W_1 + AK} \, ds = 1 \tag{8.117}$$

from which W_1 can be obtained if K is known as a function of particle size.

Notation

A	Cross-sectional area of the bed (m²)
A	Arbitrary constant in equations (8.61) to (8.70)
a_0	Distributor area per hole (m²)
B	Arbitrary constant in equations (8.61) to (8.70)
C	Total concentration (kmol/m³)
C_0	Concentration of reactant in the inlet (kmol/m³)
C_B	Concentration of reactant in the bubble phase (kmol/m³)
C_{BH}	Concentration of reactant in the bubble at the top of the bed (kmol/m³)
C_H	Concentration of reaction immediately above the top of the bed (kmol/m³)
C_P	Concentration of reactant in the particulate phase (kmol/m³)
C_S	Concentration of reactant at the particle surface (kmol/m³)
D	Bubble diameter (m)
D	Diffusivity (m²/s)
d	Particle diameter (m)
d_0	Initial particle diameter (m)
D_E	Equivalent diameter of a bubble (m)
D_T	Diameter of the bed (m)
E	Elutriation rate (kg/s)

F	Flux of reactant at the particle surface (kmol/m^{2}s)
g	Acceleration due to gravity (m/s^2)
H	Height of the bed (m)
H_0	Empirical length defined by equation (8.22) (m)
H_{mf}	Height of the bed at minimum fluidisation (m)
K	Elutriation constant (kg/m^{2}s)
k	Dimensionless reaction velocity constant defined by equation (8.50) (–)
k'	Surface reaction velocity constant (m^{-1}s^{-1})
k_2	Second-order reaction velocity constant (m^3/kmol s)
k_G	Mass transfer coefficient (m/s)
k_n	nth-order reaction velocity constant ((m^3/kmol)$^{n-1}$/s)
k_v	Effective volumetric reaction velocity constant (s^{-1})
l	Distance (m)
M	Moles of carbon within the bed (kmol)
m	Molar supply rate of carbon (kmol/s)
m_1, m_2	Roots of equation (8.62) (–)
N	Number of particles supplied to the bed per unit time (s^{-1})
N	Molar flow of oxygen (kmol/s)
n	Number of particles in the bed (–)
n	Order of reaction (–)
n^*	Number of particles per unit volume (m^{-3})
p	Pressure (N/m^2)
$p(s)$	Particle size distribution, by number, in the bed (m^{-1})
$p_0(s)$	Particle size distribution, by mass, in the feed (m^{-1})
Q	Volumetric flow rate (m^3/s)
$q(s)$	Particle size distribution in the feed (m^{-1})
R	Gas constant (J/kmol K)
R	Rate of gas production within the bed (kmol/s)
r	Rate at which particles decrease through a given diameter (s^{-1})
r_A	Rate of reaction of species A (kmol/s)
Re	Reynolds number (–)
Re_t	Reynolds number based on the terminal velocity (–)
s	Particle diameter (identical to d) (m)
Sc	Schmidt number (–)
Sh	Sherwood number (–)
T	Residence time of a bubble in the bed (s)
T	Temperature, in equation (8.77) (K)
t	Time (s)
t_B	Burnout time (s)
U	Superficial velocity (m/s)
U_A	Absolute rise velocity of the bubble (m/s)
U_B	Velocity of a bubble relative to the continuous phase (m/s)
U_G	Interstitial velocity of the gas (m/s)

U_{mf}	Superficial velocity at minimum fluidisation (m/s)
U_P	Velocity of the particles (m/s)
U_t	Terminal velocity (m/s)
V	Volume of the bubble (m^3)
W_0	Feed rate to the bed (kg/s)
W_1	Overflow rate (kg/s)
X	Cross-flow factor defined by equation (8.43)
x	Mass fraction (–)
x_0	Initial mass fraction or mass fraction in the feed (–)
x_1	Mass fraction in the bed and overflow (–)
y	Height above the distributor (m)
z	Fractional height above the distributor, $y/H(-)$
α	Parameter defined by equation (8.83)
β	Fraction of the gas passing through the bed in the bubble phase (–)
ε	Void fraction (–)
ε_{bubble}	Fraction of the bed occupied by bubbles (–)
ε_{mf}	Void fraction at minimum fluidisation (–)
μ	Viscosity (Ns/m^2)
ρ_f	Density of the fluid (kg/m^3)
ρ_m	Molar density $(kmol/m^3)$
ρ_s	Density of the solid (kg/m^3)

References

1. Carman, P.C. (1937) *Trans. Inst. Chem. Eng.*, **15**, 150.
2. Leva, M. and Gummer, M. (1952) *Chem. Eng. Prog.*, **48**, 307.
3. Davidson, J.F. and Harrison, D. (1963) *Fluidised Particles*, Cambridge University Press.
4. See, for example, Wallis, G.B. (1969) *One Dimensional Two-Phase Flow*, McGraw Hill, or Wallis, G.B. (1974) *Int. J. Multiphase Flow*, **1**, 491.
5. Whitehead, A.B. in Davidson, J.F. and Harrison, D. (1971) *Fluidzation*, Academic Press.
6. Rowe, P.N. (1976) *Chem. Eng. Sci.*, **31**, 285.
7. Baeyens, J. and Geldart, D. (1986) in reference [9].
8. See, for example, Kay, J.M. and Nedderman, R.M. (1985) *Fluid Mechanics and Transport Processes*, CUP.
9. Geldart, D. (ed.) (1986) *Gas Fluidisation Technology*, Wiley.

Further reading

Davidson, J.F. and Harrison, D. (1971) *Fluidization*, Academic Press.
Howard, J.R. (1989) *Fluidized Bed Technology*, Adam Hilger.
Kunii, D. and Levenspiel, O. (1987) *Fluidization Engineering*, Krieger.

9 Three-phase reactors
J.M. WINTERBOTTOM

9.1 General observations

Three-phase reactors have found, and are finding, many applications in the production of petrochemicals, fine chemicals/pharmaceuticals and biochemicals. The analysis and design of this type of reactor are complicated due to the many physical processes which occur during operation. These processes can include not only liquid/solid and liquid/gas mass transfer, pore diffusion, and axial mixing, but also non-linear reaction kinetics and incomplete wetting of the catalyst particles.

9.2 Applications of three-phase reactors

In this text, a three-phase reactor will be taken to mean a reactor in which gas and liquid phases are contacted with a solid catalyst. However, it is worth noting that a newly emerging area, which can be described also as three phase, is that involving the reaction of a gas with a biphasic liquid system. In this case a homogeneous catalyst may be dissolved in one solvent (usually water) and the organic reactant in another immiscible organic solvent, the reaction occurring at the liquid/liquid interface. This type of system has been reviewed by Chaudhari [1] but will not be discussed further in this chapter.

9.2.1 Triglyceride hydrogenation

The partial hydrogenation of triglyceride oils is one important example of the use of three-phase reactors. This process is generally carried out in agitated slurry reactors because heat and mass transfer rates to and from slurried particulate catalysts are better than is the case with the equivalent fixed bed reactors. However, it should be noted that the latter type of reactor has been observed to offer better selectivity for the retention of linoleate chains and, of course, its use eliminates the need for disengaging the catalyst from the product (Mukherjee [2], Boyes et al. [3]). Indeed, a related reaction, the hydrogenation of unsaturated fatty acids, is normally carried out in three-phase fixed bed reactors.

   As an example of the use of partial triglyceride hydrogenation, feedstocks such as soybean, cottonseed and rape seed oils are often hydrogenated with the objective of removing fatty acid species containing three (or more) olefinic double bonds and hence stabilising the oils against oxidation. It is very important that the hydrogenation should be selective so that fatty acid chains containing two double bonds (linoleic) and one double bond are retained. It is also desirable that these double bonds should be retained in the *cis*-isomer form and it must be admitted that the above processes do result in the conversion of some of the naturally occurring *cis*-olefinic bonds to *trans*-olefinic bonds. It has been suggested that this may present a possible health hazard. Hence, although partial hydrogenation is an important industrial process which will continue to provide very necessary synthetically manipulated edible oils for many years to come, it is quite possible that countries with more sophisticated dietary requirements may come to use synthetic oils and fats produced by other methods such as *trans*-esterification. Yet another approach is used in the manufacture of the zero calorie fat OLESTRA, produced by Procter & Gamble [4].

9.2.2 Hydrodesulphurisation

The removal of sulphur (and nitrogen) from crude oil products is of ever-increasing importance, the hydrodesulphurisation of gas oil (diesel fuel) being one example of the use of this operation. The sulphur content of many crude oils is too high to meet the statutory upper limit of 0.04% w/w sulphur. This limit was set in 1996 to limit SO_2 emissions in diesel exhaust gases. An added benefit in desulphurisation is that the lower sulphur content leads to a lower particulate level in the diesel exhaust gases.

A good example of the hydrodesulphurisation reaction is provided by the hydrodesulphurisation of thiophene according to:

$$\text{thiophene} + H_2 \xrightarrow[\text{Ni molybdate}]{\text{Co molybdate or}} C_4H_{10} + H_2S$$

The analogous reaction with nitrogen compounds produces ammonia.

9.2.3 Hydrocracking

In this three-phase process, heavy petroleum fractions are cracked in the presence of hydrogen to give lower molecular weight hydrocarbons using supported noble metals such as platinum or palladium.

9.2.4 Methanol synthesis

In many cases the production of methanol from synthesis gas ($H_2 + CO$) is carried out in the gas phase. However, there are reported advantages in carrying out this reaction in the liquid phase, comprising the product (methanol) containing the suspended catalyst (Sherwin and Frank [5]). The process is likely to command interest as methanol is a possible motive fuel as an alternative or supplement to gasoline.

9.2.5 Fine chemical/pharmaceutical production

A wide range of chemicals and intermediates is produced using three-phase catalytic reactors: the following examples are typical.

9.2.5.1 Geraniol hydrogenation (for production of fragrances)

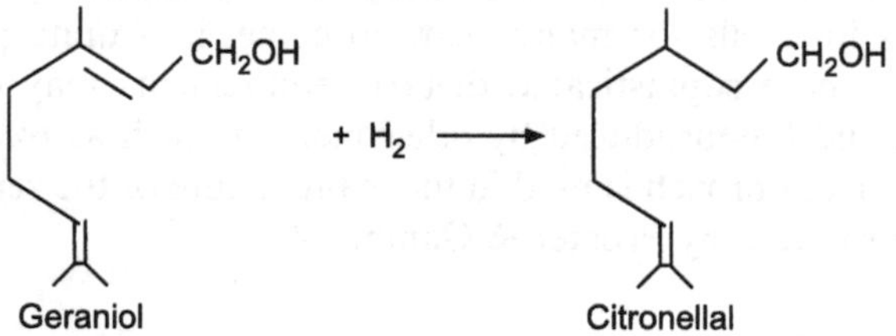

Geraniol   Citronellal

9.2.5.2 Conversion of nitriles to primary amines (used in the pharmaceutical industry)

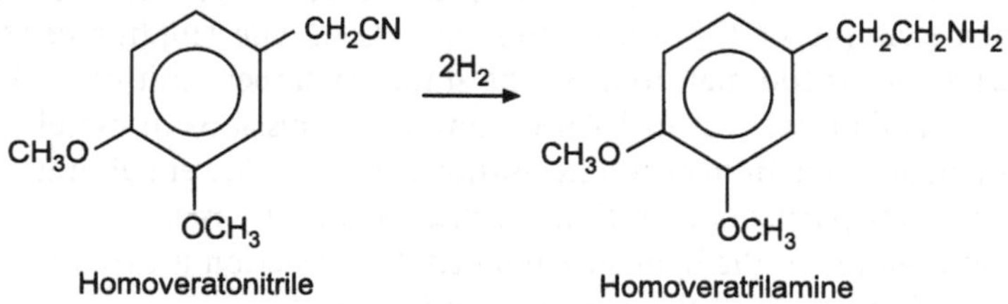

Homoveratonitrile   Homoveratrilamine

9.2.5.3 Production of dyestuffs

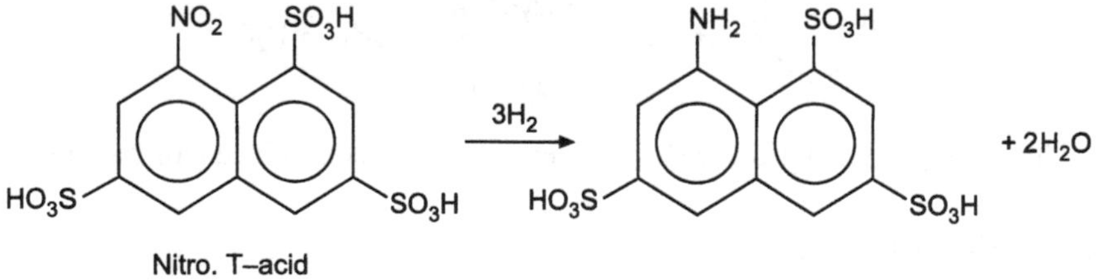

Nitro. T–acid

9.2.5.4 Production of speciality chemicals, as in the alkylation of phenol

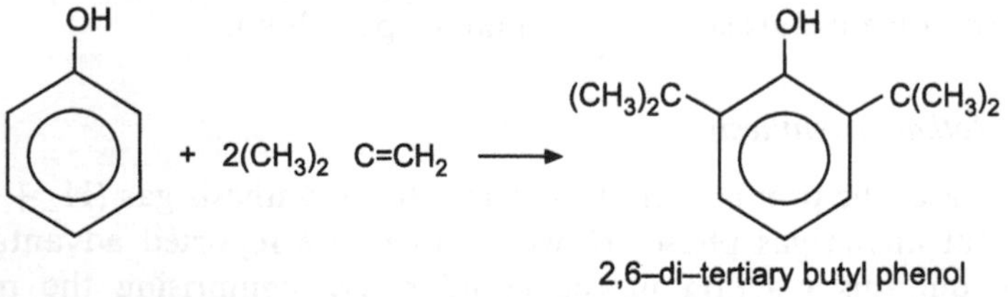

2,6–di–tertiary butyl phenol

All the above reactions are carried out in three-phase slurry reactors.

9.2.5.5 Manufacture of sorbitol from glucose. Sorbitol is an important intermediate in the production of vitamin C. It can be produced by the hydrogenation of the aldehyde groups on glucose to the corresponding $-CH_2OH$ groups

$$-CHO \xrightarrow[\text{catalyst}]{H_2} -CH_2OH$$

Raney nickel is often used to catalyse this reaction, though ruthenium catalysts are reported to operate under milder conditions with better selectivity.

9.2.5.6 Use of 'heterogenised' homogeneous catalysts in three-phase reactors. Solutions of many transition metal complexes form highly selective homogeneous catalysts for hydrogenation, hydroformylation and carbonylation operations. Hydroformylation is a process for the manufacture of aldehydes in which hydrogen and carbon monoxide react with the appropriate alkene in the liquid phase in the presence of appropriate dissolved catalyst (often a rhodium–phosphine complex). As described above, these are two-phase reactions, a typical example being

$$CH_3CH = CH_2 + H_2 + CO \longrightarrow CH_3CH_2CH_2CHO$$

For economic reasons however (rhodium-based catalysts are difficult to separate and recover from the products), the catalyst material is often attached to resin supports and then used in a three-phase reactor. Biphasic catalysis is now also being used [1].

9.3 Types of three-phase reactor

There are two main categories of three-phase reactor, namely: (a) fixed bed and (b) slurry reactors. The latter employ particulate, suspended catalysts.

9.3.1 Fixed bed reactors

In this type of reactor both the liquid and the gaseous phases flow over a stationary (fixed) bed of catalyst particles. Three flow arrangements are possible: (a) cocurrent downflow of gas and liquid, (b) liquid downflow and gas upflow and (c) cocurrent upflow of gas and liquid phases. Reactors of type (a) are designated 'trickle bed reactors' (TBR) while those of type (c) are called 'packed bubble bed reactors' (PBBR). Cocurrent operation avoids flooding problems and hence allows larger throughputs of both phases. This can be advantageous in chemical production though, if it is necessary to reduce the outlet concentration of one component to as low a value as possible (as in pollution control), countercurrent operation is better.

9.3.2   Slurry reactors

9.3.2.1   Mechanically agitated slurry reactors.   In these reactors mechanical agitation is used to keep the solid catalyst in suspension and to disperse the gas phase.

9.3.2.2   Bubble column slurry reactors.   In this case the catalyst particles are maintained in suspension by gas-induced agitation. The usual mode of operation is cocurrent with upflow of **all** phases. However there are good examples of downflow bubble columns. These are particularly suitable for fast reactions. Examples of this type are the Buss reactor [6] shown in Figure 1.5 (b) and the cocurrent downflow contactor reactor (CDCR) shown in Figure 1.5 (c) [7].

9.3.2.3   Three-phase fluidised bed reactors.   In these reactors the particles of catalyst are maintained in suspension by the combined effects of cocurrent liquid flow and bubble movement, liquid flow being the more important factor. Rather larger catalyst particles are employed with this type of reactor than is the case with bubble column slurry reactors.

The many relative advantages and disadvantages of fixed bed and slurry reactors are reviewed by Ramachandran and Chaudhari [8].

The two most common three-phase reactor types are the agitated slurry reactor and the trickle bed reactor and the remainder of this chapter will be concerned with these.

9.4   Slurry reactor theory

In the theory which follows, the reaction of a gas phase species A with a liquid phase species B on the surfaces of catalyst particles which are suspended in the liquid will be considered. The reaction may be summarised as follows:

$$A(g) \quad + \quad B(l) \quad = \quad P(l) \qquad (9.1)$$

| Reactant A fed as gas | Reactant B fed as liquid | Product P produced as liquid |

The bubble column form of slurry reactor and a mechanically agitated form in which the above type of reaction can occur are shown schematically in Figures 9.1 (a) and 9.1 (b) respectively.

Reactions of the above type involve a sequence of diffusion or reaction steps, one or more of which can be rate controlling. The following steps can be identified:

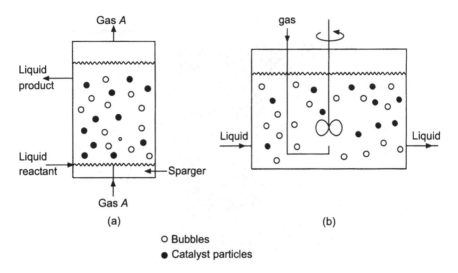

o Bubbles

● Catalyst particles

Figure 9.1 Slurry reactors: (a) bubble column type; (b) mechanically agitated type.

1. Absorption of A from the gas phase into the liquid at the gas/liquid interface enveloping each gas bubble.
2. Diffusion of A into the bulk liquid phase from the liquid side of the gas/liquid interface to the bulk liquid. The bulk liquid is taken to be perfectly mixed and of uniform composition.
3. Diffusion of A from the bulk liquid to the external surface of the solid catalyst (this can apply to species B as well, although in this case the bulk and surface concentrations are often equal).
4. Internal diffusion of reactants A and B within the catalyst pores to an active site.
5. Reaction within the porous catalyst of A with B.

The product(s) P will simultaneously diffuse out from the catalyst pores and into the bulk liquid. The steps 1 to 5 which must be taken by the reactant A before product can be formed can be regarded as resistances to the overall reaction, as illustrated in Figure 9.2.

A more detailed account of these steps and the rate at which they take place now follows.

The reactant A is initially present in the reactor as gas bubbles, its concentration in the interior of these bubbles being C_{Ag}. Before the molecules of this component can dissolve in the liquid and undergo the subsequent steps shown in Figure 9.2 leading to the formation of product, they must impinge on the gas/liquid interfaces surrounding each bubble.

The concentration of A on the gas-side of these interfaces is here denoted C_{Aig} while that on the liquid side is denoted C_{Ai}. Following usual practice we assume equilibrium between the phases at the interface so that

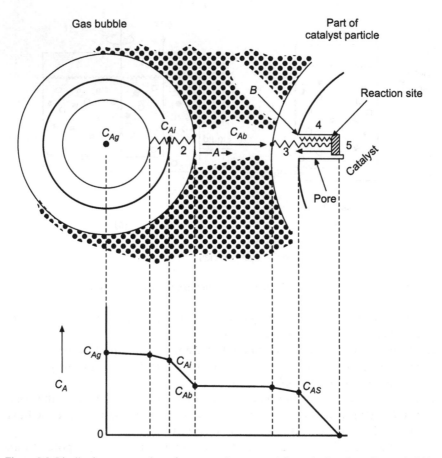

Figure 9.2 Idealised representation of passage of reactant A from the interior of a gas bubble (concentration C_{Ag}) to a reaction site within a catalyst particle. Bulk liquid (concentration C_{Ab}) is shown shaded.

$$C_{Ai} = C_{Aig}/H \qquad (9.2a)$$

where H is Henry's constant in appropriate units.

If the gas bubbles contain only small amounts of components other than A or if component A is only sparingly soluble in the surrounding liquid

$$C_{Aig} \simeq C_{Ag}$$

and

$$C_{Ai} = C_{Ag}/H \qquad (9.2b)$$

In other words, the gas absorption process is under liquid film control. This is the normal situation with slurry reactors since the feed gas is usually single component and the reacting component is usually sparingly soluble. Typical feed gases are H_2, O_2 and CO. (If the above conditions are not obeyed, C_{Aig} may be calculated from C_{Ag} in terms of the gas film mass transfer coefficient (k_{gA}) and the gas/liquid contact area per unit volume of liquid (a_b) using the expression $C_{Aig} = C_{Ag} - (R_A/k_{gA}a_b)$. R_A is the rate of transport of A across the gas/liquid interface per unit liquid volume). In the treatment below the normal situation of liquid film control will be assumed, C_{Ai} being calculated from the (known) values of Henry's constant and the concentration of A in the gas bubbles using equation (9.2b). The rate of gas absorption is determined therefore by the rate of mass transfer of A from the liquid side of the gas/liquid interface to the bulk liquid. This is given by

$$R_A = k_{lA}a_b(C_{Ai} - C_{Ab}) \qquad (9.3)$$

where k_{lA} is the mass transfer coefficient for the transport of A from the interface into the bulk liquid (taken to be well mixed) where its concentration is C_{Ab}. The rate of transport of A to the external surfaces of the catalyst pellets per unit volume of liquid is then given by

$$R_A = k_{cA}a_cm(C_{Ab} - C_{AS}) \qquad (9.4)$$

where k_{cA} is the mass transfer coefficient for the transport of A from the bulk liquid to the external surface of the catalyst. C_{AS} is the concentration of A at this surface and a_c is the external area of the catalyst pellet per unit pellet mass. m is the mass of catalyst per unit volume of slurry.

The final steps (4 and 5) involve intraparticle (pore) diffusion and reaction within the catalyst pores. The rate R_A at which A diffuses into the pores and reacts is given by

$$R_A = m(-r_{AW}) \qquad (9.5)$$

where $(-r_{AW})$ is the reaction rate **per unit mass of catalyst**. This quantity depends on the reactant concentrations at the points where the reaction occurs and is conveniently written in the form

$$(-r_{AW}) = \eta(-r_{AW}^I) \qquad (9.6)$$

where $\eta =$ catalyst effectiveness factor $(0 \leqslant \eta \leqslant 1)$ (see Chapter 3), and $(-r_{AW}^I)$ is the intrinsic reaction rate, i.e. the reaction rate which would prevail if the reactant concentrations in the pores within the catalyst particles were everywhere the same as those at the surface. This quantity can readily be calculated from the surface concentrations given a suitable kinetic equation. The effectiveness factor in equation (9.6) would be unity if diffusion of the reactants into the pores were instantaneous and tends to a very low value if pore diffusion is very slow. Methods of estimating this quantity have been discussed in Chapter 3. Equations (9.5) and (9.6) combine to give

$$R_A = m\eta(-r_{AW}^I) \qquad (9.7)$$

for the rate at which reactant A is entering the catalyst pore and reacting per unit volume of liquid.

It is now necessary to express the intrinsic reaction rate $(-r_{AW}^I)$ as a function of the reactant concentrations of the external catalyst surfaces. We shall consider two alternative ways of doing this, namely:

(a) power law expressions or

(b) expressions derived from the Langmuir–Hinshelwood–Hougen–Watson models, which have been described briefly in Chapter 3.

9.4.1   Reaction rate equation for slurry reactor based on power law expression for surface reaction rate

When adopting this approach, the rate of the chemical reaction is expressed in the form

$$(-r_{AW}^I) = \underline{k} C_{AS}^x C_{BS}^y \tag{9.8}$$

where $\underline{k}$ is the reaction rate constant based on **unit mass of catalyst**, and C_{AS} and C_{BS} are, respectively, the concentrations of A and B at the catalyst surface. The powers x and y (the reaction orders) can be non-integral and this possibility will be considered briefly in Section 9.4.2. However, we shall first consider the very simple and quite frequently occurring case where $x = y = 1$. In this situation

$$(-r_{AW}^I) = \underline{k} C_{AS} C_{BS}$$

Also, in many cases the concentration of the liquid component is large compared with that of the dissolved gas and varies very little from one position in the slurry to another, being virtually equal to the bulk concentration C_{AB} throughout. Under these conditions the intrinsic reaction rate $(-r_{AW}^I)$ can be written in the form

$$(-r_{AW}^I) = \underline{k}_1 C_{AS}$$

where

$$\underline{k}_1 = \underline{k} C_{BS}.$$

The reaction is then pseudo-first order in A.

9.4.1.1   Overall rate equation for slurry reactor based on pseudo-first-order reaction kinetics.   Insertion of the pseudo-first-order expression for $(-r_{AW}^I)$ in equation (9.7) gives, for the rates at which component A is used up in the reaction,

$$R_A = m\eta \underline{k}_1 C_{AS} \tag{9.9}$$

where $\underline{k}_1$ is the pseudo-first-order rate constant. At steady state, the rate at which reactant A is used up by the reaction (given by equation (9.9)) is equal to (a) the rate at which it penetrates the film surrounding each catalyst particle (equation (9.4)) and (b) the rate at which it is absorbed from the gas into the liquid (equation (9.3)). Hence

$$R_A = k_{lA}a_b(C_{Ai} - C_{Ab}) = k_{cA}a_c m(C_{Ab} - C_{AS}) = m\eta \underline{k}_1 C_{AS} \qquad (9.10)$$

and

$$R_A/k_{lA}a_b = C_{Ai} - C_{Ab} \qquad (9.11)$$

$$R_A/mk_{cA}a_c = C_{Ab} - C_{AS} \qquad (9.12)$$

$$R_A/\eta \underline{k}_1 = C_{AS} \qquad (9.13)$$

Adding and rearranging the above equations to eliminate the intermediate and unknown concentrations gives

$$C_{Ai} = R_A(1/k_{lA}a_b + 1/mk_{cA}a_c + 1/m\eta \underline{k}_1) \qquad (9.14)$$

or

$$C_{Ai}/R_A = (1/k_{lA}a_b + 1/m(1/k_{cA}a_c + 1/m\eta \underline{k}_1) \qquad (9.15)$$

i.e.

$$C_{Ai}/R_A = r_b + 1/m(r_c + r_r) \qquad (9.16)$$
$$= r_b + 1/m(r_c + r_{cr}) \qquad (9.17)$$

m appearing in equations (9.15) to (9.17) is the mass of catalyst per unit volume of liquid and is henceforth termed the catalyst loading. The quantities r_b, r_c, r_r and r_{cr} appearing in these equations are reaction resistances and are independent of m, being given by:

$$r_b = 1/k_{lA}a_b \qquad (9.18)$$

$$r_c = 1/k_{cA}a_c \qquad (9.19)$$

$$r_r = 1/\eta \underline{k}_1 \qquad (9.20)$$

and

$$r_{cr} = r_c + r_r \qquad (9.21)$$

The term C_{Ai}/R_A in equations (9.15) to (9.17) may be regarded as the overall resistance to the catalytic reaction between the gaseous and liquid feeds and is the sum of resistances arising from each of the mass transfer and reactive stages involved. These resistances are as below:

$r_b = 1/k_{lA}a_b$
  = gas absorption resistance $\qquad (9.22)$

$r_c/m = 1/k_{cA}a_c m$
  = resistance due to finite rate of diffusion of A from bulk slurry
  to outer surface of catalyst pellets $\qquad (9.23)$

$$r_r/m = 1/\eta \underline{k}_1 m$$

$\quad$ = resistance due to finite rates of surface reaction

$\quad\quad$ and pore diffusion within the pellets $\quad\quad\quad$ (9.24)

$$r_{cr}/m = r_c/m + r_r/m$$

$\quad$ = combined resistance due to the

$\quad\quad$ catalyst-related steps (3) (4) and (5) in Figure 9.2 $\quad$ (9.25)

r_{cr}/m will henceforth be called the 'catalyst resistance' term. It allows for finite rates of surface reaction and pore diffusion within the catalyst pellets and also for the finite rate of diffusion of A from the bulk liquid to the outer surface of these pellets. Unlike the gas absorption resistance the 'catalyst resistance' is a function of m.

It follows from equation (9.17) that in the case of first-order reactions, a plot of C_{Ai}/R_A vs $1/m$ will be linear (Figure 9.3). The intercept on the C_{Ai}/R_A axis will be $1/k_{lA}a_b (= r_b)$ and the slope will be $1/k_{cA}a_c + 1/\eta\underline{k}_1 (= r_{cr})$.

Rate-controlling steps: Referring to Figure 9.3 the ratio of the gas absorption resistance to the 'catalyst resistance' is given by

$$\frac{\text{gas absorption resistance}}{\text{'catalyst resistance'}} = \frac{r_b}{r_{cr}(1/m)} = \frac{\text{intercept} \times m}{\text{slope}} \quad (9.26)$$

i.e. the gas absorption resistance becomes rate controlling at large catalyst loadings (m) and the 'catalyst resistance' dominates at sufficiently low values of m. The latter effect is particularly marked with large catalyst particles, as may be seen from equation (9.26) as follows.

The intercept r_b in Figure 9.3 is determined only by the rate of gas absorption and so is independent of particle size. r_{cr}, on the other hand,

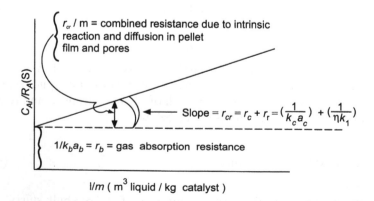

l/m (m^3 liquid / kg catalyst)

Figure 9.3 Plot to establish controlling resistances for intrinsic first-order reaction in slurry reactor.

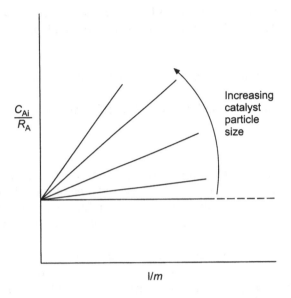

Figure 9.4 Effect of increasing catalyst particle size on performance of slurry reactor.

does depend on particle size. If this size is increased then both the effectiveness factor (η) and the mass transfer coefficient ($k_{cA}a_c$) decrease, so (from equations (9.19) to (9.21)) r_{cr} increases, as does the slope of the C_{Ai}/R_A vs $1/m$ plot (Figure 9.4). The catalyst resistance ($= r_{cr}/m$) also increases for a given value of m. The ratio of the gas absorption resistance to the catalyst resistance therefore decreases for a given catalyst loading as pellet size increases.

Factors which determine r_{cr}: From the preceding equations ((9.19) to (9.21))

$$r_{cr} = r_c + r_r = 1/k_{cA}a_c + 1/\eta\underline{k}_1 \qquad (9.27)$$

As noted above, r_{cr} may be evaluated for catalyst particles of given size by plotting C_{Ai}/R_A versus $1/m$ at given temperature and flow conditions, as in Figure 9.3. Information about the magnitudes of the quantities k_{cA}, η and $\underline{k}_1$ which determine bulk diffusion to the catalyst surface, pore diffusion and the intrinsic reaction rate can be obtained by considering the manner in which r_{cr} varies with particle size. These tests can indicate which of the three steps – namely: **External diffusion, internal (pore) diffusion** or **surface reaction** – is rate limiting.

(a) Very small particles. As the particle size falls, the contribution to r_{cr} from the term $r_c(= 1/k_{cA}a_c)$ in equation (9.27) becomes progressively smaller. This is mainly due to the progressive increase in a_c (the surface area per unit mass of pellet) which is quantified for spherical particles in equation

(9.29). Moreover, k_{cA} will also tend to increase as particle size falls (see equations (9.30) and (9.31) below). For sufficiently small particles, therefore,

$$r_{cr} \sim r_r$$

r_r is given by equation (9.20) as $r_r = 1/\eta \underline{k}_1$, where η is the effectiveness factor.

In general, for very small particles the effectiveness factor approaches unity so that

$$r_{cr} \simeq 1/\underline{k}_1 \qquad (9.28)$$

For very small particles, therefore, r_{cr} and r_r will be independent of particle size so that a plot of $\ln(r_{cr})$ vs $\ln(d_p)$ will be of zero slope.

However, some care should be exercised even for small particles. Catalysts such as Pd/charcoal or Pt/charcoal employ high area, highly porous activated charcoal. The latter material may possess a mean particle size in the region of 10–20 μm. However, if the metal particles are deposited throughout the pores rather than on the exterior surface, pore diffusion limitation can be significant despite the small particle size and will be overcome only at higher pressures [8].

(b)   Small to moderate sized particles. If the particle is porous and Thiele's modulus is sufficiently large for pore diffusion to be rate limiting, then in the case of a first-order reaction:

$$\eta = 3/\phi = 6/d_p (D_e/\underline{k}_1 \rho_p)^{1/2}$$

Hence

$$r_r = 1/\eta \underline{k}_1 = a_1 d_p$$

where η, d_p and ρ_p are the effectiveness factor and the diameter and density of the catalyst particles. D_e is the effective diffusivity of reactant A in the catalyst pores. $\underline{k}_1$ is the first-order rate constant based on unit mass of the catalyst pellet.

If the particles are still sufficiently small for r_c to be negligible, $r_{cr} = r_r = a_1 d_p$, where a_1 is constant at a given temperature.

A plot of r_{cr} vs d_p will thus be linear if pore diffusion is rate limiting.

(c)   Moderate to large size particles. As the particle size increases, resistance to the diffusion of reactant A from the bulk liquid to the external catalyst surface becomes more significant and the term $r_c(= 1/k_{cA} a_c)$ in equation (9.27) becomes more important. This is mainly because the external catalyst area per unit mass (a_c) decreases as particle size increases. For spherical particles

$$a_c = (\text{external area/mass}) = 6/\rho_p d_p \qquad (9.29)$$

The parameter k_{cA} in the expression for r_c also depends partly on particle size and this dependence must be considered before an expression for the overall dependence of r_c on particle size can be obtained.

The parameter k_{cA} depends not on particle size but also on flow conditions, and two situations will be examined.

1. *No shear stress between catalyst particle and fluid.* The catalyst particles, if sufficiently small, move with the fluid so that no shear exists between each particle and the surrounding fluid. The particles, therefore, behave as though they were suspended in a stagnant fluid. In this situation

$$Sh = 2 = k_{cA}d_p/D_{AB}$$

or

$$k_{cA} = 2D_{AB}/d_p \tag{9.30}$$

where D_{AB} is the binary diffusivity of A in liquid B. Substituting a_c from equation (9.29) gives

$$r_c = 1/k_{cA}a_c = a_2 d_p^2$$

where $a_2 = \rho_p/12D_{AB}$.

In this case, therefore, a plot by $\ln(r_{cr})$ vs $\ln(d_p)$ will have a slope of 2.

2. *Shear stress occurs between particle and fluid.* In this case an appropriate form of the Frössling equation may be employed, such as

$$Sh = 2.0 + 0.6Re^{1/2}Sc^{1/3} \tag{9.31}$$

Hence (if the 2 is ignored)

$$Sh \propto Re^{1/2}$$

and

$$Sh = k_{cA}d_p/D_{AB} \propto (d_p U/\nu_k)^{1/2}$$

where ν_k is the kinematic viscosity and U is the relative velocity of the fluid with respect to the particles, that is,

$$k_{cA} \propto U^{1/2}/d_p^{1/2} \quad \text{and} \quad k_{cA}a_c \propto U^{1/2}/d_p^{1.5}$$

so that $r_{cr} = a_3 d_p^{1.5}$.
where $a_3 = \rho_p/3.6\,Sc^{1/3}\,U^{1/2}$.

There are other correlations based on the Frössling equation. For spheres suspended in a liquid moving at a low velocity, for example,

$$Sh^2 = 4.0 + 1.21(ReSc)^{2/3} \tag{9.32}$$

so that $r_{cr} = a_4 d_p^{1.7}$.

It must be pointed out that care should be exercised in selecting suitable correlations for calculating k_{cA}. It has been pointed out by Doriaswamay and Sharma [9] that for highly turbulent systems of the type encountered in small, highly agitated stirred reactors and downflow bubble columns such as the Buss [6] and Cocurrent Downflow Contactor (CDC) reactors [3] it may be necessary to use the **liquid** properties to calculate Reynolds number. For more detail, the attention of the reader is directed to articles by Ramachandran and Chaudhari [7], Chaudhari and Ramachandran [11], Doriaswamay and Sharma [9] and Satterfield [10].

Finally, it should be remembered that the analysis given throughout the above section is based on the construction shown in Figure 9.3 and on equation (9.17), on which the construction is based. This equation applies only to reactions which are first order (or pseudo-first order) in component A. Reactions of other orders cannot be analysed in the above simple manner since r_r is not in general independent of composition (as in equation (9.20)), being given by $r_r = (C_{AS}/\eta(r^I_{AW}))$. The more general case of mth-order intrinsic kinetics is discussed in the next section.

9.4.1.2 Overall rate equation for slurry reactor based on m^{th}-order reaction kinetics.
The intrinsic reaction rate per unit catalyst mass $(-r^I_{AW})$ is in this case written in the form

$$(-r^I_{AW}) = \underline{k}_m C_{AS}^m \qquad (9.33)$$

where C_{AS} is the concentration of A in the fluid at the external surfaces of the catalyst particles.

The rate R_A at which reactant A is entering the catalyst pores and being used up in the reaction is then given by

$$R_A = m\eta(-r^I_{AW})$$
$$= m\eta \underline{k}_m C_{AS}^m \qquad (9.34)$$

(cf. equation (9.9)).

Rearranging (9.34) for $\eta = 1$ gives

$$C_{AS} = (R_A/m\underline{k}_m)^{1/m} \qquad (9.35)$$

In the steady state the rate at which A is being used up by the reactions given by equation (9.33) is equal to the rate at which it is dissolving in the bulk slurry and the rate at which it is passing by diffusion to the catalyst surface. These rates are given by

$$R_A = k_{lA}a_b(C_{Ai} - C_{Ab}) \qquad (9.3)$$

and

$$R_A = k_{cA}a_c m(C_{Ab} - C_{AS}) \qquad (9.4)$$

Combining (9.34) and (9.35) with (9.3) and (9.4) gives

$$R_A = 1/k_{lA}a_b + 1/mk_{cA}a_c \left[C_{Ai} - (R_A/m\underline{k}_m)^{1/m} \right] \qquad (9.36)$$

(i) For $m = 1/2$:

$$R_A = (m\underline{k}_{1/2})^2 / 2M_A \left[(1 + 4C_{Ai}M_A^2/(m\underline{k}_{1/2})^2)^{1/2} - 1 \right] \qquad (9.37)$$

where

$$M_A = (1/k_{lA}a_b + 1/mk_{cA}a_c)$$

(ii) For $m = 2$:

$$R_A = M_A^2 / 2m\underline{k}_2 \left[1 + 2m\underline{k}_2 C_{Ai}/M_A - (1 + 4m\underline{k}_2 C_{Ai}/M_A)^{1/2} \right] \qquad (9.38)$$

(iii) For $m = 0$ the above relationship is not applicable. However,

$$(-r_{AW}^I) = \underline{k}_0$$

and

$$R_A = m\underline{k}_0 \quad \text{(for } \eta = 1) \qquad (9.39)$$

$\underline{k}_0$ can be evaluated from tests at low catalyst loadings and small particle sizes when diffusional resistances should be negligible.

9.4.2 Rate expressions based on Langmuir–Hinshelwood–Hougen–Watson (LHHW) models for catalytic reactions

An introduction to the use of these models has been given in Chapter 3. The gas absorption and external diffusional steps occurring in the slurry reactor will not be discussed further in this section since the expressions for these have already been given in Section 9.4 (equations (9.3) and (9.4)). The treatment below is concerned only with the rate of reaction within the catalyst and comparing the rate expressions given by the power law approach with the more varied possibilities given by the LHHW model. It is assumed that both reactants chemisorb on similar sites so that the following reaction sequence could occur:

$$A + S \underset{k_{-1}}{\overset{k_1}{\rightleftharpoons}} A-S \qquad (1)$$

$$B + S \underset{k_{-2}}{\overset{k_2}{\rightleftharpoons}} B-S \qquad (2)$$

$$A-S + B-S \underset{k_{-3}}{\overset{k_3}{\rightleftharpoons}} P-S + S \qquad (3)$$

$$P-S \underset{k_{-4}}{\overset{k_4}{\rightleftharpoons}} P + S \qquad (4)$$

In step (1), reactant A is adsorbed onto a vacant site S forming A–S.
In step (2), reactant B is adsorbed onto an adjacent site forming B–S.

In step (3), the two adsorbed species A—S and B—S interact to give a vacant site and the adsorbed species A—P.

In step (4), the product P is desorbed from its site.

The net reaction rates in steps (1), (2), (3) and (4) will be written r_1, r_2, r_3 and r_4. In the following analysis it is assumed that the concentration of B at the catalyst surface is identical to its bulk concentration C_B and that step (3) above is rate determining. The rate at which this step proceeds from left to right will be written r_3. If we extend the 'elementary reaction' rate law from gas reactions to reactions between adsorbed species on a catalyst surface, we obtain

$$r_3 = k_3\theta_{(A-S)}\theta_{(B-S)} - k_{-3}\theta_{(P-S)}(\theta_S) \qquad (3R)$$

where $\theta_{(A-S)}$ is the surface concentration of adsorption sites bonded to A molecules while $\theta_{(B-S)}$ and $\theta_{(P-S)}$ are the surface concentrations of adsorption sites bonded to molecules B and P respectively. θ_S is the surface concentration of 'bare' adsorption sites.

All the other reaction steps are treated in the same way but are assumed to be rapid and at equilibrium so that from step (1)

$$k_1 C_A \theta_S = k_{-1}\theta_{(A-S)}$$

giving

$$\theta_{(A-S)} = (k_1/k_{-1})C_A\theta_S = K_A C_A \theta_S$$

where K_A is the equilibrium constant for adsorption and is identical to b_A in Chapter 3.

For step (3),

$$k_2 C_B \theta_S = k_{-2}\theta_{(B-S)}$$

therefore

$$\theta_{(B-S)} = (k_2/k_{-2})C_B\theta_S = K_B C_B \theta_S$$

Likewise for step (4),

$$\theta_{(P-S)} = (k_4/k_{-4})C_P\theta_S = K_P C_P \theta_S$$

By inserting the above expressions for $\theta_{(A-S)}, \theta_{(B-S)}$ and $\theta_{(P-S)}$, the rate-controlling equation (3R) can now be written as

$$r_3 = k_3 K_A C_A K_B C_B \theta_S^2 - k_{-3}K_p C_p \theta_S^2 \qquad (3S)$$

As in Section 3.5, it is convenient to express the surface concentrations of vacant sites and all adsorbed species as fractions of the total number of sites present. Then

$$1 = \theta_S + \theta_{(A-S)} + \theta_{(B-S)} + \theta_{(P-S)}.$$

Combining this result with the expressions for $\theta_{(A-S)}, \theta_{(B-S)}$ and $\theta_{(P-S)}$ given above gives

$$1 = \theta_S + K_A C_A \theta_S + K_B C_B \theta_S + K_P C_P \theta_S$$

and

$$\theta_S = 1/(1 + K_A C_A + K_B C_B + K_p C_p)$$

This expression for θ_S can be inserted in (3S) to give the rate of the rate-controlling step (3).

If the reaction is taken to be irreversible, as indeed many three-phase hydrogenations and oxidations are, then $k_{-3} = 0$ and

$$r_3 = k_3 K_A K_B C_A C_B/(1 + K_A C_A + K_B C_B + K_P C_P)^2 \qquad (9.40)$$

Depending on the relative strengths with which components A, B and P are adsorbed, equation (9.40) leads to a variety of different expressions for reaction rate as a function of C_{AS} and C_B. For example:

(i) If the product P is either very weakly adsorbed or, possibly, not adsorbed at all, then

$$r_3 = k_3 K_A K_B C_A C_B/(1 + K_A C_A + K_B C_B)^2$$

(ii) If, in addition, A is only weakly adsorbed, as is the case with hydrogen, and C_B is large so that its concentration has virtually no effect upon rate, then

$$r_3 = k'_3/C_A$$

where

$$k'_3 = k_3 K_A K_B C_B/(1 + K_B C_B)^2.$$

This result is similar to that from the power law approach (first order). However, other possibilities are open; for example:

(iii) If all species are only weakly adsorbed

$$1 \gg K_A C_A + K_B C_B + K_P C_P$$

and, from (9.40),

$$r_3 = k_3 K_A K_B C_A C_B$$
$$= k''_3 C_A C_B$$

where $k''_3 = k_3 K_A K_B$, so the reaction is now second order.

(iv) If B is very strongly adsorbed

$$K_B C_B \gg 1 + K_A C_A + K_P C_P$$

$$r_3 = k_3 K_A C_A/K_B C_B = k''_3 C_A/C_B$$

where $k''_3 = k_3 K_A/K_B$.

In this case, the reaction rate will be of a negative order in C_B.

It can be seen that some complex rate equations are possible. When such equations are combined with the other diffusional resistances (equations (9.3) and (9.4)) numerical solutions are usually necessary to evaluate the various kinetic parameters. Some examples of this type are given in Ramachandran and Chaudhari [7]. There is also the possibility that A may react directly from the bulk phase without adsorption in a Rideal–Eley mechanism (Chapter 3).

9.5   Trickle bed reactors

The basic transport and reaction steps for trickle bed reactors are identical to those already discussed for slurry reactors (Figure 9.2) though the mode of contact between the phases is different. In trickle bed reactors the gas flows cocurrently with the liquid through a bed of the catalyst, as described in Section 9.3.1. The main differences in the design equations arise from: (i) the different mass transfer correlations required to calculate the relevant coefficients; (ii) the greater likelihood of gas absorption resistance being significant, since it is often the case that diluent (inert) gases are used to effect quenching and to limit the effects of large reaction exotherms; and (iii) the possible necessity to consider the effects of transport resistance on the liquid component B as well as on species A.

Typical commercial trickle bed dimensions are given in Table 9.1.

When developing design equations for trickle bed reactors it is frequently assumed that the cocurrently flowing gas and liquid phases are in plug flow. This treatment is followed below.

Satterfield [10] points out that the plug flow equations should be valid for practical purposes provided the following criterion is obeyed:

$$L/d_p > (20/Pe)n \ln(1/(1 - X))$$

where L = bed height
d_p = diameter of catalyst pellets
Pe = Peclet number = $d_p U/D_{AX}$
D_{AX} = axial dispersion coefficient
n = reaction order
X = conversion of limiting reactant

If the trickle bed reactor is to be regarded as well mixed, then

$$L/d_p < 4/Pe.$$

Table 9.1 Typical dimensions of industrial scale trickle bed reactors

Bed depth	Between 3 and 6 m
Diameter	Up to 3 m
Catalyst particle size	1/8–1/32 in. (3.2×10^{-3} to 0.8×10^{-3} m)
Gas flow rate	0.013–2 kg/m^{2}s
Liquid flow rate	0.8–25 kg/m^2 s

9.5.1   Rate expressions for the transport and reaction processes undergone by reactant A

The basic transport equations parallel those for the slurry reactor, though in the present case it will not be assumed that the gas absorption process is liquid film controlled. The rate of transport of reactant A across the gas film for the trickle bed reactor is given by equation (9.42) below. It should be noted that the rates R'_A in equation (9.42) and subsequent equations for the trickle bed reactor are based on unit mass of catalyst, whereas the rates R_A for the slurry reactor were based on unit volume of liquid. One result of this difference in basis is the introduction of the term $(1/(1 - \varepsilon_b \rho_p)$ in equations (9.42) and (9.44) below. ρ_p is the density of the catalyst pellets and ε_b is the bed voidage, defined by

ε_b = (volume of 'void space' between the catalyst particles)/(volume of bed (voids + particles)).

From this definition,

$$1 - \varepsilon_b = \text{(volume of particles)/(volume of bed)} \qquad (9.41)$$

(In the reactor the 'void space' is filled with liquid.)

(a) *Rate of transport of A from bulk gas to gas/liquid interface per unit catalyst mass*

$$R'_A = k_{gA} a_i [1/(1 - \varepsilon_b)\rho_p][C_{Ag} - C_{Ai}(g)] \qquad (9.42)$$

where a = area of gas/liquid interface per unit volume of bed (m^2/m^3)

        k_{gA} = gas phase mass transfer coefficient (m/s)

        ρ_p = density of catalyst pellet (kg/m^3)

        $1 - \varepsilon_b$ = (volume of solids)/(volume of bed) (see equation (9.41))

        C_{Ag} = bulk gas phase concentration of A $(kmol/m^3)$

        $C_{Ai}(g)$ = concentration of A in gas at gas/liquid interface $(kmol/m^3)$

(b) *Equilibrium condition at gas/liquid interface*

$$C_{Ai} = C_{Ai}(g)/H \qquad (9.43)$$

where C_{Ai} = concentration of A in liquid at interface

        H = Henry's constant (cf. equation (9.1))

(c) *Rate of transport of A from gas/liquid interface to bulk liquid per unit catalyst mass*

$$R'_A = k_{lA} a (1/(1 - \varepsilon_b)\rho_p)(C_{Ai} - C_{Ab}) \quad \text{(cf equation(9.3))} \qquad (9.44)$$

where k_{lA} = liquid phase mass transfer coefficient (m/s)

        C_{Ab} = bulk liquid concentration of A $(kmol/m^3)$

(d) *Rate of transport of A from bulk liquid to external catalyst surface per unit catalyst mass*

$$R'_A = k_{cA}a_c(C_{Ab} - C_{AS}) \quad \text{(cf. equation(9.4))} \tag{9.45}$$

where k_{cA} = mass transfer coefficent for transport of solute from bulk liquid to external surface of catalyst (m/s)

a_c = external catalyst area per unit mass of catalyst (m²/g)

C_{AS} = concentration of A in the fluid at the solid/liquid interface (kmol/m³)

(e) *Diffusion and reaction in the pellet*

If the reaction is first order in A and B

$$R'_A = (-r_{AW}) = \eta \underline{k} C_{AS} C_{BS} \tag{9.46}$$

where $\underline{k}$ is the reaction rate constant based on unit mass of catalyst (cf. equation (9.9)) and C_{AS} and C_{BS} are the concentrations of A and B in the fluid at the solid/liquid interface (kmol/m³).

If the equations in (a) to (e) are combined, the following global rate equation is obtained for steady-state operation:

$$R'_A = \frac{C_{Ag}}{(1 - \varepsilon_b)\rho_p/k_{gA}a + (1 - \varepsilon_b)\rho_p H/k_{lA}a + H/k_{cA}a_c + H/\eta\underline{k}C_{BS}}$$
$$= k'_{gA} C_{Ag} \tag{9.47}$$

where k'_{gA} is the overall transfer coefficient for the gas A into the pellet.

Equation (9.47) is more complex than the corresponding equation (9.14) for the slurry reactor due to allowance for gas film resistance for the gas absorption stage and also transport limitations on component B.

9.5.2 Material balance for reactant A moving in plug flow through a catalyst bed

The trickle bed reactor may be considered as a heterogeneous plug flow reactor (Chapter 4, equations (4.18) to (4.21)).

The differential material balance then gives, for the change dF_A in the molar flow rate of A on passing through mass dW of catalyst,

$$-dF_A = (-r_{AW})dW = R'_A dW$$
$$\frac{-dF_A}{dW} = R'_A = k'_{gA} C_{Ag} \tag{9.48}$$

with k'_{gA} defined by equation (9.47).

9.5.3   Rate expressions for the transport and reaction of reactant B

9.5.3.1   Rate of transport of B from bulk liquid to external catalyst surface per unit catalyst mass.

$$R'_B = k_{cB}a_c(C_{Bb} - C_{Bs}) \qquad (9.49)$$

where C_{Bb} is the bulk liquid concentration of B and C_{BS} is the concentration of B in the fluid at the solid/liquid interface. k_{cB} is the mass transfer coefficient for transport of component B from the bulk liquid to the catalyst surface.

9.5.3.2   Rate of diffusion and reaction of B inside the catalyst pellet.

$$R'_B = (-r_{BW}) = \eta_b \underline{k} C_{AS} C_{BS} \qquad (9.50)$$

where η_b = catalyst effectiveness factor
$\underline{k}$ = reaction rate constant as in equation (9.46).
In the steady-state, equations (9.49) and (9.50) can be combined to give

$$
\begin{aligned}
R'_B &= \frac{1}{1/k_{cb}a_c + (1/\eta_b)\underline{k} C_{AS}} \cdot C_{Bb} \\
&= k'_{lB} C_{Bb}
\end{aligned}
\qquad (9.51)
$$

9.5.4   Material balance on species B moving in plug flow through a catalyst bed

A differential mass balance (see Section 9.5.2) gives, for the change dF_B in the molar flow rate of B on passing through mass dW of catalyst,

$$
\begin{aligned}
-dF_B &= (-r_{BW}) \cdot dW = R'_B \, dW \\
&= k'_{lB} C_B \, dW
\end{aligned}
$$

or

$$-dF_B/dW = k'_{lB} C_B \qquad (9.52)$$

9.5.5   Integrated equations giving catalyst mass required for a given conversion

In general, equations (9.48) and (9.52) appearing in Sections 9.5.2 and 9.5.4 must be solved simultaneously, but there are two specific limiting cases where simple analytical solutions can be obtained. These are as follows:

9.5.5.1   Mass transfer of A is rate limiting (see Section 9.5.2).   If the term $H/\eta\underline{k} C_{BS}$ in the denominator of equation (9.47) is either virtually constant

(as would be the case for a pure liquid reactant) or much smaller than the first three terms, the quantity k'_{gA} and the material balance equation for A (equation (9.48)) can be integrated to give the mass of catalyst required for a given conversion of A. (The above conditions are associated with very poor mass transfer of A compared with that of B.) For illustrative purposes the integration is carried out below for a very simple situation in which the gas flow rate remains constant down the column. This would be the case, for example, if a substantial amount of inert material was present, or conversion of the gaseous reactant was low.

Since ν_g, the volumetric flow rate of the gas stream, is taken to be constant,

$$\mathrm{d}F_A = \nu_g \, \mathrm{d}C_A.$$

Substituting in (9.48) gives

$$\nu_g \, \mathrm{d}C_{Ag}/C_{Ag} = k'_{gA} \, \mathrm{d}W$$

or

$$\mathrm{d}W = -\nu_g/k'_{gA} \int_{C_{Ag0}}^{C_{Ag}} \mathrm{d}C_A/C_{Ag}$$

and

$$W = -\nu_g/k'_{gA} \ln(C_{Ag}/C_{Ag0})$$
$$= -\nu_g/k'_{gA} \ln[1/(1 - X_A)] \qquad (9.53)$$

where k'_{gA} is given by equation (9.47).

Diffusional limitation of the transport of the gaseous reactant, as above, is found frequently in hydrodesulphurisation reactions.

9.5.5.2 Mass transfer of liquid reactant B rate limiting (see Section 9.5.4).

If the mass transport of B is rate limiting, the term k'_{lB} in equation (9.51) becomes constant and equation (9.52) can be integrated directly to give the amount of catalyst required to obtain a given conversion (X_B) of reactant B. A simple situation is considered below in which the volumetric flow rate of the liquid stream does not change appreciably as it passes down the column. This would be the case, for example, if the reacting component B were present in only small concentration in the liquid feed. Since ν_1, the volumetric flow rate of the liquid stream, is taken to be constant,

$$\mathrm{d}F_B = \nu_l \, \mathrm{d}C_B$$

and

$$\mathrm{d}W = -(\nu_l/k'_{lB}) \int_{C_{B0}}^{C_B} \mathrm{d}C_B/C_B$$

where C_{B0} is the concentration of B in the feed.

The mass of catalyst W required to achieve conversion X_B of B is now given by

$$W = (v_l/k'_{lB}) \ln(C_{B0}/C_B) = (v_l/k'_{lB}) \ln (1/(1 - X_B))$$ (9.54)

where k'_{lB} is given by equation (9.51).

9.5.5.3 Transport parameters. The various parameters k_g, k_l and k_c required for the evaluation of k'_g and k'_l in the above equations depend on a number of factors, such as packing type, flow rates, extent of particle wetting and column size. A collection of typical correlations is shown in Table 9.2.

9.6 Examples

Example 9.1
The hydrogenation of sesame seed oil is carried out on a nickel/silica catalyst. In tests using a slurry reactor hydrogen was added beneath the stirrer (Figure 9.1(b)) and initial rates of reaction were measured at 180°C, 1 atm pressure, at a stirrer speed of 750 r.p.m. and a hydrogen flow rate of 60 dm^3/h. The following reciprocal rate data were obtained, where the global reaction rate R_A is expressed as mol H$_2$/min cm^3 oil for an oil density of 0.9 g/cm^3. The concentration (C_{Ag}) of hydrogen within the gas bubbles is expressed as mol H$_2$ per cm^3 of oil.

Data for Example 9.1

m / (weight %)	0.018	0.038	0.07	0.14	0.28	1.0
$1/m$	55.6	26.3	14.3	7.1	3.6	1.0
C_{Ag}/R_A (min)	0.52	0.32	0.27	0.22	0.20	0.18

(a) Calculate $(k_b a_b)/H$ where k_b is the mass transfer coefficient for hydrogen from the liquid/gas interface to the bulk liquid, a_b is the area of this interface per unit volume of slurry and H is Henry's constant.
(b) Comment on the importance of resistance to hydrogen absorption in determining the reaction rates which can be achieved in the reactor.
(c) What would be the reaction rate for a catalyst loading (m') of 0.1% w/w if it were possible to eliminate resistance due to the hydrogen absorption process altogether?
(d) How could the gas absorption resistance be reduced?

Solution
Preliminary notes

1. Since the gas bubbles consist of pure hydrogen, gas absorption will be under liquid film control and the experimentally determined values of C_{Ag}

Table 9.2 Sample mass-transfer correlations for trickle beds[a,c]

Transport step	Correlation	Typical values	Reference
Gas-to-gas interface	$k_g a_i = 2 + 0.91\, E_g^{2/3}$ E_g is in ft $lb_f/ft^3 s$ $E_g = \left(\dfrac{\Delta P}{\Delta L}\right)_g U_g$ $k_g a_i = 2 + 0.12 E_g^{2/3}$ $E_g = kW/m^3$	$k_g\, a_i = 7.4\ s^{-1}$ for $U_g = 10\ ft/s$ $\dfrac{\Delta P}{\Delta L} 10^{-2}$ psi/ft	*IEC Proc. Des. Dev.*, **6**, 486 (1967)
Liquid interface to bulk liquid Aqueous	$\dfrac{k_l a_i}{D_l} = 8.08 \left(\dfrac{G_l}{\mu}\right)^{0.41} Sc^{1/2}$	$k_l a_i \sim 0.01 s^{-1}$ for $U_l = 0.2\,cm/s$	*Chem. Eng. Sci.*, **34**, 1425 (1979)
Organic[b]	$\dfrac{k_l a_i}{D_l} = 16.8 \dfrac{Re_l^{1/4}}{Ga^{0.22}} Sc^{1/2}\ cm^2$ $k_l a_i$ in s^{-1}, D_l in cm^2/s		*Chem. Eng. Sci.*, **36**, 569 (1981)
Bulk liquid-to-solid interface For $Re < 60$ For $Re < 20$	$Sh' = 0.815\, Re_l^{0.822} Sc^{1/3}$ $Sh' = 0.266\, Re_l^{1.15} Sc^{1/3}$	$k_c a_c \sim 0.2 s^{-1}$ for $Re_l = 50$ and $d_p = 0.5\,cm$	*AIChE J.*, **24**, 709 (1978)

$$Ga = \text{Galileo number} = \frac{d_p^3 \rho_l^2 g}{\mu_l^2},\ Re_l = \frac{G_l d_p}{\mu_l}$$

$g = 9.8\ m/s^2$

G_l = superficial mass velocity of liquid, $g/cm^2 s$

U_t = superficial velocity of liquid v_0/A_c, cm/s

$Sh' = \dfrac{k_c d_p \alpha}{D_l}$, α = fraction of external surface that is wetted

$a_s = \dfrac{6}{d_p} = \dfrac{\text{interfacial area}}{\text{volume of pellet}}$, $a_c = \dfrac{6}{p_p d_p} = \dfrac{\text{interfacial area}}{\text{mass of pellet}}$

$a_i = \dfrac{6(1 - \varepsilon_b)}{d_p} = \dfrac{\text{interfacial area}}{\text{volume of reactor}}$, ε_b = bed porosity

[a] Also see N. Midoux, B. I. Morsi, M. Purwasasmita, A. Laurent, and J. C. Charpentier, *Chem. Eng. Sci.*, **39**, 781 (1984), for a comprehensive list of correlations.
[b] In some cases this gives a low estimate of $k_l a_i$; see M. Herskowitz and J. M. Smith, *AIChEJ.*, **29**, 1 (1983); F. Turek and R. Lange, *Chem. Eng. Sci.*, **36**, 569 (1981).
[c] Table 9.2 is from Fogler, H. S., *Elements of Chemical Reaction Engineering*, 1986, p.616, and reprinted by permission of Prentice Hall, Englewood Cliffs, NJ.

will be equal to HC_{Ai} where C_{Ai} is the concentration of solute (in this case hydrogen) in the liquid at the gas/liquid interface. It is C_{Ai} which appears in the design equation (9.15) and in Figure 9.3. However, H will be independent of loading at the given temperature of 180°C so a plot of C_{Ag}/R_A versus $1/m$ will be as linear as the plot of C_{Ai}/R_A, though the intercept on the C_{Ag}/R_A axis will in this case be $H/(k_b/a_b)$ and the slope will be $(H/m)(1/(k_c a_c) + 1/(\eta \underline{k}_1)$ (use equations (9.2) and (9.15) to show this).

2. In equation (9.15) and Figure 9.3, m, the catalyst loading, is expressed as $m = $ (mass catalyst) / (volume of slurry). In the data provided, the loading is expressed as $m' = $ weight per cent of catalyst in the slurry. However, over the range of loadings considered, $m' = Am$ where A is a (dimensional) proportionality constant.

For the purpose of the present problem it is not necessary to convert the m' values given to m values. A plot of C_{Ag}/R_A versus $1/m'$ will be as linear as the plot versus $1/m$ and will make the same intercept. The slope will differ but this is not important in the present problem). A plot of C_{Ag}/R_A versus $1/m'$ is shown in Figure 9.5.

It is seen from Figure 9.5 that the plot of C_{Ag}/R_A versus $1/m'$ is linear within reasonable experimental scatter, indicating that the model on which equation (9.15) and Figure 9.3 is based is valid.

(a) **Calculation of $k_b a_b/H$**
This quantity is obtained from the intercept of the C_{Ag}/R_A vs A/m plot. This intercept is found from Figure 9.5 to be 0.184 min. Therefore,

$$k_b a_b/H = 1/0.184 = 5.4 \, \text{min}^{-1}$$

(b) **Resistance**
It is a consequence of equations (9.2) and (9.15) that the ratio of gas absorption resistance to the total resistance at a given value of m is given

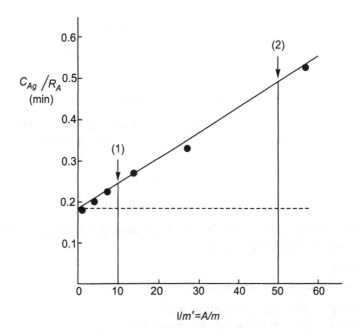

Figure 9.5 Analysis of reactor performance in Example 9.1.

by the ratio of the value of C_{Ag}/R_A at the intercept to its value at the given m (show this).

Hence, for example, the contribution of gas absorption resistance to the total resistance when the catalyst loading is 0.1% w/w Ni (condition (1) in Figure 9.5) is given by the ratio $(0.184/0.245) \times 100 = 75\%$.

For the much lower catalyst loading of 0.02 (condition (2) in Figure 9.5) the contribution to the total resistance is given by $(0.184/0.49) \times 100 = 37\%$. It can be seen from the above construction that the addition of more catalyst when the catalyst loading is already high has very little effect on reaction rate because the main resistance is then provided by gas absorption.

(c) Reaction rate

If the hydrogen absorption resistance could be eliminated at a loading of 0.1% w/w, i.e. for $1/m' = 10$, then

$$C_{Ag}/R_A = H/m(r_c + r_r) \quad = 0.245 - 0.184 \quad = 0.061 \, \text{min}$$

i.e.

$$R_A = (C_{Ag}/0.061) \, \text{mol/cm}^3 \text{min}$$

But

$$C_{Ag} = P_{Ag}/RT = 1/(0.082(273 + 180))$$
$$= 0.027 \, \text{kmol/m}^3$$
$$= 2.7 \times 10^{-5} \, \text{mol/cm}^3$$
$$= 4.4 \times 10^{-4} \, \text{mol/cm}^3 \text{min}$$

So

$$R_A = (2.7 \times 10^{-5})/0.061$$

This would be the rate if there were no gas absorption resistance. But, $C_{Ag}/R_A = 0.245 \, \text{min}$ (from Figure 9.5), so the actual global reaction rate R_A is given by

$$R_A = (2.7 \times 10^{-5})/0.245$$
$$= 1.1 \times 10^{-4} \, \text{mol/cm}^3 \text{min}$$

It therefore appears that, in the absence of gas absorption resistance, the rate would be about four times larger.

(d) Reduction

Factors which could reduce the gas absorption resistance are

(i)   smaller bubble size, i.e. better sparging
(ii) increased hydrogen flow rate for a constant bubble size, and
(iii) better agitation.

Example 9.2 Based on reference [12, p. 697]

The hydrogenation of methyl linoleate to methyl oleate was carried out in a laboratory scale reactor in which a hydrogen stream passed through a liquid containing spherical catalyst pellets. The pellet density was $2 \times 10^3 \, \text{kg/m}^3$ and the following results were obtained at 25°C.

Run	1	2	3	4
Catalyst loading (m) kg/m³	3.0	0.5	1.5	2.0
C_{Ai}/R_A (min)	0.5	3.0	1.0	3.0
Diameter of catalyst pellets (μm)	12	50	50	750

(a) Can the overall rate be enhanced by increasing the agitation, decreasing the particle size and/or improving sparger efficiency?
(b) Can the effectiveness factor (η) be determined from the above data? If so, what is its value?
(c) In order to facilitate the process, the catalyst particles used in run 4 were an order of magnitude larger than those used in earlier runs. Thiele's modulus (ϕ) for these larger pellets was 9.0. Calculate the value of the term $k_c a_c$, which determines the rate of mass transfer from the slurry into the catalyst particles. Calculate also the percentage of total resistance to the reaction which is contributed by this stage. You may assume that for Thiele's modulus values $> 3.0, \eta = 3/\phi$.

Solution

(a) C_{Ai}/R_A is plotted against $1/m$ in Figure 9.6. It is seen that the points obtained in runs 1, 2 and 3 fall on a single straight line which appears to pass virtually through the origin, though the single value of C_{Ai}/R obtained with the much larger particles falls above this line. The single line obtained for the small particles suggests that the first-order model on which equation (9.15) and Figure 9.3 are based is valid.

We first concentrate on the data for the smaller catalyst and particles;

(i) The fact that the C_{Ai}/R_A vs $1/m$ plot passes through the origin for the smaller particles indicates that $1/k_b a_b$ for these particles is very small. Since none of the quantities a_b, k_c or k_a which might respond to agitation appears to play an important role, the result of increasing the degree of agitation is going to have little effect on reaction rate. The same will be true of hydrogen sparging since the main effect of this will be to give a smaller bubble size and increased a_b. Decreasing particle size would also appear to have little effect because the point for the 12 μm particles appears to fall on the same line as the points for the 50 μm particle.

(ii) Using the data for the smaller catalyst particles and substituting the results

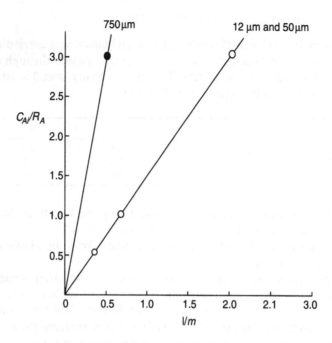

Figure 9.6 Analysis of reactor performance in Example 9.2.

(a) $1/k_c a_c \rightarrow 0$
(b) $1/k_b a_b \rightarrow 0$
(c) $\eta \rightarrow 1$ in equation (9.15)
gives $1/\underline{k}_1 = m C_{Ai}/R_A$.

For runs 1 to 3, Figure 9.6 gives $m C_{Ai}/R_A$ to be 1.5 min, or $1/k_1 = 1.5$ min.

(iii) The point for the large particle ($d_p = 750\mu$ m) falls well above those for the smaller ones, indicating that the size-dependent term ($1/k_c a_c$) can no longer be neglected and that η is no longer unity. (There is no reason why $1/k_b a_b$ should change with particle size and this will be taken as zero as before.)

We now obtain

$$C_{Ai}/R_A = 1/m(1/k_c a_c + 1/\eta \underline{k}_1)$$

or

$$3.00 = 1/2.0(1/k_c a_c + 1.5/\eta)$$

$\eta = 3/\phi = 3/9 = 0.333$ in above equation gives

$$1/k_c a_c = 6.0 - 4.5 = 1.5$$

and

$$r_{cr} = 1/k_c a_c + 1/\eta \underline{k}_1 = 6.0$$

Therefore, % resistance due to external diffusion $= (4.5/6.0) \times 100 = 75\%$.

Example 9.3

The hydrogenation of acetone to isopropanol using a Raney nickel catalyst is half-order with respect to hydrogen. Calculate the hydrogenation rate in a slurry reactor for the following conditions:

$$\text{temperature} = 14°C$$

$$\underline{k}_{1/2} = 7.4 \times 10^{-5} \, (m^3 \, kmol)^{1/2} \, kg^{-1} \, s^{-1}$$

$$C_{Ai}(H_2) = 2.75 \times 10^{-2} \, kmol/m^3$$

$$k_{lA}a_b = 0.0186 \, s^{-1}$$

$$mk_{cA}a_c = 0.266 \, s^{-1}$$

$$\text{catalyst loading } (m) = 25 \, kg/m^3$$

The nomenclature in the above table is that of Section 9.4.1.1 in the text. Calculate the surface concentration of hydrogen (C_{AS}) and state whether or not the external mass transfer is significant in this case.

Solution (see Section 9.4.1.2)

For a half-order reaction

$$R_A = (m\underline{k}_{1/2})^2 / 2M_A [(1 + 4C_{Ai}M_A^2/(m\underline{k}_{1/2})^2)^{1/2} - 1] \qquad (9.37)$$

where

$$M_A = (1/k_{lA}a_b + 1/mk_{cA}a_c) = 1/0.0186 + 1/0.266 = 0.0174 \, s^{-1}$$

$$R_A = \frac{(2.5 \times 7.4 \times 10^{-5})^2}{2 \times 0.0174} \left\{ \left(1 + \frac{4 \times 2.75 \times 10^{-2} \times (1.74 \times 10^{-2})^2}{(25 \times 7.4 \times 10^{-5})^2} \right)^{1/2} - 1 \right\}$$

$$= 2.23 \times 10^{-4} \, kmol/m^3 \, \text{slurry s}$$

The concentration of A (H_2) at the external surface of the catalyst is given by

$$C_{AS} = C_{Ai} - R_A/M_A$$

$$= (2.75 \times 10^{-2}) - (2.23 \times 10^{-4})/1.74 \times 10^{-2}$$

$$= 1.47 \times 10^{-2} \, kmol/m^3$$

Hence $C_{AS} \ll C_{Ai}$ and the external mass transfer resistance is significant.

Example 9.4

The hydrogenation of crotonaldehyde is carried out at 1 atm pressure in a trickle bed reactor under differential conditions using a catalyst consisting of palladium in porous alumina. The reaction is first order with respect to hydrogen, the rate constant being $\underline{k}_1 = 3.69 \times 10^{-1} \, m^3/s \, kg \, Pd$ at 51°C.

The catalyst contains 0.5% w/w Pd on $\gamma-Al_2O_3$. The solubility of hydrogen in the liquid medium at the reaction conditions is $2.8 \times 10^{-3}\,kmol/m^3$.

Other data are as follows:

Particle radius $= 2.5 \times 10^{-3}\,m$

Catalyst density $= 1.5 \times 10^3\,kg/m^3$

$$k_{lA}a_b = 2 \times 10^{-2}\,s^{-1}$$

$$k_{cA} = 2 \times 10^{-4}\,m/s$$

D_e (effective diffusivity in catalyst for $H_2 = 2 \times 10^{-9}\,m^2/s$

catalyst hold-up $(1 - \varepsilon_b) = 0.6$

Calculate the rate of reaction.

Solution
(i) **Rate constant**
The rate constant $\underline{k}_1$ is given as:

$$\underline{k}_1 = 3.69 \times 10^{-1}\,m^3/s\,kg\,Pd\,at\,51°C$$

If this is converted to unit mass of catalyst (which is only 0.5% w/w Pd)

$$\underline{k}_1 = 3.69 \times 10^{-1} \times 0.5 \times 10^{-2}$$
$$= 1.84 \times 10^{-3}\,m^3/s\,kg\,catalyst$$

(ii) **Catalyst effectiveness factor (η)**
We first evaluate Thiele's modulus ϕ_1 (see Chapter 3) which in this case is given by

$$\phi_1 = R\left(\frac{\rho_p \underline{k}_1}{D_e}\right)^{1/2}$$
$$= 0.25 \times 10^{-2}\left(\frac{1.5 \times 10^3 \times 1.84 \times 10^{-3}}{2 \times 10^{-9}}\right)^{1/2}$$
$$= 93$$

This is a sufficiently large value of ϕ_1 for the diffusion limited result $\eta = 3/\phi$ to be used.
In the present case,

$$\eta = 3/\phi_1 = 3/93 = 0.032$$

(iii) The required reaction rate R_A can now be calculated using the first-order model:

$$C_{Ai}/R_A = 1/\underline{k}_b a_b + 1/m(1/\underline{k}_c a_c + 1/\eta \underline{k}_1)$$

with

$$C_{Ai} = 2.8 \times 10^{-3} \text{ kmol/m}^3$$

$$m = \rho_p(1 - \varepsilon_b) = 1.5 \times 10^3 \times 0.60 = 0.9 \times 10^3 \text{ kg/m}^3$$

$$k_{cA} = 2 \times 10^{-4} \text{ m/s}$$

$$\underline{k}_1 = 1.84 \times 10^{-3} \text{ cm}^3\text{s kg catalyst}$$

$$a_c = 6/\rho_p d_p = 6/1.5 \times 10^3 \times 5 \times 10^{-3} = 0.8 \text{ m}^{-1}$$

Substituting the above values into the first-order equation gives

$$2.8 \times 10^{-3}/R_A = 1/2 \times 10^{-2} + (1/0.9 \times 10^3)\Big((2 \times 10^{-4} \times 0.8)^{-1}$$

$$+ (0.03 \times 1.84 \times 10^{-3})^{-1}\Big)$$

$$= 50 + 1/0.9 \times 10^3(1/1.6 \times 10^{-4} + 1/5.5 \times 10^{-5})$$

$$= 50 + 1/0.144 + 1/0.05$$

$$= 50 + 6.9 + 20 = 76.9$$

Therefore,

$$R_A = (2.8/76.9) \times 10^{-3} = 3.6 \times 10^{-5} \text{ kmol/m}^3 \text{ s}$$

References

1. Chaudhari, R.V., Bhattacharya, A. and Bhanage, B.M. (1994) *Catalysis in Multiphase Reactors*, Lyon (7/9 Dec.).
2. Mukherjee, K.D. (1975) *J. Amer. Oil Chem. Assoc.*, **52**, 282.
3. Boyes, A.P., Chugtai, A., Khan, Z., Raymahasay, S., Sulidis, A.T. and Winterbottom, J.M. (1995) *J. Chem. Tech. Biotechnol*, **64**, 55.
4. Lemonick, M.D. (1996) *Time* (8 Jan), 31.
5. Sherwin, M.B. and Frank, M.E. (1976) *Hydrocarbon Proc.*, **55**, 122.
6. Greenwood, T.S. (1986) *Chem. Ind.*, **3**, 94.
7. Ramachandran, P.A. and Chaudhari, R.V. (1983) *Three Phase Catalytic Reactors*, Gordon & Breach Science Publications, New York.
8. Acres, G.J.K. and Cooper, J. (1972) *J. Appl. Chem. Biotechnol.*, **22**, 769.
9. Doriaswamay, L.K. and Sharma M.M. (1984) *Heterogeneous Reactions*, Vol. 2, Wiley Interscience, New York.
10. Satterfield, C.N. (1970) *Mass Transfer in Heterogeneous Catalysis*, MIT Press, Massachusetts.
11. Chaudhari, R.V. and Ramachandran, P.A. (1980) *AIChEJ*, **26**, 177.
12. Fogler, H.S. (1986, 1992) *Elements of Chemical Reaction Engineering*, 1st edition 1986, 2nd edition 1992, Prentice Hall.

10 Bioreactors

A.N. EMERY

10.1 Introduction

Before constructing design equations for bioreactors (Section 10.4) it is necessary to outline first the background biology, biochemistry and bioreaction kinetics on which they are based.

Two principal factors distinguish bioreactions from other reaction systems. The first concerns the biological nature of the reactions themselves – these involve either living materials or at least entities, even unicomponent biocatalysts, derived from such living material. The second concerns the type of products. With some few exceptions, the products of biotechnology are for use, directly or indirectly, in closely regulated markets such as human or veterinary clinical treatment and/or diagnosis or, alternatively, they are for use, again directly or indirectly, for food or beverages. In such cases the licence to produce will specify in detail both the product and the process which produces it, including the raw materials, their sourcing and handling, the processing conditions and the means of assuring both process and product integrity. These will be defined much more stringently than in most chemical reaction processes. It is also characteristic of such products that the time for development and testing prior to licensing is long; 10–12 years is typical for pharmaceuticals. Commercial success is often determined by the ability to get to the market first and it follows that, since technical innovation must by its nature demand closer and longer scrutiny, there are strong pressures for conservatism in design.

In considering the **biological** nature of these processes we need to distinguish between two processes. In the one, product is being produced as a result of the **living processes** of cells, and progress therefore depends on a cell state (the **physiological state**). In the other, the production process is brought about using, for example, **biocatalysts** isolated from or fashioned from (now dead) cellular material. Of course this begs the question as to what is living and what is dead, but we shall duck the more philosophical questions by taking as a simple definition: living = capable of division and multiplication; dead = incapable of ever being revived to do so.

Processes that involve living cells therefore must in most cases cope with the fact that the **biomass** (i.e. the living cell material) is actually increasing throughout the process. It follows that its demand rates for raw materials and also its production of products are also increasing, i.e. the process is essentially **autocatalytic**. If the biocatalyst is not 'living' the process is **catalytic** and will continue at a single rate solely fixed by the environment – although the biocatalyst may also decay in its activity. Figure 10.1 illustrates the contrast between the progress of typical reactions of the two types. For the most part this chapter concerns living autocatalytic processes because it is these that particularly distinguish the practice of bioreaction from chemical reaction. By and large the theoretical treatment underlying non-living biocatalysis is not significantly different in principle from other multi-phase catalytic processes and we shall only discuss here some fundamental kinetics of enzyme catalysis and their application.

We shall therefore consider below the kinetics of biomass multiplication (which we shall normally call **growth**), consumption of raw materials (which we shall usually call **substrates**), and production of **products** (which may be either desired or not). To do so, we shall take a grossly simplistic view of the biomass – which we shall consider simply as an unstructured mass without any regard to the many chemical components of which it is comprised. Likewise we shall not be concerned with **all** the substrate and product components which are present in the reacting environment, but only with those which have a major effect (particularly one that is rate-limiting) on the progress of growth and product formation. Having identified these, we shall

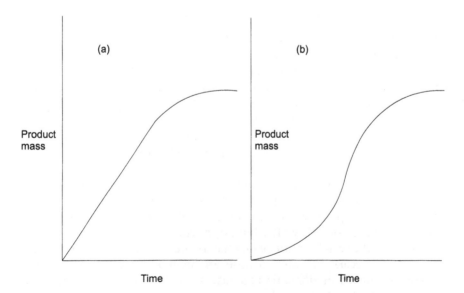

Figure 10.1 The progress of simple batch (a) catalytic and (b) autocatalytic bioreactions compared. (It is assumed that the reaction initially proceeds at a rate solely determined by the concentration of biocatalyst, but subsequently the rate becomes limited by the exhaustion of a particular reactant.)

REACTOR DESIGN FOR CHEMICAL ENGINEERS

examine their influences on reactor configuration and operating mode, concentrating particularly on those features which are unique to bioreactors.

Water is an essential requirement for living processes. Therefore all bioreactions take place in a wet environment. Some bioreactions may take place in a largely **solid-phase environment** (for example, ensiling and composting) and some in the presence of an **organic phase** (an area of rapidly increasing interest in biotechnology) but by far the majority of those we are interested in take place in a **continuous aqueous phase**. Particularly in the latter the concentrations of the principal participants in bioreactions may often be very low indeed. A few may be measured in $kg\,m^{-3}$, more usually it is in $g\,m^{-3}$, sometimes as little as $mg\,m^{-3}$. A consequence of this, and of the fact that the **characteristic time scale** of biological growth is in minutes and hours rather than seconds, is that bioreactors are often very large indeed. Many industrial bioreactors have capacities of hundreds of tonnes, a few even of thousands. **Scale-up** of such facilities is therefore by no means straightforward – indeed it will involve compromise between competing requirements in mass and heat transfer, homogeneity, controllability and practicality in construction and operation.

10.2 Biological reactions

10.2.1 First: a little microbiology – and its applications

The living cells that we are concerned with in bioreactor design fall into various categories each of which may present different problems to the design engineer. The classification and naming of the biological world (**taxonomy**) and the science of the functions and phenomena of living things (**physiology**) are well covered both in standard texts: [1] and [2] are very readable and well illustrated, and also now in computer-based packages such as [3] and [4]. At this stage all we need is a rather broad-brush picture. We shall start with the smallest living organisms and work upwards:

10.2.1.1 Viruses. Viruses are sub-microscopic ($< 0.1\mu m$ equivalent diameter) entities, largely made up of DNA surrounded by protein/lipid coats, which exist and reproduce only by infecting other organisms or tissues (the **host**). In bioprocessing they offer both opportunities and threats.

Viruses – either whole or attenuated (i.e. rendered incapable of infection) – are used as **vaccines** in order to raise an immune response to infectious disease in humans or domestic animals. Viruses which specifically infect certain insects (**baculoviruses**) have also been used as effective insecticidal agents. In either case we therefore have to produce host cells and then infect them and allow the virus to replicate in the hosts. The bioreaction problems are therefore primarily those of producing the host cell – but the engineering problems are largely those of preventing release to the environment of what can be extremely infectious agents. The standards of containment required in this type of operation (e.g. for the production of vaccines against human infectious diseases as well as foot and mouth disease and rabies) are among the highest in biotechnology.

Viruses are also a threat to bioprocesses in that they can be agents of inadvertent and unwanted infection. For example, viruses that multiply in bacteria (known as **bacteriophage** or just **phage**) are the nightmare of every bacterial fermentation operator, who therefore designs equipment and operating procedures specifically to exclude them.

10.2.1.2 Bacteria. Bacteria are characterised by their small size (typically $\sim 1\ \mu\text{m}$ equivalent diameter), ability to form large colonies of single cells and relatively (compared to higher life-forms) simple metabolism. Some (the **aerobes**) require 'free-form' oxygen as in air, some (the **anaerobes**) do not; others can do both and are said to be **facultative**. There is great diversity in their shape (round, rod-like, spiral or even filamentous), and in their growth requirements and replication rates. Many replicate by fission; i.e. the parent cell divides into two daughter cells of similar mass, thus providing a relatively simple organisation for modelling purposes. In suspended culture they can grow to high concentrations (even $> 100\ \text{kg m}^{-3}$), provided that the bioreactor engineering allows adequate transfer of nutrients, products and heat to be achieved. The growth rate of bacteria can be extremely rapid (doubling in as little as 20 minutes) and such fermentations provide the biochemical engineer with significant design and control challenges.

The number of industrial bacterial processes is extensive (at least 80% of those listed in Table 10.1 for example) and is continuously increasing, particularly since bacteria are the main vehicles for application of the results of recent developments in recombinant DNA technology and protein engineering. By these means, **heterologous** genes (i.e. from another source) may express their activity (i.e. direct the synthesis of their specific product) in bacteria which are simple to grow rapidly to high concentrations. There are manifest advantages when the original source organism or tissue for the gene is itself difficult (or toxic) to grow. Moreover, **amplification** techniques can allow such genes to express their product at far higher levels than are found in the source. Public concern for the control of genetic engineering has ensured that the regulatory requirements for such processes are most severe. However, it is possible (indeed necessary under statute law) to complete sophisticated hazard assessments of both small- and large-scale operations [5, 6], and the appropriate level of process containment can then be set.

10.2.1.3 Fungi. Fungi are strictly plants that lack chlorophyll (and they cannot therefore use radiant energy from the sun). Most can use rather simple nutrients, though growth is often enhanced by specific growth factors

Table 10.1 Some industrial products obtained using bacteria, yeasts and filamentous fungi

1. Whole cells as products	Bacterial insecticides
	Starter cultures for yoghurts, cheeses and other dairy products
	Inocula for veined cheeses
	Starter cultures for fermented food products, Koji, Tempeh, Salami, pickles, yeasts for baking
	Cells (biomass) as complete foods or as food and feed ingredients
	Modification of ice-nucleation (preventing frost damage in crops or snow creation)
2. Low molecular weight products **Primary metabolites**	Alcoholic beverages
	Ethanol and other solvents and fuels
	Amino acids (feed supplements, flavour enhancers and chemical intermediates)
	Vitamins (Riboflavin B_2, B_{12}) and vitamin precursers (β-carotene)
	Nucleotides (meat flavour enhancers)
	Citric, Gluconic, Itaconic and other organic acids
Secondary metabolites	Antibiotics (Penicillins, Streptomycins, Cephalosporins, Tetracyclines, Cephamycins, Rifamycins, Griseofulvin, etc.)
	Cyclosporin (immunosuppressant)
	Gibberellins (acceleration of barley germination)
Products of microbial transformation	Steroids (11-α-hydroxylation, 1-dehydroxylation, 16α-hydroxylation and other reactions to produce therapeutically active steroids)
	Vitamin C (D-Sorbitol to L-Sorbose)
	L-Phenylalanine (precursor of sweetener manufacture) from phenylpyruvic acid
	Copper, uranium (bioleaching of low-grade mineral ores)
3. High molecular weight products **Polysaccharides**	Xanthan (food gelling agent/stabiliser, potentially an oil recovery agent)
	Dextran (blood extender, pharmaceutical carrier)
	Pullulan (biodegradable packaging)
Lipids	Poly (3-hydoxybutyrate) PHB (biodegradable thermoplastic)
	Poly (3-hydroxyvalerate) (biodegradable thermoplastic)
Proteins *Enzymes*	Amylases (starch hydrolysing for manufacture of carbohydrate syrups)
	Proteases (acid (chymosin replacement for cheese manufacture), alkaline (detergent additive) and neutral (brewing aid))
	Glucose isomerase (sweet high-fructose syrups from starch)
	Pectinase (clarification of fruit juices)
	Lactase (aid to digestion of dairy products)
	Lipase (cheese processing, vegetable-based oil manufacture)
	Penicillin acylase (manufacture of 6-APA, precursor of 'semi-synthetic' penicillins)
	Urokinase (aid to recovery from heart attack)
Therapeutic proteins	Human insulin (diabetes)
	Human growth hormone (treatment of dwarfism)
	Erythropoetin (treatment of anaemia)
	Human serum albumin
	Ricin
Vaccines	Hepatitis B vaccine
Others	Bovine somatotrophin (increase of milk yield)

Source: Based on information given in Primrose [1].

(B vitamins, etc.). We may divide them into two categories: the **yeasts** and **filamentous** or **mycelial fungi**. The former grow essentially as colonies of single oblate cells. Many are **facultative** with respect to oxygen. They are however larger than bacteria (2–12 μm equivalent diameter), grow slightly more slowly (generation times > 1.2 h) and most (but not all) replicate by a **budding** process rather than by fission. They can be grown to cell concentrations as high as $100\,\mathrm{kg\,m^{-3}}$. The filamentous fungi grow as highly branched networks of microtubes, known as **hyphae**, to form a **mycelium**. The minimum generation times in practice rarely fall below 3–4 h and growth rates are often much slower than this. When grown in suspended culture, this mycelium can become very dense – in volume if not mass terms (maximum cell concentrations above $\sim 15\,\mathrm{kg\,m^{-3}}$ are hard to achieve) – and can impart significant viscosity to the culture, including non-Newtonian, particularly shear-thinning, characteristics. All that are of interest to us are strictly aerobic. While individual hyphae may only be of the order of 1 μm in diameter, aggregated mycelia take us well into the size range visible with the naked eye. This mycelium is only the vegetative form of the organism. Filamentous fungi have complex reproduction cycles, including asexual reproduction through spore formation in fruiting bodies – at their most obvious in mushrooms or on mouldy foods. In bioreaction processes we are generally trying to avoid spore formation in the large-scale production process, although the use of a spore inoculum is common. Many of the antibiotics, organic acids and enzymes shown in Table 10.1 are produced by filamentous fungi as are some food (e.g. cheese, Koji and Tempeh) starters and ingredients as well as the 'mycoprotein' food known in the UK as *Quorn*®

By far the largest bioprocessing application for yeasts is in the production of alcoholic beverages, a largely anaerobic process. Even though most such processes are now carried out with defined pure cultures on well defined raw materials (compared with traditional processes which may involve a succession of naturally occurring flora), nonetheless the process aims to generate flavour and other organoleptic characteristics rather than simply to maximise the concentration of alcohol. The latter does, however, become a major objective, if not the only one, where alcohol is being produced for spirit, solvent or fuel manufacture. These are classic examples also of processes whose rate is limited by increasing concentrations of the product. Few alcohol fermentations can operate to concentrations beyond 12–15% v/v – many stop long before that.

Aerobic growth is used on a large scale to produce yeasts for baking and to a much smaller extent for the manufacture of other products. During the 1960s and 1970s there was considerable development of processes for the large-scale production of yeast biomass (and also of bacteria) for animal feed purposes but subsequent changes in agricultural structures and prices have led to the winding-up of virtually all such projects, at least in the

Western world. While it lasted, however, this development led to some outstanding biochemical engineering achievements, including high-rate operation of very large (up to $800\,m^3$ capacity) bioreactors in continuous mode with recycle streams and integrated separations. More recently yeast-based processes for the expression of foreign genes (for which yeasts have some advantages over bacteria) have also been widely developed.

Processes using filamentous fungi originally used **surface culture** with the fungi grown as mats on the surface of solid or liquid media. The development of the culture of *Penicillium* species suspended in liquid media for antibiotic production during World War 2 led to the first use of the term 'biochemical engineering'. Large-scale manufacture of many antibiotics is representative of processes for the production of **secondary metabolites** (see below) and requires operation in at least two time phases, one to maximise cell growth and one to maximise product titre, each of which may be operated in batch mode but with continuous feed of nutrient – hence **fed-batch** operation. Many of these characteristics of fungal bioreactions are shared by processes using a particular class of bacteria – the **actinomycetes** – particularly *Streptomyces* species.

All the filamentous fungi (and actinomycetes) replicate vegetatively by linear extension of the hyphae and by creation of new branch points in the network; both detailed and global morphology may change with time and operating conditions. Consequently, simple modelling of their replication is not always adequate – or indeed so simple – even without the complication of spore formation!

10.2.1.4 Animal cells. Such cells have originally been excised from animal tissue (mammalian, avian, insect or fish) which has been broken up into individual cells by enzymic action. With care, continuously replicating cell lines that are genetically identical (**clones**) can be established and maintained over many generations. Cell lines can be developed using fusion and transformation processes to produce required characteristics, in particular the ability to produce specific products, to grow in suspension (rather than requiring a surface to attach themselves to) and to replicate without limitation (non-transformed cells will not continue to replicate beyond typically 45–50 generations).

The first use of animal cell culture processes was to produce hosts for subsequent viral infection – the manufacture of foot and mouth disease vaccine is still, in capacity terms, the largest application of animal cell culture. The development of hybridoma technology for the production of **monoclonal antibodies** (antibodies that recognise only a single structure and which are derived from a single clone of cells) gave a major impetus to new cell culture technology in the 1980s as did the use of animal cells for the expression of heterologous genes, particularly where the protein product was for human **biopharmaceutical** use (e.g. growth factors, blood clotting factors, interferons, interleukins and other cytokines). For these applications the detailed structure of the protein must approach the natural form in the patient (to avoid immune response as well as to ensure optimum activity). Only production in animal cells can achieve all the stages of protein processing that are required to produce such fidelity, particularly the addition of sugar residues to the protein structure (**glycosylation**) and the final structural folding of the completed protein. These post-translational modifications can only be properly achieved in animal cells. There is no such capability in bacteria and it is limited in fungi.

This is a pity because animal cells are by no means easy to grow. They are quite large (12–20 μm diameter) and rather fragile, as they have no structural cell wall as have bacteria and fungi, have complex nutrient requirements and, since they grow rather slowly (generation times 12–25 h or more), their culture is highly susceptible to contamination by faster growing bacteria and fungi. Moreover, simple batch culture will only give rather low cell densities (typically, $< 4 \times 10^6$ cells per ml compared, for example, to tissue density of $\sim 10^9$ cells per ml) and consequently both final product titre and bioreactor productivity are usually very low. Animal cell culture has become a fertile field for innovative process intensification, for example using combined bioreaction/separation processes.

10.2.1.5 Plant cell culture. Culture of plant cells, including freshwater and marine microalgae, shares many of the problems of animal cell culture. Cells can be even larger (20–150 μm diameter), delicate (though they do have a cell wall), slow growing, and difficult to maintain in pure culture. The major difference is that plant cells can use light energy, although they are in fact only rarely cultured in this manner. Plant cell culture should have become more widespread – after all many prescribed drugs are plant-derived – but it has been beset by technical problems which have allowed only a few processes to be competitive with whole plant processes. These are mostly outside the scope of an introductory text such as this so our reference to plant cell culture will be brief. Mention should, however, be made to the importance of the technology of **immobilisation** which, experimentally at least, has been widely applied to plant cells. The products of interest are all secondary metabolites and the cells are slow and difficult to grow and prone to aggregation, so it makes sense to apply a technology which allows cells to be 'immobilised' by attachment or entrapment within the bioreactor in a slow-growing but high-producing state. Such systems introduce potential problems of mixing and mass transfer limitation and these have been extensively reviewed [7]. A subset of plant cell culture is plant organ culture, particularly the descriptively named '**hairy-root**' culture for bacterially transformed plant cells, for which a number of novel designs have been developed to try to overcome problems of nutrient transport to and within a distributed organised structure.

With these latter examples we have come to the point where the complexity of the biological response is of a quite different order to that in simpler organisms – and this provides us with a legitimate if arbitrary point to halt a review primarily concerned with reactor engineering. It would, however, be incomplete to do so without at least passing reference to the animal and plant as bioreactors. Recombinant DNA techniques afford the opportunity to express heterologous genes in, for example, sheep [8] or fast-growing plants [9] and these may perhaps in the future provide an agricultural alternative to the industrial reaction technology with which we are concerned here – but not for many years yet. Indeed, given the ethical concerns raised, particularly regarding the former, the acceptance of these technical directions is not by any means assured.

10.2.2   A (very) little biochemistry

The **elemental composition** of biomass will vary with the cell type and the environment in which it is grown, but we can start to understand the metabolism of cells in culture with a much simplified proximate analysis. We can approximate the elemental (ash-free) composition of, for example, a yeast by $CH_{1.8}O_{0.5}N_{0.2}$. It follows that, if we wish such biomass to increase, we must supply these elements. In addition, we shall need phosphorus, sulphur, mineral ions and, in many cases, specific growth factors which the cells cannot or will not synthesise themselves. Not only will we have to supply the raw materials for new biomass synthesis (and, where appropriate, expressed product synthesis) but also we shall have to supply the means for the cell to derive the energy needed to drive not only this synthesis but also its continued maintenance. With some assumptions we can calculate the energetic requirements for the cell from a fundamental thermodynamic basis [10].

By and large, only small molecules can pass across the boundary between the cell and its environment, but many cells are capable of secreting **enzymes** into their surroundings which break down, generally by **hydrolysis**, macromolecules such as **proteins** and **polysaccharides** into components small enough to pass into the cell. We often stimulate this activity in cells in order to manufacture such enzymes both for industrial applications (as in the manufacture of glucose) and for domestic use (as in biological washing powders).

Therefore, the basal nutrients of microbial growth can be supplied either as monomers (e.g. glucose, ammonia, amino acids) or as polymers (e.g. starch, proteins), providing in the latter case that the cells have the capability to break them down. If polymers have to be broken down, one result may be that the kinetics of growth or bioproduct synthesis are determined by the rate of breakdown, particularly if the polymer is only partially soluble, rather than those of intrinsic biomass synthesis.

For most of the situations that we need to consider, the **carbon** substrate is that from which cells derive not only the material for **biomass and product**

synthesis but also their **energy**. In many cases the carbon substrate is a **carbohydrate** but it can be, for some micro-organisms, an alcohol, carboxylic acid, short-chain hydrocarbon, fatty acid, etc. Animal cells in culture derive much of their energy from the metabolism of certain amino acids, particularly glutamine.

Proteins are both structural and catalytic components in biomass and their synthesis requires C, N, H, O, S and other trace elements. **Nitrogen** can be supplied in simple forms (as **ammonia** for example), to many bacteria and fungi. Plant cells and some bacteria can also use **nitrate**. At the opposite extreme, we may use complex nitrogen sources such as **extracts** from yeast, cereals and other plants or animal sera. Some of these may be important in supplying other essential nutrients. Animal cells require the supply of certain amino acids in free form for protein synthesis, cell growth and energy metabolism.

The principal internal energy currency of the cell is in the form of **high-energy phosphate** bonds. For example, glucose passes first through phosphorylation reactions and thence on to the triose, pyruvate. Figure 10.2 shows how this can lead to entry into the **citric acid cycle** and thence through **oxidative phosphorylation** to complete oxidation to CO_2 and water. These **catabolic** reactions result in the formation of high energy bonds in ATP (Adenosine triphosphate). This is just one example (from aerobic metabolism) of the complex pathways involved in such degradative, energy-yielding reactions. Many of the others will result in the production of partially oxidised products, for example alcohols, ketones, fatty acids, etc., many of which are important bioproducts. The biosynthetic **anabolic** reactions are driven by the Gibbs free energy released by the hydrolysis of ATP to produce the diphosphate (ADP) and/or monophosphate (AMP) which are in turn phosphorylated back to ATP via the catabolic reactions. All this, together with the fact that P is required for synthesis of both DNA (the building block of the cell's genetic code) and RNA (the driver of protein synthesis) and also of phospholipid cell membranes, means that P must be supplied, usually as **inorganic phosphate**, for new cell synthesis, albeit in rather small quantities – a C/P ratio of 100/1 might be appropriate.

Next, we should consider the needs of bioproduct synthesis. Biotechnologists seek to squeeze the cells of their choice into a corner from which there is no escape except to produce the desired product. Of course, if the product is to be just more biomass, then both cell and biotechnologist have the same objective. More often, the biotechnologist uses the knowledge of the control of the cellular metabolism at the molecular level to drive the cell, through genetic construction and/or environmental control, to over-produce a desired product. Some products arise from regulation of the primary energy pathways of the biomass. Such products as ethanol, acetic acid, citric acid, etc., are called **primary metabolites**. Other important microbial products, for example antibiotics, are not essential products of the

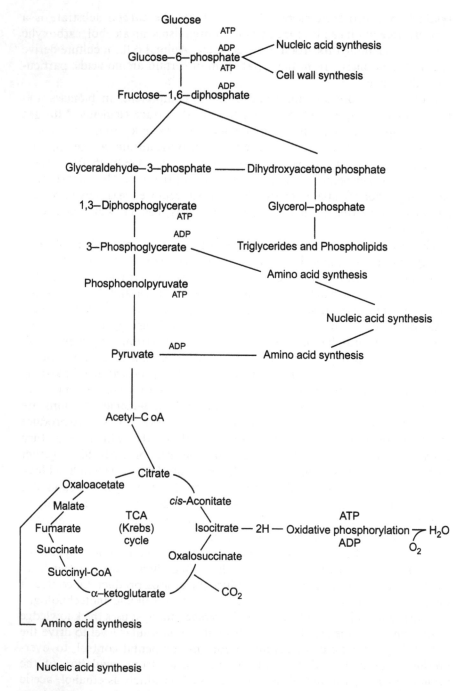

Figure 10.2 A highly simplified metabolic map to show the aerobic catabolism of a carbohydrate and the generation of energy for use in biosynthetic anabolic reactions.

central energy and biomass producing metabolism of the cell. They are called **secondary metabolites** and, in order to promote their overproduction, we will often need to supply **precursors** for their synthesis. An example is the use of phenylacetic acid or phenoxyacetic acid to drive production of Penicillins G and V respectively. Other reactions involve more overtly direct **bioconversions**. The conversion of a **sorbitol** feedstock to **sorbose** is an essential step in vitamin C manufacture. The **hydroxylation** and **dehydrogenation** of **steroids** at very specific sites are important industrial processes. The **specificity** of bioconversions makes bioreaction often an integral part of the synthesis of new **chiral** products.

We have seen that the enzymes and other functional proteins produced by cells are important products. Some are **homologous** but we make the cell over-express them. Others result from the expression of **heterologous** genes in recombinant cells. The latter is often regulated so that the expression phase is separated from the growth phase by avoiding addition of an **inducer** until the latter is complete. There are good reasons for doing this – but not the space to describe them now. Nor will we describe the important differences between **non-secreted** and **secreted** protein products (i.e. those that are retained within or exported from the cell respectively) except to mention that inadequate secretion in recombinant bacteria can lead to such gross intracellular accumulation that protein aggregates (**inclusion bodies**) are formed within the cells.

If we take an **unstructured** view of the biomass, we need go no further into the internal reactions of the cell. We are only concerned here to know that any of the **extracellular** substrates or products that we have referred to above may limit, through their availability or accumulation, the rate at which a desired reaction proceeds and we shall examine in the next section the means to describe that limitation.

Nonetheless we must recognise the limitation of such an approach. We are assuming that complicated changes in the internal structure and composition of the biomass, and their regulation, can be described by simple empirical functions. Implicitly we are accepting that steady states are the norm. For the prediction of operating conditions in continuous processes this may be adequate though, even here, there may be limitations in their ability to predict responses to transients such as shifts in flow rate, nutrient concentration, etc. Batch processes are inherently unsteady state and we will only be able to apply unstructured models if we make the assumption usually known as **balanced growth**: that is to say, as growth proceeds, all the biomass components (and the reactions in which they are involved) are changing at the same rate relative to each other. This approach has many and obvious weaknesses but in many cases it is the only one available. It is adopted in the following sections. It must be recognised, however, that these first steps in unstructured modelling should not be applied beyond their capabilities.

Nor will we consider, at this stage, the fact that a cell population will at any instant contain cells at different stages in the replication process (i.e. it will be **segregated**): some will be larger cells about to divide, others will be new smaller daughter cells proceeding through the phases of synthesis and rest which make up the **cell cycle**. We can justify this assumption on the basis that we are dealing with large numbers of events (cell concentrations in the range 10^6–10^{12} per ml) and the error in taking a mean value for any property is small, even when the individual standard deviation is quite large.

10.3   Biological reaction kinetics

10.3.1   Enzyme biocatalysis

Enzymes are the **catalysts** that drive all biological reactions. They are true catalysts in that:

(i) they participate in the reaction but are recovered unchanged at its completion

(ii) they are required at very low concentrations compared to those of the reactants

(iii) while they may change the rate of reaction by many orders of magnitude they do not change the reaction equilibrium.

However, as has been pointed out in Chapter 1, enzymes exhibit much greater levels of specificity than do present-day synthetic catalysts; i.e. they only catalyse a specific reaction of a single substrate – or a group of rather similar substrates. They may be inactivated by many physical or chemical conditions, for example extremes of pH or temperature or the presence of chemical **denaturants**.

We can take a mechanistically based approach to the description of the kinetics of enzymically catalysed reactions which is based on the assumption that enzyme E and substrate S form reversibly an enzyme–substrate complex ES which then dissociates irreversibly to give product P and uncombined E again. Thus:

$$S + E \underset{k_1}{\overset{k_1}{\rightleftharpoons}} ES \overset{k_2}{\rightarrow} P + E$$

The more rigorous of a number of derivations starting from this model is that of Briggs and Haldane which is based on the further key assumption that the concentration of ES (denoted, as are other concentrations, by square brackets, []) does not change, i.e.

$$\frac{\mathrm{d}[ES]}{\mathrm{d}t} = 0 \tag{10.1}$$

This approximation, the so-called 'quasi-steady state' approximation, is discussed also in Section 3.1.4.5 where it is applied to the concentration of an activated complex. The result is demonstrably valid in a closed system, provided that the concentration ratio of enzyme and substrate at time zero $[E]_0/[S]_0$ is small. Of course $[E]_0$ is going to be the total enzyme present at all times through the reaction, though some will be free and some will be complexed so that:

$$[E] + [ES] = [E]_0 \qquad (10.2)$$

With material balances on S and ES and equations (10.1)–(10.2) it is then possible (and a useful exercise) to define the reaction rate v in terms of:

$$v = -\frac{d[S]}{dt} = \frac{k_2[E]_0[S]}{[(k_{-1} + k_1)/k_1] + [S]} \qquad (10.3)$$

At high values of $[S]$ then, v clearly takes a maximum value given by:

$$v_{max} = k_2[E]_0 \qquad (10.4)$$

We can also represent the ratio in the denominator in equation (10.3) by a single term:

$$(k_{-1} + k_1)/k_1 = K_m$$

so that:

$$v = \frac{v_{max}[S]}{K_m + [S]} \qquad (10.5)$$

which is known as the **Michaelis–Menten** rate expression, the form of which is illustrated in Figure 10.3; K_m is known as the Michaelis–Menten constant and may be calculated as the value of $[S]$ at $v = v_{max}/2$.

v should clearly have units of kmol s^{-1} and K_m should have the same kmol units as $[S]$ – however, the definition of a mole is not always simple, particularly, for example, in the hydrolysis of macromolecules such as polysaccharides and proteins, and the literature is therefore full of such expressions using mass concentration units.

Our introduction has indicated that reaction rates may be influenced by other factors than just the substrate concentration. Can we account for the effects, for example, of temperature, pH, inhibition of the reaction by the accumulation of product, or by the presence of other chemical inhibitors, even high concentrations of the substrate itself? In fact, mechanistically based derivations can account for all these and for reversible reactions (such as isomerisation of glucose–fructose). More advanced texts [11] give both derivations and functional forms; here we shall summarise only the following:

1. The effect of an inhibitory product such as alcohol in yeast fermentations (it is termed a reversible non-competitive inhibitor) can be described by

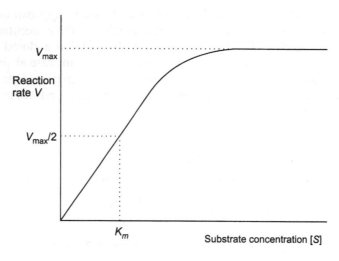

Figure 10.3 Michaelis–Menten enzyme reaction kinetics (following equation (10.5)).

$$v = \frac{v_{max}[S](K_P)}{(K_m + [S])(K_P + [P])}$$ (10.6)

where $[P]$ is the concentration of the (inhibitory) product and K_P is an inhibition constant. This is by no means the only form of inhibition and/ or reversibility for which we may derive rate equations. It does need to be borne in mind, though, that as models become more complex so the number of parameters needed to fit the model may become unrealistically large for practical application.

2. Enzyme reaction rates plotted against pH show a characteristic 'bell-shaped' curve (see Figure 10.4) which can be derived from simple models of enzyme–hydrogen ion interactions in which protonation and de-protonation of its active site lead to reversible inactivation of the enzyme. As the pH moves further away from the optimum, so the effects of drastic denaturation of the protein structure are reflected in irreversible loss of activity.

3. Temperature effects on enzyme reactions generally follow classical Arrhenius forms such that a reaction rate constant k is related to absolute temperature T by:

$$k = A\exp(-E_a/RT)$$ (10.7)

so that a plot of log k vs $1/T$ gives a straight line. A is the frequency factor, E_a the activation energy for the reaction and R the gas law constant.

What we have to bear in mind is that we are dealing with two functions. One is the intrinsic effect of temperature on the enzyme reaction rate; as

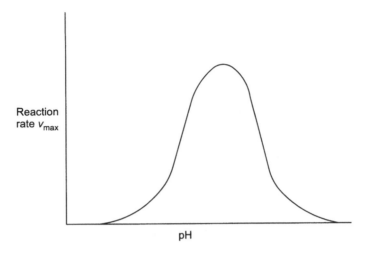

Figure 10.4 The characteristic 'bell-shaped' relationship between enzyme reaction rate and pH.

the temperature increases we should expect an increasing rate of reaction. However, at higher temperatures the effects of denaturation of the enzyme predominate and we see an increasingly rapid loss of activity. One model [10] for this situation is of a form directly analogous to the Hougen and Watson expression in catalyst kinetics:

$$\nu_{max} = \frac{A \exp(-E_a/RT)}{1 + B \exp(-\Delta G_d/RT)} \qquad (10.8)$$

where the additional term in the denominator represents the denaturation reaction having a free energy change ΔG_d. This is represented in Figure 10.5 which is typical of observed behaviour.

Our interest in biocatalysis is not confined to simple enzyme–substrate reactions. It should be appreciated that much of the above will also apply in situations in which biomass is maintained in an effectively non-growing state in the bioreactor. This is usually done by **immobilising** the biomass in some manner. This can be achieved either by retaining the biomass with a semi-permeable membrane which allows the reactant solution to **perfuse** through the active biomass or by associating the biomass with a solid phase (by, for example, adsorption or entrapment) in a packed or fluidised bed reaction system. Whether the biomass in such reactors is viable in terms of being able, given favourable conditions, to resume replication, may also be a process choice. This depends on the ability of the proposed reaction system to maintain the desired activity without either unwanted side-reactions or the breakdown of the biomass structure. Limitation of reaction rates by transport of nutrients and products in such heterogeneous systems must always

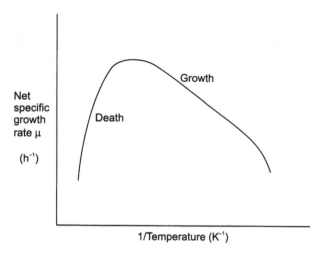

Figure 10.5 Enzyme reaction rate and enzyme inactivation as inverse functions of temperature (following equation (10.8)).

be a potential problem – it is examined in a substantial body of literature (e.g. [12, 13]) but not any further here.

Such **immobilisation** can also be applied even to single enzymes extracted from an appropriate source. At one time this technology appeared to have many potential applications, affording as it did the opportunity to achieve many of the reaction control, as well as productivity, targets of synthetic catalysis. While there has been some commercial use, particularly in amino acid manufacture, practical manufacturing applications have rather failed to match the considerable research effort in this area.

For our purposes, we need only to appreciate that the kinetic relationships set out above form a useful starting point for analysis of such systems. We have seen that the v_{max} term is proportional to the total enzyme catalytic activity $[E]_0$ and therefore, when this is present as non-growing biomass, we should need only to incorporate a **biomass concentration term** x in place of $[E]_0$. In the next section we examine the auto-catalytic situation when the biomass is allowed, indeed encouraged, to grow!

10.3.2   Basic unstructured microbial growth kinetics

If we inoculate with the desired micro-organisms a medium adequately furnished with nutrients for their growth, and control the reaction environment (pH, temperature, etc.), we shall then observe the increase of the biomass to pass through a number of phases, as are illustrated in Figure 10.6.

10.3.2.1   The lag phase (1).   **Phase 1** is clearly one in which there is negligible apparent growth. Physiologically, what is happening is that the cells,

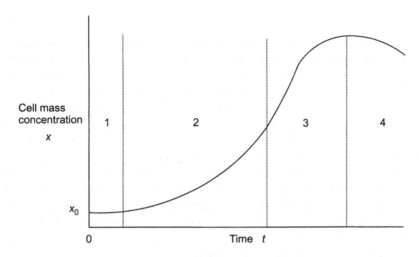

Figure 10.6 A typical microbial growth cycle showing the phases of growth: 1, lag phase; 2, unlimited exponential growth phase; 3, limited exponential growth phase; 4, death phase.

which have been taken from one growth environment, are readapting to a new one. Usually the **inoculum** has been taken from a culture in a vigorous growth phase – in which the substrate concentrations are different to those of the new medium in which they are now placed. The cells then have to synthesise a complement of enzymes able to cope with this new condition and during this time little growth will occur. We simply call this the **lag phase**.

In seeking to model this phase we must therefore have some measure of the previous history of the inoculum, and this is virtually impossible on any generic basis. There are empirical models relating the length of the lag phase to specific aspects of the previous history but these have no value outside their immediate context. What is interesting is that even the simplest of structured models, which divide the cells into structural-genetic and synthetic components, can model the lag phase – and changes in cell size during batch growth – in a way that unstructured models cannot (see, for example, the model of Williams [14] as described (and modified) by Bailey and Ollis [15] and by Nielsen and Villadsen [16]).

Fortunately, we can usually manage our bioreactions well enough so that the lag phase is reduced to rather short times compared to those of the other phases.

10.3.2.2   The unlimited exponential growth phase (2).   In **phase 2** the cells will be observed to divide at a rate that is proportional to their own number. In its simplest sense, 1 cell becomes 2, becomes 4, 8, 16 and so on. Each division takes the same time. Essentially we have a first-order growth process such that the growth rate $r_x(\text{kg m}^{-3}\,\text{h}^{-1})$ is given by:

$$r_x = \mu x \tag{10.9}$$

where μ is the first-order rate constant (h^{-1}) – which is called the **specific growth rate constant**. x is the biomass concentration $(kg\ m^{-3})$. We should strictly write equation (10.9) in terms not of mass concentration but of **cell numbers** n (cells m^{-3}). Whether we do or not depends on the means that we use to measure cell growth. In most microbial applications, biomass is measured directly or indirectly as dry mass. In animal cell culture it is more usual to count cells and therefore to express quantities in terms of cell numbers. Throughout this work we shall use only mass units. This does, of course, require the assumption that the cells do not change their **specific mass**. As far as the individual cell is concerned, we know that this cannot be true since a cell will double its mass before dividing into two, but, again, we fall back on the fact that, since we have such large numbers of cells involved in the typical bioreactor application, the assumption of a single mean specific cell mass is statistically justifiable.

If we consider a **batch growth** process at constant volume with no feed or product streams, then a material balance on x over a time t will give (accumulation = reaction)

$$\frac{dx}{dt} = r_x = \mu x \tag{10.10}$$

and this can be integrated between limits of $x = x_0$ at $t = t_0$ (initial concentration at initial time) and $x = x_t$ at time $t = t$ to give:

$$\ln \frac{x_t}{x_0} = \mu t \quad \text{or} \quad x_t = x_0 \exp(\mu t) \tag{10.11}$$

Hence the terms **exponential** or **logarithmic growth**. We can see that, if we put $x_t = 2x_0$, we can derive the relationship between the **doubling** or **generation time** t_{gen} and the **specific growth rate constant** μ to be:

$$t_{gen} = \frac{\ln 2}{\mu} = \frac{0.693}{\mu} \tag{10.12}$$

[Try calculating typical values for μ from the doubling times quoted in Section 10.2.1.]

Since we have said that the generation time in this phase is constant, it follows that μ is also constant.

10.3.2.3 The limited exponential growth phase (3).

Phase 2 cannot go on for ever, and eventually the growth rate becomes limited by something – the exhaustion of a nutrient and/or the accumulation of a toxic product for example – or the inability of the bioreactor to achieve adequate transfer processes to match the demands of the bioreaction. Doubling still takes place exponentially (8 will become 16, 32, etc.) but we see that the doubling time in this phase becomes longer and longer; i.e. μ is no longer a true constant. There have been many ways of modelling this limitation but, for now, the most important is that description of the relationship between the growth rate and the concentration of a limiting substrate $[S]$ known as the **Monod growth equation**, expressed as:

$$\mu = \frac{\mu_{max}[S]}{K_S + [S]} \tag{10.13}$$

μ_{max} is now the **maximum specific growth rate constant** and is truly constant. Equation (10.13) is of the same form as the Michaelis–Menten rate equation for enzyme reactions and there must be a temptation to assign a mechanistic basis to Monod's equation analogous to that used in Section 10.3.1. That would be incorrect. Is it after all logical to believe that the division and growth of the cell is controlled at all stages by the activity of a single enzyme? Moreover, we can of course arrive at the same form of equation from a mechanism of adsorption and first-order reaction (see Chapter 3). The Monod equation is no more than an acceptable empirical description of the growth observed under various limitations; carbon, nitrogen, growth factors, oxygen, etc. Note that we use K_S to denote the **saturation constant** rather than K_m; also that its value is given by that of $[S]$ at $\mu = \mu_{max}/2$. We also see that the value of μ in phase 2 when there was no substrate limitation (all $[S]$ were well above K_s) was actually what we now recognise to be μ_{max}.

10.3.2.4 The decline phase (4). We have up to now assumed that cells only grow; they do not die. Of course this cannot be true. Cells starved of nutrients and growth factors will first cease to divide and then they will die. They may do this because they are no longer able to maintain their **membrane functions** (particularly osmotic balance) and they simply break up structurally – a process known in animal cells as **necrosis**. Animal cells switch on their own 'suicide' programmes – genetically programmed sequences of controlled breakdown reactions known as **apoptosis** [17]. The final destruction of the cell, known as **lysis**, is accomplished by a battery of released enzymes, particularly broad-spectrum proteases and carbohydrases.

This is again clearly a complex process, but an empirical approach – largely supported by experiment – describes cell death as a first-order process, i.e.

$$r_d = k_d x \tag{10.14}$$

where k_d is a death rate constant (h^{-1}). In most practical growth situations $k_d \ll \mu$. There are important exceptions to this; in waste water treatment, substrate concentrations are usually very low, hence so is μ; and in batch animal cell culture k_d is observed often to be of the same order of magnitude as μ_{max}, although death does not here become significant until after the exhaustion of key substrates (k_d cannot therefore have a constant value in this latter situation). Usually, it is not as if growth stops and then death starts. What is happening is that the disappearance of substrate is progressively reducing the value of μ to the level of and below k_d so that, in this phase (phase 4), we see a maximum value of x achieved before death becomes predominant. This balance of growth and death can in some situations be maintained over a period of time in batch culture – an important condition known as **stationary growth**.

Our problem with stationary growth, and indeed generally with phase 4, is that experimentally it is rather difficult, unless one uses sophisticated radioactive or other tagging techniques, to determine more than the **net growth rate**, $r_{g\,net}$, represented by:

$$r_{g\,net} = r_x - r_d = x(\mu - k_d) \tag{10.15}$$

For one thing, it is difficult to measure cell **viability** in bacterial cultures except by rather tedious and/or complex methods. Moreover, if there is rapid onset of lysis, dead cells may be disappearing before you can register them.

In all the further analyses below we shall for simplicity ignore cell death effects.

10.3.2.4   Growth as a function of product concentration, temperature and pH.   By arguments similar to those used above for enzyme biocatalysis we could derive μ as a function of inhibitory product concentration in a manner analogous to that expressed in equation (10.6):

$$\mu = \frac{\mu_{max}[S](K_P)}{(K_S + [S])(K_P + [P])} \tag{10.16}$$

We should note that the same reservations must exist about the assigning of a simple mechanistic basis to a complex process as we expressed about the Monod vs Michaelis–Menten equations. Therefore it is not surprising that other empirical functionalities are often used, for example:

$$\mu = \frac{\mu_{max}[S]}{(K_S + [S])} \exp(-\boldsymbol{K}_P[P]) \tag{10.17}$$

(recognising that the first terms of the series expansion of the exponential in equation (10.17) given by equation (10.16) where $K_P = 1/\boldsymbol{K}_P$).

An alternative model is based on the assumption that there is a **limiting concentration** of P, given by $[P]_m$, beyond which growth effectively ceases:

$$\mu = \frac{\mu_{max}[S]}{(K_S + [S])}(1 - [P]/[P]_m) \tag{10.18}$$

Likewise an analogous approach to that for enzyme biocatalysis can be taken to the functionality of growth and death on temperature. Both μ_{max}

and K_d can be Arrhenius functions of absolute temperature, similar to that expressed in equation (10.7).

pH dependence of growth may also be described by similar bell-shaped functions to those that can be derived for single enzyme kinetics, though again with the reservation that this is a gross simplification of a complex functionality.

10.3.3   Unstructured models of microbial product formation

It should already be appreciated that the relationship of **product formation** to growth and substrate consumption in cells is so complex that simple unstructured models having generic application are of strictly limited use. Much progress has now been made to relate production dynamics to functional metabolic models of the biomass, or indeed to other determinants of what may be termed the **biological state** of the system – for example, the **morphological state** of fungal mycelium may be an important indicator of its capacity to produce an antibiotic. However, these are largely outside the scope of this text so we confine ourselves here to the simplest unstructured production models, while appreciating the limitations that this implies. Those interested in structured models of growth and product formation should start with those described in [16].

The simplest distinction that can be made is between those products that may be termed **growth-associated** and those that are **non-growth-associated**. The former might be expected to be products of the **primary metabolism** of the cell or **constitutive** enzymes (i.e. those that are present in more-or-less constant amounts in the cell regardless of changes in the cell's environment).

The description of the rate of product formation for growth-associated products $r_{P(g)}$ might then be expected to be given by:

$$r_{P(g)} = \alpha r_x \tag{10.19}$$

Non-growth-associated products would be those not linked directly to the primary metabolic pathways of the cells, including those we have already recognised as secondary metabolites such as antibiotics, or those proteins whose production is induced by, for example, addition of a chemical inducer or by a temperature shift, in a slowly or non-growing biomass. In these circumstances, the production rate $r_{P(n)}$ is going to be proportional only to the mass of biomass, thus:

$$r_{P(n)} = \beta x \tag{10.20}$$

Of course, not all products are going always to fall neatly into one of these categories, so we may be more general by using:

$$r_P = \alpha r_x + \beta x \tag{10.21}$$

Equation (10.21) was first applied by Luedeking and Piret [18] to describe
the pH dependence of lactic acid production (the constants α and β varied
with pH) and their names are now often used to describe this most general of
production rate models. The general use of α and β as the constants of
proportionality also derives from their original work. These different
dynamics of production are illustrated in Figure 10.7.

We may also find that α and β can be related to substrate and product
concentrations, temperature and pH, by functions analogous to those that
we have seen above for enzyme biocatalysis. Some reactions may be other
order functions of the reactant concentrations. Note, therefore, that the
limiting reactant of the product formation reaction may not be that of
growth and vice versa. Models can be assembled to cope with these situations but it will be obvious that the complexity, and the number of parameter
values to be determined, will be increased, perhaps to the point where the
utility of the models becomes limited.

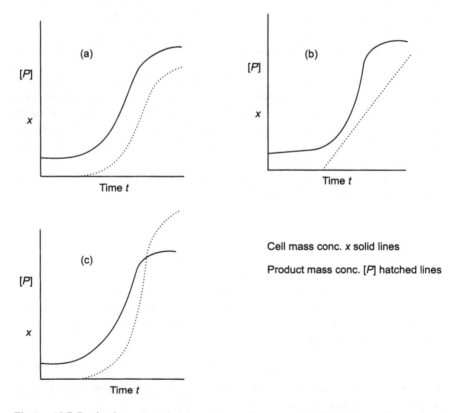

Figure 10.7 Production rate models: (a) growth-associated; (b) non-growth-associated;
(c) Luedeking and Piret [18] kinetics (equation (10.21)).

10.3.4 Unstructured substrate consumption models

Obviously, substrates are being consumed to produce biomass and products and also to drive the energy metabolism of the cells as well. Again, we take a short cut to the unstructured description of an essentially complex relationship but we do so using what are certainly the weakest assumptions in these simple models of bioprocess dynamics. We assume that constant **yield coefficients** $Y_{x/S}$ and $Y_{P/S}$ (kg/kg or kmol/kmol) may be defined in the following manner:

$$r_x = -Y_{x/S}r_{S(x)} \qquad (10.22)$$

and

$$r_P = -Y_{P/S}r_{S(P)} \qquad (10.23)$$

where $r_{S(x)}$ and $r_{S(P)}$ represent rates of substrate utilisation for growth and product formation respectively. These are additive to give a total rate of substrate utilisation r_S so that:

$$-r_S = \frac{1}{Y_{x/S}}r_x + \frac{1}{Y_{P/S}}r_P \qquad (10.24)$$

Of all the assumptions made in unstructured modelling these are the least well-supported experimentally. It is in fact widely observed that $Y_{x/S}$ and $Y_{P/S}$ do vary both with the environmental conditions and with r_x. Indeed, we might say that, if they did not, there would be rather little point in fermentation development. The bioengineer and bioscientist after all seek to manipulate the bioreaction environment to maximise one and/or the other for their particular purposes.

Nonetheless we shall stick with equations (10.22) to (10.24) as, at least for the moment, the best we have got. Note that the S referred to can be any substrate, though obviously within equation (10.24) both terms must refer to the same substrate. We could therefore define (taking growth only for example purposes) values of:

$Y_{x/C}$ for growth on the carbon substrate
$Y_{x/N}$ for growth on the nitrogen substrate (ammonia for example)
$Y_{x/O}$ for the yield on oxygen utilised

Moreover we can define other yield coefficients based, for example, on ATP yields related to utilisation of various substrates. Theoretical values for these can be calculated and one can therefore consider the energy efficiency of metabolism.

10.4 Choices in bioreactor configuration and operating mode

In the following sections we examine the choices available to us for both biomass and product manufacture. We illustrate how the kinetic models discussed above can be applied in ideal reactor design – that is to say, in the absence of potential limitations through reactant transfer and of any effects of non-ideal flow in reactors. To avoid further complication at these stages we ignore cell death and assume that all the biomass x that is present is active and reproducing.

Our analyses will be based always on application of the fundamental **balance equations** formulated on:

$$\text{accumulation} = \text{input} - \text{output} + \text{reaction}$$

so that we may write for biomass x, product P and substrate S:

$$\frac{\mathrm{d}(Vx)}{\mathrm{d}t} = Q_F X_F - Q_P X_P + Vr_x \tag{10.25}$$

$$\frac{\mathrm{d}(V[P])}{\mathrm{d}t} = Q_F[P]_F - Q_P[P]_P + Vr_P \tag{10.26}$$

$$\frac{\mathrm{d}(V[S])}{\mathrm{d}t} = Q_F[S]_F - Q_P[S]_P + Vr_S \tag{10.27}$$

where V is the reaction volume, Q_F and Q_P are the volumetric flow rates in and out of the bioreaction system, x_F, $[P]_F$ and $[S]_F$ are the concentrations of x, P and S in the inflow, and x_P, $[P]_P$, $[S]_P$ those in the outflow.

10.4.1 Simple batch reactors

There are no inflow and outflow terms and V is assumed to be constant. Hence:

$$V\frac{\mathrm{d}x}{\mathrm{d}t} = Vr_x \tag{10.28}$$

$$V\frac{\mathrm{d}[P]}{\mathrm{d}t} = Vr_P \tag{10.29}$$

$$V\frac{\mathrm{d}[S]}{\mathrm{d}t} = Vr_S \tag{10.30}$$

We first assume that no significant product formation takes place and therefore, if the initial conditions are defined by x_0 and $[S]_0$ at $t = 0$, then at time t, $x = x_t$ and $[S] = [S]_t$ so that a yield coefficient $Y_{x/S}$, following from equation (10.22), is given by:

$$x_t - x_0 = -Y_{x/S}([S]_t - [S]_0) \tag{10.31}$$

This enables us to express equations (10.28) and (10.30), together with the appropriate rate expressions (we use simple Monod kinetics {equations (10.10) and (10.13)} without cell death) in terms only of x_t or $[S]_t$ respectively:

$$\frac{\mathrm{d}x_t}{\mathrm{d}t} = \frac{\mu_{\max}\{[S]_0 - (1/Y_{x/S})(x_t - x_0)\}x_t}{K_S + \{[S]_0 - (1/Y_{x/S})(x_t - x_0)\}} \tag{10.32}$$

$$\frac{d[S]_t}{dt} = \frac{\mu_{max}[S]_t\{x_0 + Y_{x/S}([S]_0 - [S]_t)\}}{K_S + [S]_t} \qquad (10.33)$$

Integrating the two equations (10.32) and (10.33) between the limits described above gives:

$$\ln x_t = \mu_{max}t + \ln x_0 - \frac{Y_{x/S}K_S}{x_0 + Y_{x/S}[S]_0} \ln \frac{(x_t/x_0)(Y_{x/S}[S]_0)}{Y_{x/S}[S]_0 + x_0 - x_t} \qquad (10.34)$$

and

$$\ln[S]_t = \ln\{[S]_0 + Y_{x/S}([S]_0 - [S]_t)([S]_0/x_0)\}$$
$$+ \left(\frac{x_0 + Y_{x/S}[S]_0}{Y_{x/S}K_S}\right)\ln\left(\frac{x_0 + Y_{x/S}([S]_0 - [S]_t)}{x_0}\right) - \mu_{max}t\left(\frac{x_0 + Y_{x/S}[S]_0}{Y_{x/S}K_S}\right) \qquad (10.35)$$

Neither gives an explicit analytical solution but, using numerical methods, we can generate curves corresponding to phases 2 and 3 of the growth cycle. It is relatively simple to extend this approach to include cell death terms, and Nielsen and Villadsen [16] extend the analytical solution (and its derivation) to include growth-associated product formation. They also give a nice example of the use of a simple structured model in the analysis of batch reactor performance.

Biocatalytic systems not involving biomass growth are rather simpler to analyse, although the analytical solutions are still not explicit, since the only appropriate balance equations are now those for P and S. r_x is zero and $x_0 = x_t$. If we assume that the reaction rate r_S is given by the Michaelis–Menten form of equation (10.5) and a simple yield coefficient $Y_{P/S}$ can be defined according to equation (10.23), then $[S]_t$ and $[P]_t$ can be shown to be given by

$$[S]_t - K_m \ln([S]_t/[S]_0) = [S]_0 - v_{max}t \qquad (10.36)$$

and

$$[P]_t = [P]_0 + Y_{P/S}([S]_0 - [S]_t) \qquad (10.37)$$

v_{max} is a product of the biomass concentration x_t and a rate constant.

Analytical solutions of varying complexity are also possible for other bioreaction rate equations involving reversibility, inhibitors, etc.

10.4.2 Ideal plug flow reactors

For reasons that will be apparent later (see Section 10.5.2), such conformations are actually rather unrealistic in most bioreaction systems involving biomass growth. An exception is waste water treatment – though even here the departure from ideality is in practice so profound that the ideal case solution does not necessarily give us a useful prediction of practical performance. We do however, include them here because their analysis is a logical extension of that of the batch reactor. Effectively, as introduced in Chapter 4, the plug-flow reactor is a batch reaction happening in space rather than time. Using the concept of space time t defined by:

$$t = \frac{\text{axial distance along the reactor}}{\text{mean axial flow velocity}} = \frac{z}{\nu} \qquad (10.38)$$

and, given a constant reactor cross-section area A, then $\nu = Q_F/A$, and we can simply replace the time terms in equations (10.34), (10.35) and (10.36) by zA/Q_F to give the conversion as a function of axial distance or, for an ideal PFR of volume V, just V/Q_F.

10.4.3   Completely mixed stirred-tank reactors

The CSTR is not only, at least in some situations, a feasible reactor choice as a manufacturing bioreactor but it is also, as we shall see, a widely used experimental tool for the examination of biological response and for the determination of many of the parameters which we have defined in the previous sections. We shall in our analyses assume always that our reactors are indeed completely mixed and that, therefore, the inlet stream is changed instantly to what is a uniform condition throughout the reactor and that this latter must then be the condition of the outlet. Therefore, starting again from the balances given by equations (10.25)–(10.27), and assuming steady-state operation, we may write in this case:

$$V\frac{dx}{dt} = 0 = Q_F x_F - Q_P x_P + V r_x \qquad (10.39)$$

$$V\frac{d[P]}{dt} = 0 = Q_F[P]_F - Q_P[P]_P + V r_P \qquad (10.40)$$

$$V\frac{d[S]}{dt} = 0 = Q_F[S]_F - Q_P[S]_P + V r_S \qquad (10.41)$$

where r_x, r_P and r_S are all functions of the outlet conditions x_P, $[P]_P$ and $[S]_P$. Again, we initially consider the situation in which $r_P = 0$, the growth rate is given by equations (10.10) and (10.13) (with $[S] = [S]_P$ and $x = x_P$), and we define the yield constant $Y_{x/S}$ by:

$$x_P - x_F = -Y_{x/S}([S]_P - [S]_F) \qquad (10.42)$$

Therefore, taking the x balance (10.39) with the growth rate equation (10.10) we can write:

$$Q_P x_P = Q_F x_F + V \mu x_P \qquad (10.39\text{a})$$

Normal operation of a CSTR would require a period of batch operation in order to establish an initial biomass concentration followed by the initiation of a feed of a substrate solution. Equation (10.39a) demonstrates that, in steady-state continuous operation, we do not need then to have a feed of x. Indeed, this is the normal practice for most, but by no means all, applications. So, assuming $x_F = 0$ and $Q_P = Q_F$:

$$Q_F x_P = V \mu x_P \quad \text{or} \quad Q_F/V = \mu \qquad (10.43)$$

We normally call Q_F/V the **dilution rate** D (h^{-1}). It will be recognised that $\mu = D = 1/\tau$, where τ is the **bioreactor residence time**. This is an important result, because it shows us that we can determine the specific growth rate of cells in a given bioreactor solely by the substrate feed rate. This makes the CSTR useful to the experimental physiologist who can use it to examine the biological responses of cells grown at different specific growth rates. From the initial studies carried out by Herbert and others in the 1950s [19], two means of operation have been available. In one, the **turbidistat**, the outlet concentration x_P is monitored and its value maintained by controlling Q_F but this is rarely used. Much more usually Q_F (and $[S]_F$) are set at fixed values and the response of x_P monitored – this is called **chemostat** operation. We can calculate the resulting values of x_P and $[S]_P$ from:

$$D = \mu = \frac{\mu_{\max}[S]_P}{K_S + [S]_P} \qquad (10.44)$$

which can be rearranged to give:

$$[S]_P = \frac{D K_S}{\mu_{\max} - D} \qquad (10.45)$$

and, applying the yield coefficient definition, we derive:

$$x_P = Y_{x/S} \left\{ [S]_F - \frac{D K_S}{(\mu_{\max} - D)} \right\} \qquad (10.46)$$

These two functions take the form illustrated in Figure 10.8 and it is obvious that D is restricted to a maximum value. This value, which we term the **washout rate**, is seen by putting $[S]_P = [S]_F$ in equation (10.44). Above this dilution rate the biomass cannot grow rapidly enough to match the loss in the outflow.

We can also define a **reactor biomass productivity** in terms of $D x_P$ (i.e. kg biomass produced per m^3 bioreactor volume per hour), the functionality of which is shown in Figure 10.9. It is possible (and a useful exercise) to show that the maximum reactor productivity is then obtained by operation at:

$$D_{\max \text{ prod}} = \mu_{\max} \left[1 - \frac{(K_S)^{1/2}}{(K_S + [S]_F)^{1/2}} \right] \qquad (10.47)$$

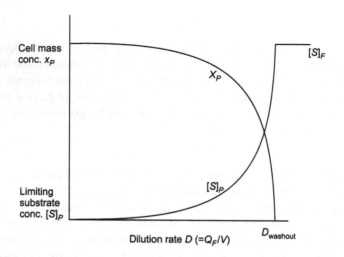

Figure 10.8 Substrate and biomass concentrations ($[S]_P$ and x_P) as functions (equations (10.45) and (10.46) respectively) of the dilution rate ($D = Q_F/V$) in CSTR bioreactor operation.

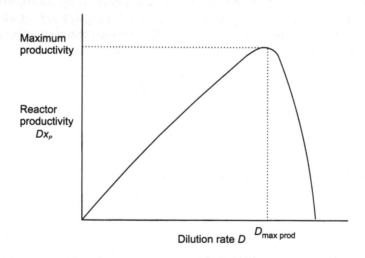

Figure 10.9 Biomass productivity Dx_P as a function of dilution rate D in a CSTR bioreactor. ($D_{\text{max prod}}$ is given by equation (10.47).)

Comparison of Figures 10.8 and 10.9 will show, however, that selection of the practical optimum operating condition is not so simple. Maximum productivity would be obtained by fixing $[S]_F \gg K_S$. However, we should then find that the value of $[S]_P$ rises. While $Y_{x/S}$ is fixed in this theoretical treatment, the **conversion** is lowered and we would effectively be throwing good substrate down the drain – a mode of operation both environmentally and economically unsound. The underlying lesson of the CSTR as bioreactor must therefore be that a compromise will usually need to be made between high-rate operation and high-conversion operation – you cannot have both.

One feature of CSTR operation that is attractive to the regulatory agencies is of course that, barring mismanagement, it should be possible to maintain constant steady-state operation, producing product at a defined concentration and condition (theoretically) *in perpetuum*. This attraction is exploited by some manufacturers, in spite of the compromises in rate and conversion that have to be made.

10.4.3.1 Parameter estimation using the chemostat. It should follow that, since we can fix the specific growth rate of cells in a CSTR of fixed volume simply by regulating the feed flow rate, this conformation may readily be used for the determination of the parameter values in equations (10.42) and (10.44). A range of flow rates may be taken, each of which defines a value of μ (remember $\mu = D$), at each of which determinations are made of x_P and $[S]_P$. Equation (10.44) may be rearranged in such forms as:

$$\frac{1}{\mu} = \frac{1}{\mu_{max}} + \frac{K_S}{\mu_{max}} \cdot \frac{1}{[S]_P} \tag{10.48}$$

or

$$\frac{[S]_P}{\mu} = \frac{K_S}{\mu_{max}} + \frac{[S]_P}{\mu_{max}} \tag{10.49}$$

or

$$\mu = \mu_{max} - \frac{K_S \mu}{[S]_P} \tag{10.50}$$

There are statistical arguments for the use of each of these in different circumstances but the first approach would normally be through equation (10.48), which is known as the **Lineweaver–Burk** plot, because the most accurately determined values (those near to μ_{max}) will be near to the intercept, which we should determine first (hence μ_{max}). The disadvantage is that the slope (and hence the value of K_S) will be strongly influenced by those values for $[S]_P$ which will be less accurately determinable. Bailey and Ollis [15] suggest the use of a plot of μ vs $[S]_P$ and determination of K_S as the value of $[S]_P$ at $\mu = \mu_{max}/2$. $Y_{x/s}$ can be determined from simultaneous measurements of x_P and $[S]_P$.

10.4.3.2 Prediction of CSTR performance from batch data. Even if we have not determined values for the kinetic constants in equation (10.44), it is still possible to make at least a first estimate of CSTR performance from a **single batch experiment**. If we determine x_t over a batch time t, then, for each value of x_t we can evaluate the slope dx_t/dt. Using the x balance (equation

(10.28) with $r_x = \mu_x$) we see that, on a plot of dx_t/dt vs x_t, the slope of the line from the origin to any point on the curve should give a value of μ. Since $\mu = D = Q_F/V$ for the CSTR, it follows that, for a sterile feed ($x_F = 0$), the slope of this line is also D. Therefore, if we fix any two out of x_P, Q_F and V, the other can be determined, as shown in Figure 10.10. There is, however, a considerable assumption involved – after all, the batch data have come from an unsteady-state process in which the conditions at any time are a product of the previous history of the culture. To use them to predict performance in a steady-state process, in which the cells see no such history, suggests some caution in their use. Nonetheless the method is useful because it shows up certain features; early in the batch, D has a maximum value (this is of course $\cong \mu_{max}$). It nicely demonstrates the limitation in CSTR operation that we have referred to above – we cannot have both high D and high x_P. Finally, we shall see in the next section that it can serve to introduce the analysis of multiple CSTRs.

10.4.4   Operations with multiple CSTRs

We can extend the approach we have used above to the case of multiple CSTRs in series. The system is illustrated in Figure 10.11, from which it can be seen that an x balance on the ith reactor would give

$$Q_F(x_{Pi} - x_{P(i-1)}) = V_i \mu x_{Pi} \qquad (10.51)$$

from which, using our earlier argument that $\mu = (dx_{Pi}/dt)/x_{Pi}$, then the dilution rate Q_F/V_i in the ith reactor will be given by $(x_{Pi} - x_{P(i-1)})/(dx_{Pi}/dt)$

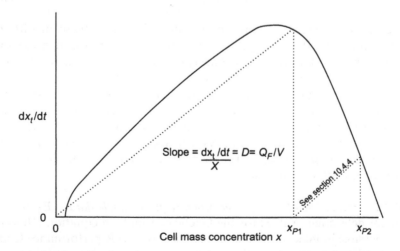

Figure 10.10 A short-cut method for the prediction of CSTR bioreactor performance from batch data. See Sections 10.4.3.2 (single CSTR) and 10.4.4 (CSTRs in series).

which is the slope of the line in Figure 10.10 from $x_{P(i-1)}$ on the x axis to the curve at $x_P = x_{Pi}$. This is illustrated for the case of a second reactor $(i = 2)$. If the reactors are of the same size, then the slopes of the lines are all constant and this affords a means for the prediction of the x_P value for each reactor. This may be used, incidentally, to demonstrate that, as the number of reactors increases, so the performance more nearly approaches that of a plug flow reactor.

We can expand equation (10.51) to

$$Q_F(x_{Pi} - x_{P(i-1)}) = \frac{V_i\,\mu_{\max}[S]_{Pi}x_{Pi}}{K_S + [S]_{Pi}} \tag{10.52}$$

and, by using a yield coefficient such that

$$x_{Pi} - x_{P(i-1)} = -Y_{x/S}([S]_{Pi} - [S]_{P(i-1)})$$

we can set up the solutions for $[S]_P$. For two reactors we can obtain an analytical solution based on a quadratic in $[S]_{Pi}$. Beyond this, numerical solutions are more practical.

10.4.5 CSTR and/or PFR: a simplified exercise in optimisation

The following exercise, first described by Bischoff [20], affords an interesting means for the comparison of PFR and CSTR performance and, given an objective function based on the minimisation of the total reactor volume

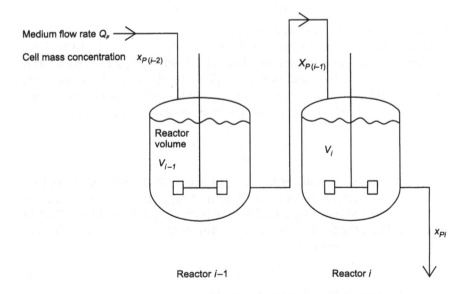

Figure 10.11 Multiple CSTRs in series (see Section 10.4.4).

required to achieve a given concentration of x_P (hence minimising τ_{total}), it also allows a simple means of optimising reactor configuration and size. For ease of use we express x_P and K_S in non-dimensional terms and r_x as a specific rate, thus:

$$x^* = \frac{x}{x_F + Y_{x/S}[S]_F} \tag{10.53}$$

and

$$K_S^* = \frac{K_S}{[S]_F + (x_F/Y_{x/S})} \tag{10.54}$$

and

$$r_x^* = \frac{r_x}{x_F + Y_{x/S}[S]_F} \tag{10.55}$$

We can then show (another useful exercise) that

$$r_x^* = \frac{\mu_{max} x^* (1 - x^*)}{K_S^* + 1 - x^*} \tag{10.56}$$

If we now consider an x^* balance over a CSTR, we obtain:

$$Q_F(x_P^* - x_F^*) = V r_x^* \tag{10.57}$$

and the residence time $\tau_{CSTR} (= V/Q_F)$ is given by

$$\tau_{CSTR} = (x_P^* - x_F^*)/r_{xP}^*$$

where r_{xP}^* is the value of r_x^* at $x^* = x_P^*$.

Given then a plot of $1/r_x^*$ as a function of x^* (from equation (10.56)), we can then see that the residence time in a CSTR achieving a final concentration x_P^* from an initial condition x_F^* is given by the area of the rectangle shown as ABCD in Figure 10.12. Likewise, we may show also that the same conversion in a PFR would be achieved in a residence time τ_{PFR} given by

$$\tau_{PFR} = \int_{x_F^*}^{x_P^*} dx^*/r_x^* \tag{10.58}$$

which is clearly the area under the curve in Figure 10.12 between x_F^* and x_p^*. If indeed the objective function is to minimise τ_{total}, then it follows that this would be achieved by using a CSTR to achieve x_{min}^* followed by a PFR in series to achieve the final concentration. Moreover, it can readily be demonstrated that x_{min}^* is given by

$$x_{min}^* = K_S^* + 1 - \{(K_S^*)(K_S^* + 1)\}^{1/2} \tag{10.59}$$

The combination is illustrated in Figure 10.12.

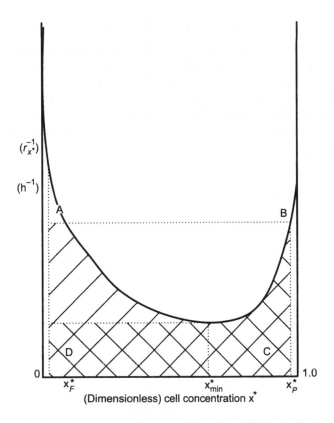

Figure 10.12 Optimising reactor configuration (after Bischoff [20]). See Section 10.4.5.

10.4.6 CSTR with recycle: improving productivity and conversion

We had earlier indicated that operation of the CSTR usually involved a sterile feed stream (i.e. $x_F = 0$). Moreover, we have emphasised the compromise that must be made between achieving high conversion (low final $[S]_P$ and high x_P) **and** high rates. This is particularly acute where our aim is to approach very low $[S]_P (\ll K_S)$ as, for example, in the treatment of waste water, without impracticably long reactor residence times. Another practical example can be given from high-rate alcohol production, in which the product inhibition effects may slow reaction rates down even more drastically. A solution is to separate and recycle a concentrated biomass stream to the bioreactor. The flow sheet is shown in Figure 10.13, from which it will be seen that, to obtain the biomass balance across the reactor, it is first necessary to complete the balance about point A in order to determine the inlet x to the reactor. If this is done it will follow that the equality previously expressed through equations (10.43) and (10.44) is now modified as

$$\mu = D[1 + R - R(x_R/x_P)] \qquad (10.60)$$

where R is a 'recycle ratio' given by the ratio of recycle flow stream flow rate to the 'fresh feed' flow rate (Q_R/Q_F) and x_R is the biomass concentration in this recycle stream. As usual it is assumed that x_F in the fresh feed is zero. x_P is the biomass concentration leaving the reactor, not the separator.

Figure 10.14 illustrates the effect of varying the 'modifier' on the right-hand side of equation (10.60) and shows clearly that dilution rates can be

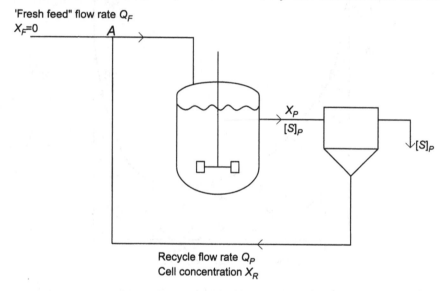

Figure 10.13 CSTR bioreactor with recycle: flow sheet.

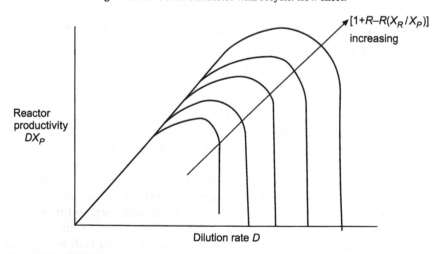

Figure 10.14 CSTR bioreactor with recycle – the effects of varying recycle ratio and concentration. See Section 10.4.6.

obtained well above the maximum specific growth rate μ_{max} together with appropriately increased reactor productivities.

10.5 Batch and continuous reactor modes and configurations: practical comparisons

Before examining other options for reactor operation it is well to pause and consider some of the more **practical** influences affecting our choice. For ease of access we can set these out in terms of tables of advantages and disadvantages below.

10.5.1 Batch fermentations: advantages and disadvantages

ADVANTAGES

- *Batch identity* – important for pharmaceuticals, food, etc., for regulatory purposes.
- *Flexible* – it is relatively simple to switch from one product to another in successive batches (though regulatory pressure may limit this in some cases) and there are no losses in the switch.
- *Identifiable optimum time* for harvesting, though compromise may be needed (or careful control) balancing the needs of maintaining product quality and recoverability and the scheduling of upstream and downstream operations.
- *Simple* – charge, inoculate and leave.
- *Operations* (other than supervisory cover) can be concentrated in *single shifts*.

DISADVANTAGES

- *Labour and service intensive* – in short bursts, producing extreme demands on services such as steam and cooling water (during sterilisation prior to fermentation).
- *Other service demands* (e.g. agitation and compression energy) may vary over the time course of the batch, because oxygen demand changes, viscosity develops, etc.
- *Harvest optimum time* may not fit scheduling requirement.
- *Inadequate/incomplete control of metabolism* – for example, high initial concentrations of sugars may lead to repression of the desired activity, or the cells may simply 'waste' substrate, taking a pathway which yields little energy, and of course less raw material, for biomass synthesis than is desired.
- *Product synthesis and biomass synthesis may be incompatible* in a single batch reaction. Many products (secondary metabolites in particular)

require conditions for their synthesis quite different from those for optimal growth.

- *Model-based control* is often still very problematical because the modelling of the inherently unsteady-state responses is inadequate.

10.5.2   Continuous fermentations: advantages and disadvantages

ADVANTAGES

- *Steady-state operation* – CSTR simple to control, the state of the system is given by that measured at any point; likewise, any control action acts on the whole reaction volume. PFR difficult to control having varying conditions, demands and responses over the length of the reactor.
- *Steady demand on services.*
- *CSTR has simple kinetic basis for model-based control strategies.*
- *Productivity high* – no down-time between batches.
- *Regulatory bodies* like guarantee of quality in steady-state operation.
- Good for *single product dedicated operation*, particularly at high production rates to closely regulated specification.

DISADVANTAGES

- *24 hour a day operation* – requires full three-shift cover. *Upstream and downstream* operations needed to match (or large intermediate holding capacity).
- *Compromise* needed between high productivity and high conversion in CSTR.
- *Lacks flexibility* – effectively single product campaigns only. Moreover, changes in dilution rate change growth rate, hence metabolism – effectively may be one flow rate only in a closely regulated situation.
- Single CSTR *unsuitable for secondary metabolite* production or for products which are inhibitory or which require induction. Multiple CSTRs too complex – compromises to control and reliability generally unacceptable.
- *PFR difficult to achieve in practical operation* – energy input to achieve adequate mass/heat transfer induces excessive fluid movement for maintenance of true plug flow. Control extremely difficult.

10.5.3   Fed-batch operation: advantages and disadvantages

ADVANTAGES

- *All the advantages* (except the last) of *batch fermentation.*
- *High cell/product concentrations* are achievable at high yields.
- *Successive phases* optimised for growth and product synthesis can be operated.

- *Conditions can be matched to the phase* (e.g. temperature, pH, limiting nutrient).
- *Precursers and/or inducers* can be fed at the right time.
- *Avoid catabolite repression.*
- *Avoid Crabtree and similar* effects.
- *Much previous development experience* – cell strains and operational strategies have been developed within the context of fed-batch operation.

DISADVANTAGES

- *Still labour intensive* – and requires greater technical skills on hand.
- *Control (particularly model-based control)* can still be problematical. Problem of process identification (what is the biological state of the system at any defined time). This could be at least partly overcome if the control system could embody the knowledge base and decision-making capacity of the 'expert'. Hence, this is an active area of research and development for the application of 'expert system' based identification and control.

10.6 Fed-batch operation of bioreactions

There are, therefore, many examples of practical **fed-batch** processes. Recombinant bacteria can be grown to high concentrations with glucose feeds prior to switching on the product expression process; baker's yeast yields on molasses can be maximised; growth of filamentous fungi such as *Penicillium* can be maximised prior to induction of antibiotic production with feed of a precursor; animal cell concentrations can be maximised by feeding nutrients which would degrade rapidly in simple batch culture. One example illustrating just why fed-batch operation can be so useful is given below.

10.6.1 The Crabtree effect and how it lowers the yield of Bakers' yeast

Figure 10.15 is the key to understanding what is going on here. If we examine the top two plots we shall see that (looking first at the left-hand top plot (a)), at sugar concentrations above a particular level, the relationship between rate of growth and sugar concentration follows a path significantly below that which we would expect it to follow with simple Monod type kinetics (illustrated by the heavy-dotted line). Looking across at plot (b) we see that we not only suffer this penalty but the yield constant also declines rapidly. Commercially this would be a disaster – yeast is a cheap commodity product made from a cheap raw material and margins depend absolutely on high yield. So why does it happen? Plot (c) shows that in this same sugar concentration range the cells actually start to produce ethanol, even when

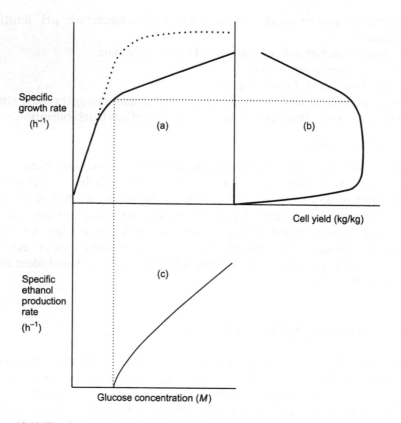

Figure 10.15 The Crabtree effect: loss of yield in yeast fermentations at high sugar substrate concentrations (see Section 10.6.1).

conditions are fully aerobic. This is the **Crabtree** effect. The cells would eventually use the ethanol they have produced as a carbon source, once they had used up the sugar, but, in doing so, they would burn more of the carbon off as CO_2 and reduce the overall yield. The solution must clearly be to feed the sugar in at such a rate that the consumption matches the supply and the concentration of sugar is maintained at levels below that at which the Crabtree effect comes into play.

10.6.2 Fed-batch bioreaction models

Consider the balances on the substrate, the biomass and the product. Taking a balance on M_x, the **total biomass** in the reactor ($= Vx$):

$$\frac{d(M_x)}{dt} = Vr_x + Q_F x_F \qquad (10.61)$$

where x_F is the biomass concentration in the feed (usually zero). Also

$$\frac{d(M_x)}{dt} = \frac{d(Vx)}{dt} = \frac{xdV}{dt} + \frac{Vdx}{dt} \tag{10.62}$$

So that, equating (10.61) and (10.62) and noting that $dV/dt = Q_F$,

$$\frac{Vdx}{dt} = Vr_x + Q_F(x_F - x)$$

and, on a unit volume basis with $Q_F/V = D$,

$$\frac{dx}{dt} = r_x + D(x_F - x) \tag{10.63}$$

Similar arguments can be used for S and P to derive

$$\frac{d[S]}{dt} = r_S + D([S]_F - [S]) \tag{10.64}$$

and

$$\frac{d[P]}{dt} = r_P + D([P]_F - [P]) \tag{10.65}$$

where $[S]_F$ and $[P]_F$ are the substrate and product concentrations in the feed solution.

These are the **general balance** equations, subject to the various restrictions and reaction rates appropriate to the operating conditions chosen. Their application can be illustrated using two typical scenarios for fed-batch bioreaction operation.

10.6.2.1 Product formation in a non-growing system. This would be the situation in, for example, the production of an antibiotic with the precursor and a carbon source fed at such a rate that the concentrations of each would be maintained effectively at zero. Obviously, this would be the second phase of a process, the first phase of which had maximised the yield and concentration of the biomass. Now, in this phase, the growth rate of the biomass is constrained to be effectively zero by the absence of nutrient. We now want to predict the biomass and product concentrations over the time that the solution is fed. Clearly the feed rate of substrate will have to vary in order to maintain our required zero substrate concentration. In this case, therefore, $r_x = 0$ and $x_F, \frac{d[S]}{dt}, [S]$ and $[P]_F$ are also all zero.

So, based on equation (10.63)

$$\frac{dx}{dt} = -Dx \tag{10.66}$$

and on equation (10.64)

$$0 = r_S + D[S]_F \tag{10.67}$$

The reaction rate term r_S can be found from the product production rate, assuming the substrate is only being consumed for product formation since

$$r_P = -Y_{P/S}r_S$$

The Luedeking and Piret equation (10.21) for product formation can be used but, in this case, $r_x = 0$, so:

$$r_S = \beta x$$

and hence

$$r_P = -\frac{\beta x}{Y_{P/S}}$$

so that, from equation (10.67):

$$D = \frac{\beta x}{Y_{P/S}[S]_F} \tag{10.68}$$

Substituting (10.68) into (10.66) then gives:

$$\frac{dx}{dt} = -\frac{\beta x^2}{Y_{P/S}[S]_F} \tag{10.69}$$

which, on integration between limits $x = x_0$ at $t = 0$ and $x = x_t$ at $t = t$, leads to

$$\frac{1}{x_t} - \frac{1}{x_0} = \frac{\beta t}{Y_{P/S}[S]_F} \tag{10.70}$$

(x_0 is the biomass concentration **at the start** ($t = 0$) **of this production phase**). Equation (10.70) can be rearranged as

$$x_t = x_0\left(1 + \frac{\beta x_0 t}{Y_{P/S}[S]_F}\right)^{-1} \tag{10.71}$$

The P balance (equation (10.65)) can in this case be expressed as

$$\frac{d[P]}{dt} = \beta x - D[P]$$

and substituting for D from equation (10.68) above gives

$$\frac{d[P]}{dt} = \beta x - \frac{\beta x[P]}{Y_{P/S}[S]_F} \tag{10.72}$$

which, in this case, can be expressed in integral form as

$$\int_0^{[P]_t} \frac{d[P]}{1 - [P]/Y_{P/S}[S]_F} = \int_0^t \beta x\,dt \tag{10.73}$$

which, on integration and substitution for x, leads to

$$\ln\frac{1}{(1 - [P]_t/Y_{P/S}[S]_F)} = \frac{\beta}{Y_{P/S}[S]_F} \cdot \int_0^t x_0\left(1 + \frac{\beta x_0 t}{Y_{P/S}[S]_F}\right)^{-1} dt \tag{10.74}$$

and hence

$$\ln \frac{1}{(1 - [P]_t / Y_{P/S}[S]_F)} = \frac{\beta x_0}{Y_{P/S}[S]_F} \cdot \frac{Y_{P/S}[S]_F}{\beta x_0} \cdot \ln \left(1 + \frac{\beta x_0 t}{Y_{P/S}[S]_F}\right)$$

so that:

$$\frac{1}{1 - [P]_t / Y_{P/S}[S]_F} = \left(1 + \frac{\beta x_0 t}{Y_{P/S}[S]_F}\right)$$

and hence

$$[P]_t = \beta x_0 t \left(1 + \frac{\beta x_0 t}{Y_{P/S}[S]_F}\right)^{-1} \tag{10.75}$$

Actually, there are two very much easier ways to calculate $[P]_t$ as the result above should suggest:

(i) Since the total cell mass $M_x = Vx$ does not change and therefore (because there is no cell growth)

$$V_t x_t = V_0 x_0$$

we can calculate V_t, knowing both V_0 and x_0 at $t = 0$ and calculating x_t at $t = t$ using equation (10.71).

Clearly, the total product produced, $[P]_t V_t$, will depend only on the mass M_x and the time of reaction. So,

$$[P]_t V_t = \beta M_x t = \beta V_t x_t t = \beta V_0 x_0 t$$

All the values on the right-hand side are known, and so is V_t, hence $[P]_t$.

(ii) The volume change $(V_t - V_0)$ represents the volume of feed solution fed to the batch over this time, and hence, given the feed substrate concentration $[S]_F$, the total substrate fed can be calculated. All this has been converted to product with a given yield coefficient $(Y_{P/S})$. Hence the total product produced can be found and when this is divided by the volume V_t at $t = t$ you have the concentration $[P]_t$ at this time.

10.6.2.2 Biomass production at a fixed specific growth rate (avoiding the Crabtree effect).

This is a situation characteristic of the production of baker's yeast that was mentioned above, but applicable also to any process in which maximisation of biomass yield is the desired objective – for example, in the first phase of the process just referred to above. In this case $[S]$ in the bioreactor is controlled at a fixed but non-zero value by varying the feed flow rate to meet the demands of the increasing total biomass M_x. From equation (10.13) it follows that the specific growth rate μ is therefore also constant. It is assumed that no significant product formation takes place at this condition.

Again our objectives are to predict the values of x and V at time t. However, given that our aim is to maximise the total biomass $M_x(= xV)$

we shall base the analysis on this objective and, on the way, be able to show how the substrate feed rate Q_F varies with time as well. Since a fixed $[S]$ is going to be maintained, μ will be constant also. Call this μ_0.

Taking the total biomass balance (equation (10.61)) and assuming the feed x_F is zero:

$$\frac{d[M_x]}{dt} = \frac{d(xV)}{dt} = Vr_x \tag{10.76}$$

which can be integrated between limits of $x_0 V_0$ (where x_0 is the biomass concentration in the volume V_0 at inoculation) at $t = 0$ and $x_t V_t$ at $t = t$ to give:

$$[M_x] = x_t V_t = x_0 V_0 \exp(\mu_0 t) \tag{10.77}$$

The substrate balance (equation (10.64)) gives (remembering that $d[S]/dt = 0$):

$$0 = -Vr_x/Y_{x/S} + Q_F([S]_F - [S])$$

and hence

$$Q_F = \frac{\mu_0 x V}{Y_{x/S}([S]_F - [S])} \tag{10.78}$$

so that, substituting (10.77) in (10.78) for $xV = x_t V_t$, the feed rate needed to maintain the defined $[S]$ at time t is given by:

$$Q_F = \frac{\mu_0 x_0 V_0}{Y_{x/S}([S]_F - [S])} \exp(\mu_0 t) \tag{10.79}$$

To separate x_t and V_t in equation (10.77) one way is to derive x_t starting from the x balance (equation (10.63)) and assuming $x_F = 0$, so that:

$$\frac{dx}{dt} = \mu_0 x - Dx \tag{10.80}$$

recognising that $D = Q_F/V$, equations (10.78) and (10.80) can be combined to give

$$\frac{dx}{dt} = \mu_0 x - \frac{\mu_0 x^2}{Y_{x/S}([S]_F - [S])} \tag{10.81}$$

which, on integration between $x = x_0$ at $t = 0$ and $x = x_t$ at $t = t$, yields:

$$x_t = \frac{x_0 \exp(\mu_0 t)}{1 - Kx_0(1 - \exp(\mu_0 t))} \tag{10.82}$$

where $K = (Y_{x/S}([S]_F - [S]))^{-1}$.

Alternatively, the ratio V_t/V_0 may be determined by accounting for all the biomass and substrate at times zero and t in terms of 'actual' and 'equivalent' biomass using the yield constant $Y_{x/S}$. Thence, by substituting the value of x from equation (10.82), one arrives at:

$$V_t = V_0[1 - Kx_0(1 - \exp(\mu_0 t))] \qquad (10.83)$$

These are not the only operating strategies that might be used. Another might involve a feed rate controlled to meet the demand imposed by a limitation such as that imposed by oxygen transfer capability. Many processes may simply use a fixed or stepped feed rate, in which case all the concentrations and the volume vary with time.

10.7   Other reaction rate limitations in bioreaction systems

Until now it has been assumed that reactor configuration and operating strategy can be determined solely by the biological reaction kinetics. In the first steps of analysis this may be reasonable, but it will soon be appreciated that any more realistic analysis must account for other system rate determinants. Here we can give only outline reference to some of them – for in-depth treatment, including, for example, the effects of **non-ideal flow regimes** (see Chapter 4), reference should be made to [15], [21] and [22], or other comprehensive bioreactor texts. Nonetheless, it is necessary at least to introduce what is in many practical bioreaction situations the most important reaction rate limitation – **transfer** of poorly soluble substrates, particularly **oxygen**.

'Free' oxygen is a requirement of all aerobic growth and production systems. It is usual to supply it from air bubbles **sparged** into the bioreactor. In some designs, the action of the rising bubbles alone is used to provide convective flow within the bioreactor, for example in **bubble columns, air-lifts** and related devices [21]. In other designs, additional energy input is made through **mechanical agitation** [22], the purposes of which are generally both to enhance the rates of interphase transfer and to mix the reactor contents in respect of all three phases – gas bubbles, liquid medium and solid microorganisms. Figure 10.16 illustrates a number of bioreactor designs.

The energy inputs in such systems may be very considerable (up to $1 kW\ m^{-3}$ in microbial fermentations, though up to two orders of magnitude less in animal and plant cell systems) and these represent significant process costs as well as additional load on the cooling system. Moreover, the point inputs of both aeration and mechanical agitation may also impose significant local energy dissipation rates which may have deleterious effects on the structure and integrity of the biological systems in suspension. There are, for example, clear indications (if not yet complete understanding) of the interactions (in both directions) between mechanical agitation inputs and the morphology (and hence bioactivity) of filamentous fungi. While, in another example, it is generally accepted that the energy dissipation rates in mechanical agitation of animal cell cultures are not themselves sufficient usually to *(a)*   *Reactors with mechanical energy input*

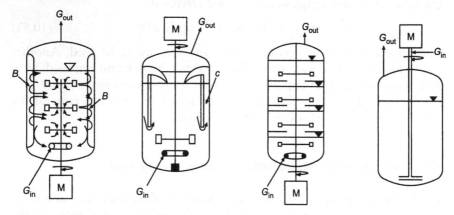

(b)   *Reactors with energy input from external pump*

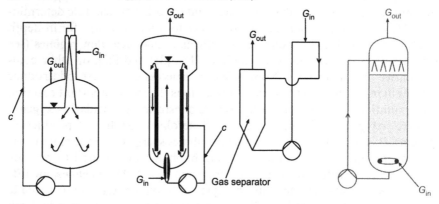

(c)   *Reactors in which energy input is from compressed gas supply*

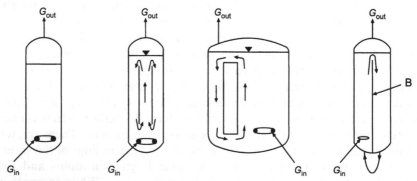

Figure 10.16 Typical bioreactor configurations (after Schügerl [23]): G_{in} = gas inlet; G_{out} = gas outlet; B = baffle; C = conduit; M = motor; ▼ = liquid level.

cause loss of integrity or death of the cells, it is now recognised that the intense energy dissipation around bubbles bursting at the surface of the medium can and often does lead to cell death and reduced cell and product yield.

10.7.1   Oxygen transfer in aerated and agitated bioreactors

Application of classical **two-film theory** to the transfer of oxygen between air bubbles and liquid medium can be simplified by the (justifiable) assumptions that gas film and interfacial resistances are negligible by comparison with that of the liquid film. We may then consider that the overall mass transfer coefficient k_L (which is all that we can measure in practice) is effectively the liquid film mass transfer coefficient $k_L(\text{m h}^{-1})$ so that we express the volumetric oxygen transfer rate $Q_{O_2}(\text{mol m}^{-3}\text{h}^{-1})$ as

$$Q_{O_2} = k_L a([O_2]^* - [O_2]) \qquad (10.84)$$

where $a(\text{m}^{-1})$ is the specific interfacial area, $[O_2]^*(\text{mol m}^{-3})$ is the oxygen concentration in equilibrium with the gas phase composition (to which it is linked by **Henry's law**) and $[O_2](\text{mol m}^{-3})$ is the actual liquid phase oxygen concentration.

We can then set up the two balances on x and $[O_2]$ assuming that the specific growth rate μ varies with $[O_2]$ according to a Monod type saturation function, and no other nutrient is limiting:

$$\frac{dx}{dt} = r_x = \mu x = \frac{\mu_{max}[O_2]}{K_{S0} + [O_2]}x \qquad (10.85)$$

$$\frac{d[O_2]}{dt} = \frac{-r_x}{Y_{x/0}} + k_L a([O_2]^* - [O_2]) \qquad (10.86)$$

Direct transfer between gas phase and cells can be neglected – cells effectively only use oxygen in solution. $[O_2]^*$ varies with the medium and the gas phase composition and, for air bubbles in contact with medium, will typically be about $0.25\,\text{mol m}^{-3}$. Oxygen consumption and recirculation of the gas phase ensure that the mean value in the bioreactor is often much lower than this. For most purposes $[O_2]^*$ would be assumed to be that concentration in equilibrium with the outlet gas composition.

In a batch growth process then, as x increases with time, so the required transfer rate Q_{O_2} will increase. If the maximum oxygen transfer capability represented by $k_L a[O_2]^*$ is fixed by the equipment and its operating conditions then numerical solution of equations (10.85) and (10.86) will demonstrate the phenomenon of so-called **linear growth**, as illustrated in Figure 10.17. Indeed, such an observation in the laboratory or pilot plant is a good indication of inadequate oxygen transfer capacity.

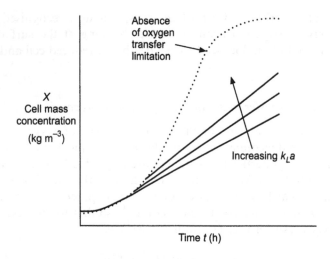

Figure 10.17 Showing how oxygen mass transfer limitation may lead to so-called 'linear growth' in batch bioreactions.

The question follows, therefore, as to how the oxygen transfer capability, characterised by $k_L a[O_2]^*$, can be practically enhanced.

10.7.2   Improving oxygen transfer in bioreactors

For most practical purposes, it is difficult and unnecessary to separate the 'a' and 'k_L' terms (although they may be handled separately in more mechanistically based approaches) and therefore many empirical correlations link the so-called 'absorption coefficient' $k_L a$ to the reactor operating parameters. There are many such correlations that claim application in bioreactor operating regimes, most of which are based on

$$k_L a = k(P_g/V)^a u_S^b \mu^c \tag{10.87}$$

Values for the constant k and the exponents a, b and c are widely available in the literature. Broadly k tends to be sensitive to the fluid and geometry of the system while a and b are rather less sensitive, both tending to fall within the range 0.3 to 0.7. Experimentally the strict dependence of c (typically < 0.2) is difficult to distinguish from effects due to the viscosifier also being a solute. The correlations may be used in both air-driven reactors and stirred reactors but limitations in their application must be recognised. These are illustrated below for impeller-agitated vessels. Schügerl [21] and Chisti [24] review air-driven bioreactors and their design and scale-up.

The constants and exponents in equation (10.87) are all vessel and impeller geometry-specific – they will vary, for example, with the design of impeller used and with the number and spacing. The **ungassed specific power input** P/V may be calculated from

$$P = P_0 \rho N_i^3 D_i^5 \tag{10.88}$$

where P_0 is the dimensionless **power number** specific to the type, number and spacing of the impeller(s), ρ the fluid density $(\mathrm{kg\ m^{-3}})$, N_i is the impeller rotational speed $(\mathrm{s^{-1}})$ and D_i the impeller diameter (m). In the turbulent regime P_0 is independent of the impeller Reynolds number.

However, to use equation (10.87) we need to employ the **gassed** power input P_g and here particular difficulties arise. Nienow [25] has demonstrated the different gas-handling regimes in the operating envelope of both single and multiple impellers, the existence of which cast doubt on the applicability of any simple prediction for P_g/P. Figure 10.18 (based on Nienow [25]) shows data for a number of impellers expressed as plots of measured P_g/P vs the gas flow rate Q_g at equivalent impeller power dissipation rates. In the case of those fermentations in which significant non-Newtonian character-istics are generated, a further difficulty is faced with the viscosity term (μ) in equation (10.87). The question arises as to what shear rate should be used for assessing apparent viscosity – should it be a maximum or an average and how should it be calculated? Given all the uncertainties, the safest ground is to use a measured power input at the appropriate conditions. Nonetheless, we may at least conclude from all the above analyses that, given a need to increase oxygen transfer in a given vessel, the following steps could be taken:

(i) Increase impeller speed – limited by mechanical and power draw (hence cost) considerations. Moreover, all the energy put in by agitation has to be taken out again in the cooling water.

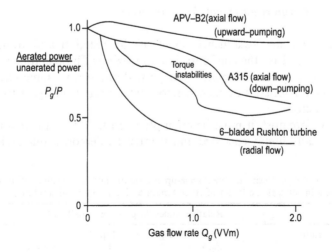

Figure 10.18 The effects of sparged air flow rate Q_g (volume of air per volume of fluid per minute) on the gassed power drawn (shown as P_g/P) by a range of impellers in agitated tanks (after Nienow [25]).

(ii) Increase impeller diameter – limited as above and also by the starting torque needed to get the impeller rotating.
(iii) Change the impeller configuration/number/spacing (and hence the power number).
(iv) Increase the gas throughput – may be limited by foaming of the medium. Moreover, the increased superficial velocity (u_S and hence the aeration number) may significantly reduce the power input through the impeller (depending on the impeller design). This should have the additional benefit of increasing the partial pressure of oxygen in the bioreactor outlet and therefore also increasing $[O_2]^*$.
(v) Increase the total pressure in the bioreactor (hence increasing $[O_2]^*$) – limited again by mechanical design.
(vi) Supplementing the air supply with oxygen (also to increase $[O_2]^*$ but limited by the costs of pure oxygen – though this has been used in some effluent treatment processes.)

Finally it should be concluded that many **scale-up** approaches are based on the maintenance of a constant $k_L a$ with changing scale. Yet it should clearly be appreciated that many other factors must also be accommodated on scale-up, not the least being the need to maintain adequate circulation of the contents in order to assure homogeneity of the reactor contents. Table 10.2 from Amanullah et al. [26] shows clearly, for a number of commonly used scale-up criteria, how the circulation time increases as the vessel size increases.

10.8   Other design requirements in bioreactors

While we have now established some of the reaction and transfer rate parameters that drive the configuration and sizing of bioreactors, we should not leave the subject of bioreactor design without at least some reference to the other characteristics of such reactors which set them apart from other chemical reactors.

First and foremost, it must be re-emphasised that both process and product are usually highly specified and it must be demonstrable (**validatable**)

Table 10.2 The effects of using different scale-up criteria on the relative values of the circulation time t_c using a linear scale-up factor of 10 and maintaining geometrical similarity.

	Relative values: large scale : small scale				
Scale-up criterion	Equal P_g/V	Equal N_i	Equal Re	Equal Re and vvm*	Equal $k_L a$ and U_s
Circulation time $t_c (\propto N_i^{-1})$	4.55	1.0	100	5.88	4.55

*vvm = volume of air per volume of medium per minute.
Source: Based on Amanullah et al. [26].)

that operation cannot allow other than the desired process to take place. In particular, the design, construction and operation must be such that no other biological growth or production process can take place other than that specified. To achieve this, the vessel and its contents and all the flow-streams into the vessel must have the numbers of any other organisms reduced to levels that have no impact on the integrity of the process and product. Heat processing is almost invariably used to kill all organisms in the vessel before inoculation with the desired cells and it is often used on the liquid substrate solutions as well. Often the vessel and its batched contents will be treated together. A typical regime would consist of heating to 121°C, followed by a 'hold' at this temperature for between 0.3 and 2 hours and cooling down to the bioreaction temperature (typically in the range 25–37°C). Many medium components, e.g. sugars and proteins, will exhibit a propensity to burn onto surfaces unless close attention is paid to smoothness of design – austenitic stainless steels (usually in the 304, 316 and 321 series) will be specified not only because of their corrosion resistance but also because they can take on a very smooth surface finish which provides little 'key' for such materials, or indeed for micro-organisms, to attach to.

Air entering (and leaving) the bioreactor will be filtered in deep-fibre or (more usually) membrane filters. All other feedstreams into the system – for example, of nutrients fed to the reaction or acid/alkali used for pH control – will have to be either filtered or heat treated to the same standards. In the case of animal cell processes, the sensitivity of many of the medium components dictates that media are generally filter sterilised.

Close monitoring and control of bioreaction parameters will be continuous, examples of the measurable variables being given in Table 10.3. Although much development is being expended, it is as yet difficult to estimate the **biological state** of the system using in-line **bioprobes**, and state estimates must be inferred from other parameters (bioreactor **off-gas analysis** being particularly useful and fast in response) and from off-line measurements. Given that one is blowing large amounts of air into media containing proteins and other surface-active materials, it may be expected that **foaming** will often be a severe process problem which will have to be detected and controlled.

Our bioreactor is therefore rather a complex vessel. It is a container, the absolute integrity of which is essential both to processor and process, a mixer, often of rheologically complex gassed suspensions of solids in surface-active fluids, a mass transfer enhancer, and a heater/cooler/temperature controller. In it are carried out reactions, the complexities of which are often poorly understood, but the closely specified control of which is totally essential. The contents of a single vessel will often have a potential value of many hundreds of thousands of pounds or dollars – so it is worth time and care in design and operation.

Table 10.3 Variables measurable on-line or near-on-line in fermentation processes (after Bull [27])

On-line
Temperature of vessel contents
Vessel contents (and gas hold-up) by mass or level
Foam level
Temperature(s), pressure(s) and flow rate(s) of process cooling streams
Agitator rotation speed
Mechanical (strain gauge) and/or electrical power (wattmeter) input for agitation
Vessel head pressure
Temperature, pressure, flow rate(s) and composition(s) of inlet air (and other gases)
Temperature, pressure, flow rate and composition of exhaust gas
Concentration (tension) of dissolved oxygen and other volatiles
Redox potential
pH and amounts and/or rates of acid and base addition for pH or redox control
Ammonia and other specific ions (specific ion electrodes)
Amounts and/or rates of addition of nutrients, substrates and other control reagents

Near-on-line (e.g. using automatic sampling and flow injection analysis)
Metabolite/product concentrations and/or activities (HPLC, NIR spectroscopy, etc.)
Cell concentration by mass, count or volume
Cell cycle and cell size distributions (flow cytometry)
Broth viscosity (with or without biomass)

Determined from measured variables
Oxygen uptake rate, carbon dioxide evolution rate and respiratory quotient
Oxygen transfer coefficient
Substrate, intermediate and product turnover rates and concentrations
Biomass concentration and specific growth rate
Yield and maintenance coefficients
Heat balances and heat transfer coefficients
Mean and point energy dissipation rates
Apparent broth viscosity and/or flow behaviour indices
Process identification and control parameters, e.g.
– kinetic rate constants and other process model parameters
– dynamic process response times
– optimal process paths

References

1. Primrose, S.B. (1991) *Molecular Biotechnology*, 2nd edition, Blackwell Scientific Publications, Oxford.
2. Glazer A.N. and Nikaido H. (1995) *Microbial Biotechnology*, W.H. Freeman & Co., San Francisco.
3. The BIOTOL (Biotechnology by Open Learning) packages published by Open Universiteit, Netherlands and Greenwich University Press, London, UK.
4. Williams G. (1997) *Hyper CELL 1997*, Garland Publishing, Hamden, CT.
5. ACGM/HSE/DOE Note 6 (1987) *Guidelines for the Large-Scale Use of Genetically Manipulated Organisms*, HSE, London. Currently under revision as part of an *ACGM Compendium of Guidance*.
6. ACGM/HSE/DOE Note 7 (1993) *Guidelines for the Risk Assessment of Operations Involving the Contained Use of Genetically Modified Micro-organisms*, HSE, London. Currently under revision as part of an *ACGM Compendium of Guidance*.

7. Payne, G., Brigi, V., Prince C. and Shuler M. (1994) *Plant Cell and Tissue Culture in Liquid Systems*, John Wiley Inc., New York.
8. First N. (1991) *Transgenic Animals*, Butterworth-Heinemann, Oxford.
9. Cunningham C. and Porter A. (1997) *Recombinant Protein Production in Plants: Production and Isolation of Clinically Useful Compounds*, Humana Press, UK.
10. Roels J.A. (1983) *Energetics and Kinetics in Biotechnology*, Elsevier Medical Press, Amsterdam.
11. Cornish-Bowden A. (1995) *Fundamentals of Enzyme Kinetics*, Portland Press, UK.
12. Mavituna F. (1986) Activity of immobilised cell particles. In: *Process Engineering Aspects of Immobilised Cell Systems* (eds C. Webb, G.M. Black and F. Mavituna), I. Chem. E., Rugby, pp. 134–150.
13. Fukuda H. (1994) Immobilized microorganism reactors. In: *Bioreactor System Design*, (eds J.A. Asenjo and J.C. Merchuk), Marcel Dekker, New York, pp. 339–375.
14. Williams F.M. (1967) A model of cell growth dynamics. *J. Theoret. Biol.*, **15**, 190–207.
15. Bailey J.E. and Ollis D.F. (1986) *Biochemical Engineering Fundamentals*, 2nd edition, McGraw-Hill, New York.
16. Nielsen, J. and Villadsen, J. (1994) *Bioreaction Engineering Principles*, Plenum Press, New York.
17. Singh, R.P., Al-Rubeai, M. Gregory C.D. and Emery A.N. (1994) Cell death in bioreactors: a role for apoptosis. *Biotechnol. Bioeng.*, **44**, 720–726.
18. Luedeking R.L. and Piret E.L. (1959) A kinetic study of the lactic acid fermentation batch process at controlled pH. *J. Biochem. Microbiol. Technol. Eng.*, **1**, 393–412.
19. Herbert, D., Elsworth R. and Telling R.C. (1956) The continuous culture of bacteria. A theoretical and experimental study. *J. Gen. Microbiol*, **14**, 601–622.
20. Bischoff K.B. (1966) Optimal continuous fermentation reactor design. *Can. J. Chem. Eng.*, **44**, 281–284.
21. Schügerl, K. and Lübbert, A. (1994) Pneumatically agitated bioreactors. In: *Bioreactor System Design* (eds J.A. Asenjo and J.C. Merchuk), Marcel Dekker, New York, pp. 257–303.
22. Reuss M. (1994) Stirred tank bioreactors. In: *Bioreactor System Design* (eds J.A. Asenjo and J.C. Merchuk), Marcel Dekker, New York, pp. 207–255.
23. Schügerl, K. (1982) New bioreactors for aerobic processes. *Int. Chem. Eng.*, **22**, 591–610.
24. Chisti, M.Y. (1989) *Airlift Bioreactors*, Elsevier Science Publ., Amsterdam.
25. Nienow A.W. (1990) Gas dispersion performance in fermenter operation. *Chem. Eng. Prog.*, **86**, 61–71.
26. Amanullah, A., Baba, A., McFarlane, C.M. Emery A.N. and Nienow A.W. (1993) Biological models of mixing performance in bioreactors. In: *Bioreactor and Bioprocess Fluid Dynamics* (ed. A.W. Nienow), Mech. Eng. Publ. London, pp. 381–400.
27. Bull D.N. (1985) Instrumentation for Fermentation process control. In: *Comprehensive Biotechnology* (ed. M. Moo-Young), Vol. 2 (eds C.L. Cooney and A.E. Humphrey), Pergamon Press, Oxford, pp. 149–163.

Symbols

A large number of symbols are used in this book but most of them are defined close to the point of use. The list given below is not exhaustive but is intended to include those for which confusion might arise particularly in Chapters 1 to 4. Symbols for Chapter 8 are given at the end of that Chapter.

$A, B, C, D \ldots i \ldots$	(as subscripts) denote components $A, B, C, D \ldots$ $i \ldots$ respectively.
A	limiting reactant (section 2.1.4.4)
A	constant in reaction rate equation (Example 4.19)
A	pre-exponential term in the Arrhenius equation for reaction rates (sometimes called the frequency factor)
a, b, p, q	stoichiometric numbers (Eq. 3.1)
b_A	adsorption coefficient (or equilibrium constant) for adsorption of component A (Eq. 3.84)
c	(as subscript) denotes critical value
C_i	concentration of component i
C_A	concentration of reactant A (mol m^{-3} unless otherwise stated)
C_{A0}	concentration of A at inlet to flow reactor or initial concentration of A in batch reactor
C_{An}	concentration of A in outlet from reactor n
C_p	total heat capacity of contents of batch reactor
$\underline{C}_p$	total heat capacity of mixture (J K^{-1}) leaving flow reactor (or element of flow reactor) over specified period in steady state
C_{pmi}	molar heat capacity of component i (J mol^{-1} K^{-1})
$\bar{C}_{pmi}$	average molar heat capacity of component i over temperature range of interest (J mol^{-1} K^{-1})
D	reactor diameter (m unless otherwise specified)
D_{AB}	binary diffusivity of A with B (m^2 s^{-1}, Eq. 3.115)
D_{KA}	Knudsen diffusion coefficient of component A (m^2 s^{-1})
$D_{A,pore}$	net diffusivity of A into pore in binary case (m^2 s^{-1}, Eq. 3.123a)
D_A^M	net diffusivity of A into pore in multicomponent case (m^2 s^{-1}, Eq. 3.124)
$D_{A,eff}$	effective diffusivity of A into catalyst pellet (m^2 s^{-1}, Eq. 3.125)
e	(as subscript) denotes equilibrium value

E	activation energy for reaction (J mol^{-1} unless otherwise stated)
f_i	fugacity of component i in mixture under specified conditions (bar)
F_i	fugacity of pure component i under specified conditions (bar)
F_A	molar flow rate of reactant A
F_{A0}	molar flow rate of reactant A at entry point to reactor (mol s^{-1} unless otherwise stated)
F_t	total molar flow rate
F_{An}	molar flow rate of reactant A leaving reactor n
G	Gibbs energy (i.e. the Gibbs Function) (J)
$\Delta_r G^\theta(T)$	standard Gibbs energy change for a reaction (Eq. 2.71)
$\Delta_r G^{\theta,g}(T)$	standard Gibbs energy change based on ideal gas standard states at temperature T
$\Delta_f G_i^\theta(T)$	standard Gibbs energy of formation (J mol^{-1} unless otherwise stated, Eq. 2.72)
H	enthalpy
H_0	initial enthalpy of contents of batch reactor
H_F	enthalpy of contents of batch reactor when conversion is X_{AF} and temperature T_F
$\underline{H}_{\text{in}}$	enthalpy of fluid entering flow reactor over specified period at steady state
$\underline{H}_{\text{out}}$	enthalpy of fluid leaving flow reactor over same period
ΔH	increase in enthalpy of batch reactor
$\Delta H_{R,A}(T)$	or ($\Delta H_{R,A,T}$) increase in enthalpy per mole of reactant A consumed at temperature T (J mol^{-1} unless otherwise stated, Eq. 2.36)
$\Delta_r H^\theta(T)$	standard enthalpy change for reaction (Eq. 2.54)
$\Delta_f H_i^\theta(T)$	standard enthalpy of formation of component i at temperature T (Eq. 2.58)
j_A	flux of component A into pore due to bulk flow and 'bulk diffusion' combined (mol s^{-1} m^{-2}, Eq. 3.116) (bulk diffusion is sometimes described as 'mutual diffusion' or 'normal diffusion')
j_{KA}	flux of A in a Knudsen diffusion process (mol s^{-1} m^{-2})
$J_{A,\text{pore}}$	flux of A into pores due to combined effects of bulk diffusion, Knudsen diffusion and bulk flow (mol s^{-1} m^{-2})
k	rate constant (units depend on order of reaction)
k_1	first order rate constant (s^{-1} unless otherwise stated)
$k(n)$	first order rate constant for reactor n (s^{-1} unless otherwise stated)
k_{CA}, k_{cA}	film mass transfer coefficient for component A (m s^{-1})
$k_v^{apparent}$	apparent first order rate constant based on unit volume of catalyst pellet (s^{-1})
k_v^1	intrinsic first order rate constant based on unit volume of catalyst pellet (s^{-1})

k_B	Boltzmann's constant (J K^{-1})
K	thermodynamic equilibrium constant at given temperature (dimensionless, Eq. 2.74)
K_A	adsorption coefficient (or equilibrium constant) for adsorption of component A. This is denoted as b_A in Chapter 3.
$\underline{K}$	thermodynamic equilibrium constant based on ideal gas states at given temperature (dimensionless, Eq. 2.74a)
$\underline{K}_p$	(sometimes written K_p) thermodynamic equilibrium constant in ideal gas system expressed in terms of partial pressure/standard pressure ratios (dimensionless, Eq. 2.77): pressure dependent in non-ideal systems
K_p^1	'kinetic' equilibrium constant expressed in terms of partial pressures (may have dimensions)
K_c	'kinetic' equilibrium constant expressed in terms of concentrations (may have dimensions, Eq. 2.70)
l	axial distance from reactor inlet (m)
m	mass (kg)
m	(as subscript) denotes molar property
M	molar mass in SI system
n	number of moles
n_i	number of moles of component i in batch reactor
n_{i0}	number of moles of component i initially present in batch reactor
n_{A0}	number of moles of limiting reactant A initially present in batch reactor
$\underline{n}_i$	number of moles of component i leaving flow reactor in period during which $\underline{n}_{i0}$ moles of i and $\underline{n}_{A0}$ moles of A enter
Δn	increase in moles per mole of A reacting
N_i	number of molecules of component i
N	Avogadro's number
p_i	partial pressure of component i in gas mixture ($p_i = P y_i$) units are Pa in SI system. The bar (10^5 Pa) is often a convenient unit to use.
p_i^0	vapour pressure of pure component i at specified temperature, units as above
p^θ	standard state pressure (usually 10^5 Pa or 1 bar)
P, p	absolute pressure (units as for p_i)
$\dot{Q}$	rate of heat input (W)
$d\dot{Q}$	rate of heat input to element dl of reactor (W)
ΔQ	heat entering batch reactor (J, Eq. 2.30)
$\Delta \underline{Q}$	heat entering flow reactor in specified interval at steady state (J, Eq. 2.33)
r	radial co-ordinate in spherical catalyst pellet (m)

$(-r_A)$	reaction rate = net rate of consumption of limiting reactant A per unit volume of (homogeneous) reaction mixture (mol s^{-1} m^{-3} unless otherwise specified)
$(-\overrightarrow{r_A})$	rate of consumption of A in forward reaction per unit volume
$(\overleftarrow{r_A})$	rate of generation of A in reverse reaction per unit volume
$(-r_{AV})$	rate of consumption of limiting reactant A per unit volume of catalyst pellet (mol s^{-1} m^{-3} unless otherwise specified)
$(-\underline{r}'_A)$	rate of consumption of limiting reactant A per unit volume of catalyst bed (mol s^{-1} m^{-3} unless otherwise specified)
$(-\underline{r}_{AW})$	rate of consumption of limiting reactant A per unit mass of catalyst bed (mol s^{-1} kg^{-1} unless otherwise specified)
$(-r^1_{AV})$	intrinsic value of $-(r_{AV})$, i.e. rate which would occur if external surface conditions of temperature and composition prevailed throughout the catalyst pellets (mol s^{-1} m^{-3} unless otherwise specified)
$(-\underline{r}^1_{AW})$	intrinsic value of $(-\underline{r}_{AW})$ (mol s^{-1} kg^{-1} unless otherwise specified)
$(-r_{AS})$	rate of consumption of A per unit area of catalyst surface (mol s^{-1} m^{-2} unless otherwise specified)
R	effective radius of catalyst pellet (m)
R_m	(sometimes abbreviated to R) molar gas constant (J mol^{-1} K^{-1})
RMM	relative molecular mass or "molecular weight"
S	cross section area of tubular reactor (m^2 unless otherwise specified)
S	entropy (J K^{-1})
STP	standard temperature and pressure (0°C and 1.0133×10^5 Pa)
t	residence time (s unless otherwise stated) (defined in Section 4.2 and Chapter 5)
t_0	time required to empty batch reactor and prepare it for next batch
t_r	reaction time required to achieve specified conversion in batch reactor
t_T	$= t_0 + t_r$
T	temperature (K)
T_0	temperature at inlet to flow reactor or initial temperature of batch reactor (K)
T_F	temperature at outlet from flow reactor or terminal temperature in batch reactor
u	average fluid velocity within tubular reactor
U	internal energy (J)
V_r	volume of fluid contained in reactor (m^3 unless otherwise stated)
V_b	volume of catalyst bed
V'	molar volume (m^3 mol^{-1} unless otherwise stated)
ΔW_s	'Shaft work' done on batch reactor (J, Eq. 2.30)

$\Delta \underline{W}$ Shaft work done on flow reactor

x axial distance from mouth of pore (m)

X_A conversion of limiting reactant A (Eq. 2.19)

X_{Ae} equilibrium conversion of limiting reactant A

X_{A0} initial conversion of A

X_{AF} conversion of A at outlet from flow reactor or terminal conversion of A in batch reactor process

y_A mole fraction of reactant A in gas stream

y_{A0} mole fraction of A at inlet to reactor

z axial distance from feed point to catalyst bed (m unless otherwise specified)

Z compressibility factor (Eq. 2.12)

Z quantity defined by Eq. (2.63) or dimensionless measure of radius (Eq. 3.130)

ε_A the 'expansion factor', a measure of volume change on reaction defined by Eq. (2.63)

η effectiveness factor (Eqs. 3.113, 3.143)

η_{Sa} effectiveness factor for nth order reaction in spherical pellets

θ_i fraction of adsorption notes occupied by component i (sometimes written $\theta_{(i-S)}$)

θ_S fraction of adsorption sites which are bare

ν volumetric flow rate (m^3 s^{-1} unless otherwise stated)

ν_0 volumetric flow rate of fluid entering reactor

ν_i stoichiometric coefficient for component i (taken to be negative for reactants and positive for products, Eq. 3.2)

τ 'space time' $= (V_r/\nu_0)$ (s unless otherwise stated)
      for ideal plug flow reactor with no volume change during reaction, $\tau = t$

τ_p space time for plug flow reactor (s)

τ_C space time for CSTR reactor (s)

ϕ_{S1} Thiele's modulus for first-order reaction in spherical pellets (Eqs 3.131, 3.132)

ω acentric factor

Index

adiabatic operation

Milton Keynes UK
Ingram Content Group UK Ltd.
UKHW021850071024
449327UK00021B/1564